Deep Learning for Internet of Things Infrastructure

Deep Learning for Internet of Things Infrastructure

Edited by

Uttam Ghosh, Mamoun Alazab, Ali Kashif Bashir,
and Al-Sakib Khan Pathan

CRC Press is an imprint of the
Taylor & Francis Group, an **informa** business

First edition published 2022
by CRC Press
6000 Broken Sound Parkway NW, Suite 300, Boca Raton, FL 33487-2742

and by CRC Press
2 Park Square, Milton Park, Abingdon, Oxon, OX14 4RN

ISBN: 978-0-367-45733-4 (hbk)
ISBN: 978-1-032-06470-3 (pbk)
ISBN: 978-1-003-03217-5 (ebk)

Typeset in Times
by codeMantra

Dedicated to

---

my loving wife Pushpita and darling children
Upadhriti and Shriyan – Uttam Ghosh

my family – Mamoun Alazab

my parents, wife, and children – Ali Kashif Bashir

my family – Al-Sakib Khan Pathan

Contents

Acknowledgments

First and foremost, praises and thanks to God, the Almighty, for His showers of blessings throughout my life to complete the book successfully. It is another incredible experience of book editing, and our sincere gratitude goes to the contributing authors, copyeditors, and the publisher for facilitating the process. During this book editing journey, I learned a lot which also enhanced my patience, communication, and tenacity. Last but not least, our very best wishes are for our family members whose support and encouragement contributed significantly to complete this book.

Uttam Ghosh
Vanderbilt University, USA

Mamoun Alazab
Charles Darwin University, Australia

Ali Kashif Bashir
Manchester Metropolitan University, UK

Al-Sakib Khan Pathan
Independent University, Bangladesh

Editors

Uttam Ghosh is an Assistant Professor of the Practice in the Department of Electrical Engineering and Computer Science, Vanderbilt University, Nashville, Tennessee, USA. He earned a PhD in electronics and electrical engineering at the Indian Institute of Technology Kharagpur, India, in 2013, and has post-doctoral experience at the University of Illinois in Urbana-Champaign, Fordham University, and Tennessee State University. He has been awarded the 2018–2019 Junior Faculty Teaching Fellow (JFTF) and has been promoted to a graduate faculty position at Vanderbilt University. He has published and presented fifty papers in reputed international journals, including IEEE Transaction, Elsevier, Springer, IET, Wiley, InderScience, and IETE, and in international conferences sponsored by IEEE, ACM, and Springer. He has conducted several sessions and workshops related to Cyber-Physical Systems (CPS), SDN, IoT, and smart cities as co-chair at top international conferences including IEEE Globecom, IEEE MASS, SECON, CPSCOM, IEMCON, and ICDCS. He has served as a technical program committee (TPC) member at renowned international conferences, including ACM SIGCSE, IEEE LCN, IEMCON, STPSA, SCS SpringSim, and IEEE Compsac. He is an associate editor of the *International Journal of Computers and Applications*, Taylor & Francis, and a reviewer for international journals, including IEEE Transactions, Elsevier, Springer, and Wiley. Dr. Ghosh is a guest editor for special issues with *IEEE Transaction on Industrial Informatics* (TII), *IEEE Journal of IoT, IEEE Transaction on Network Science and Engineering* (TNSE), and *ACM Transactions on Internet* as well as ten edited volumes on Emerging CPS, Security, Machine/Machine Learning with CRC Press, Chapman Hall Big Data Series, and Springer. His main research interests include cybersecurity, computer networks, wireless networks, information-centric networking, and software-defined networking.

Mamoun Alazab earned a PhD in computer science at the Federation University of Australia, School of Science, Information Technology and Engineering. He is an Associate Professor in the College of Engineering, IT and Environment at Charles Darwin University, Australia. He is a cyber-security researcher and practitioner with industry and academic experience. His research is multidisciplinary, with a focus on cyber security and digital forensics of computer systems, including current and emerging issues in the cyber environment, such as cyber-physical systems and the Internet of Things. His research takes into consideration the unique challenges present in these environments, with an emphasis on cybercrime detection and prevention. He has a particular interest in the application of machine learning as an essential tool for cybersecurity, examples of which include detecting attacks, analyzing malicious code, and uncovering vulnerabilities in software. He is the Founder and Chair of the IEEE Northern Territory Subsection (February 2019–present), a Senior Member of the IEEE, and a Member of the IEEE Computer Society's Technical Committee on Security and Privacy (TCSP). In addition, he has collaborated with government and industry on many projects, including work with IBM, Trend Micro, Westpac, the Australian Federal Police (AFP), and the Australian Communications.

Ali Kashif Bashir is a Senior Lecturer/Associate Professor and Program Leader of BSc (H) Computer Forensics and Security at the Department of Computing and Mathematics, Manchester Metropolitan University, United Kingdom. He also enjoys positions as Honorary Professor and Chief Advisor of Visual Intelligence Research Center, School of Information and Communication Engineering, University of Electronics Science and Technology of China (UESTC), China, and Adjunct Professor of School of Electrical Engineering and Computer Science, National University of Science and Technology (NUST), Islamabad, Pakistan. His past assignments include Associate Professor of ICT, University of the Faroe Islands, Denmark; Osaka University, Japan; Nara National College of Technology, Japan; the National Fusion Research Institute, South Korea; Southern Power Company Ltd., South Korea; and the Seoul Metropolitan Government, South Korea.

He earned a PhD in computer science and engineering at Korea University, South Korea. He is a senior member of IEEE, a member of IEEE Industrial Electronic Society, a member of ACM, and a distinguished speaker of ACM. He has authored over 150 research articles; received funding as PI and Co-PI from research bodies of South Korea, Japan, EU, UK, and the Middle East; supervised or cosupervised several graduate (MS and PhD) students. His research interests include Internet of Things, wireless networks, distributed systems, network/cyber security, network function virtualization, and machine learning. He is the editor-in-chief of the IEEE Future Directions Newsletter. He is also serving as area editor of *KSII Transactions on Internet and Information Systems* and associate editor of *IEEE Internet of Things Magazine*, *IEEE Access*, *IET Quantum Computing*, and *Journal of Plant Disease and Protection*. He has also organized 24 guest editorials in the top journals of IEEE, Elsevier, Springer, and Elsevier. He has delivered more than 20 invited/keynote talks. He is leading many conferences as a chair (program, publicity, and track) and had organized workshops in flagship conferences like IEEE Infocom, IEEE Globecom, and IEEE Mobicom.

Al-Sakib Khan Pathan is an Adjunct Professor of Computer Science and Engineering, Independent University, Bangladesh. He earned a PhD in computer engineering in 2009 at Kyung Hee University, South Korea, and a BSc in computer science and information technology at Islamic University of Technology (IUT), Bangladesh, in 2003. In his academic career, he has worked as a faculty member at the CSE Department of Southeast University, Bangladesh, during 2015–2020; at Computer Science Department, International Islamic University Malaysia (IIUM), Malaysia; at BRACU, Bangladesh during 2009–2010; and at NSU, Bangladesh 2004–2005. He was a guest lecturer for the STEP project at the Department of Technical and Vocational Education, Islamic University of Technology, Bangladesh, in 2018. He also worked as a researcher at the Networking Lab, Kyung Hee University, South Korea, from September 2005 to August 2009 where he completed an MS leading to a PhD. His research interests include wireless sensor networks, network security, cloud computing, and e-services technologies. Currently, he is also working on some multidisciplinary issues. He is a recipient of several awards and best paper awards, and has several notable publications in these areas. So far, he has delivered over 22 keynotes and invited speeches at various international conferences and events. He has served as a general chair, organizing committee member, and technical program committee (TPC) member in numerous top-ranked international conferences/workshops such as INFOCOM, GLOBECOM, ICC, LCN, GreenCom, AINA, WCNC, HPCS, ICA3PP, IWCMC, VTC, HPCC, and SGIoT. He was awarded the IEEE Outstanding Leadership Award for his role in IEEE GreenCom'13 conference and the Outstanding Service Award in recognition for the service and contribution to the IEEE 21st IRI 2020 conference. He is currently serving as the editor-in-chief of *International Journal of Computers and Applications*, Taylor & Francis, UK. He is also editor of *Ad Hoc and Sensor Wireless Networks*, Old City Publishing; *International Journal of Sensor Networks*, Inderscience Publishers; and *Malaysian Journal of Computer Science*. He is associate editor of *Connection Science*, Taylor & Francis, UK, and *International Journal of Computational Science and Engineering*, Inderscience; area editor of *International Journal of Communication Networks and Information Security*; guest editor of many special issues of top-ranked journals; and editor or author of 22 books. One of his books has been included twice in Intel Corporation's Recommended Reading List for Developers, 2nd half 2013 and 1st half of 2014; three books were included in IEEE Communications Society's (IEEE ComSoc) Best Readings in Communications and Information Systems Security, 2013; several other books were indexed with all the titles (chapters) in Elsevier's acclaimed abstract and citation database, Scopus, in February 2015; at least one has been approved as a textbook at NJCU, USA, in 2020; and one book has been translated to simplified Chinese language from English version. Also, two of his journal papers and one conference paper were included under different categories in IEEE Communications Society's (IEEE ComSoc) Best Readings Topics on Communications and Information Systems Security, 2013. He also serves as a referee of many prestigious journals. He received some awards for his reviewing activities such as one of the most active reviewers of IAJIT several times; Elsevier Outstanding Reviewer for Computer Networks, Ad Hoc Networks, FGCS, and JNCA in multiple years. He is a Senior Member of the Institute of Electrical and Electronics Engineers (IEEE).

Contributors

Asad Abbas
Department of Information Technology
University of Central Punjab
Lahore, Pakistan

Arthi B.
Department of Computer Science and Engineering
College of Engineering and Technology, SRM Institute of Science and Technology
Kattankulathur, India

Rajakumar Arul
Department of Computer Science and Engineering
Amrita School of Engineering, Amrita Vishwa Vidyapeetham University
Bengaluru, India

Aruna M.
Department of Computer Science and Engineering
College of Engineering and Technology, SRM Institute of Science and Technology
Kattankulathur, India

Shunxing Bao
Department of Electrical Engineering and Computer Science
Vanderbilt University
Nashville, Tennessee

Yogesh Barve
Department of Electrical Engineering and Computer Science
Vanderbilt University
Nashville, Tennessee

Shakila Basheer
Department of Information System
College of Computer and Information Sciences
Princess Nourah Bint Abdulrahman University
Riyadh, Saudi Arabia

Ali Kashif Bashir
Department of Computing and Mathematics
Manchester Metropolitan University
Manchester, United Kingdom

Anirban Bhattacharjee
Department of Electrical Engineering and Computer Science
Vanderbilt University
Nashville, Tennessee

Somenath Chakraborty
Computational Science and Computer Engineering Department
University of Southern Mississippi
Hattiesburg, Mississippi

Thomas Damiano
Advanced Technology Labs
Lockheed Martin
Cherry Hill, New Jersey

Moses Garuba
Data Science and Cybersecurity Center (DSC2), Department of Electrical Engineering and Computer Science
Howard University
Washington, District of Columbia

Uttam Ghosh
Department of Computer Science
Vanderbilt University
Nashville, Tennessee

Aniruddha Gokhale
Department of Electrical Engineering and Computer Science
Vanderbilt University
Nashville, Tennessee

Nitin Gupta
Department of Computer Science and Engineering
National Institute of Technology, Hamirpur
Hamirpur, India

Md Enamul Haque
School of Medicine
Stanford University
Stanford, California

Mehedi Hassan
Center for Advanced Computer Studies
University of Louisiana at Lafayette
Lafayette, Louisiana

Muhammad Usama Islam
Center for Advanced Computer Studies
University of Louisiana at Lafayette
Lafayette, Louisiana

Rajesh Kaluri
School of Information Technology and Engineering
Vellore Institute of Technology
Vellore, India

Zhuangwei Kang
Department of Electrical Engineering and Computer Science
Vanderbilt University
Nashville, Tennessee

Shweta Khare
Department of Electrical Engineering and Computer Science
Vanderbilt University
Nashville, Tennessee

Ananda Kumar S.
School of Computer Science Engineering
Vellore Institute of Technology
Vellore, India

Kuruva Lakshmanna
School of Information Technology and Engineering
Vellore Institute of Technology
Vellore, India

Guy M. Lingani
Data Science and Cybersecurity Center (DSC2), Department of Electrical Engineering and Computer Science
Howard University
Washington, District of Columbia

Aaisha Makkar
Department of Computer Science and Engineering
Chandigarh University
Chandigarh, India

Bighnaraj Naik
Department of Computer Application
Veer Surendra Sai University of Technology
Burla, India

Janmenjoy Nayak
Department of Computer Science and Engineering
Aditya Institute of Technology and Management
Tekkali, India

Suzann Pershing
School of Medicine
Stanford University
Stanford, California

G. M. Sai Pratyusha
Department of Computer Science and Technology
Dr. L. Bullayya College of Engineering (for Women)
Visakhapatnam, India

Riya
Department of Computer Science and Engineering
National Institute of Technology, Hamirpur
Hamirpur, India

P. Muralidhara Rao
School of Computer Science and Engineering
Vellore Institute of Technology
Vellore, India

Danda B. Rawat
Data Science and Cybersecurity Center (DSC2), Department of Electrical Engineering and Computer Science
Howard University
Washington, District of Columbia

Prishita Ray
School of Computer Science and Engineering
Vellore Institute of Technology
Vellore, India

Dukka Karun Kumar Reddy
Department of Computer Science and Technology
Dr. L. Bullayya College of Engineering (for Women)
Visakhapatnam, India

Thippa Reddy G.
School of Information Technology and Engineering
Vellore Institute of Technology
Vellore, India

Praveen Kumar Reddy M.
School of Information Technology and Engineering
Vellore Institute of Technology
Vellore, India

P. Saraswathi
Department of Computer Science Engineering
Gandhi Institute of Technology and Management
Visakhapatnam, India

R. Sivaranjani
Department of Computer Science Engineering
Anil Neerukonda Institute of Technology and Sciences
Visakhapatnam, India

Qin Xin
Faculty of Science and Technology
University of the Faroe Islands
Vestarabryggja, Faroe Islands

S.M. Zobaed
Center for Advanced Computer Studies
University of Louisiana at Lafayette
Lafayette, Louisiana

1 Data Caching at Fog Nodes under IoT Networks

Review of Machine Learning Approaches

Riya and Nitin Gupta
National Institute of Technology, Hamirpur

Qin Xin
University of the Faroe Islands

CONTENTS

1.1 INTRODUCTION

In recent years, small devices embedded with sensors produce a large amount of data by sensing real-time information from the environment. The network of these devices communicating with each other is recognized as the Internet of Things (IoT), sometimes called as Internet of Everything [1]. The data produced by the IoT devices need to be delivered to the users using IoT applications after processing and analyzing. Further, data produced by the IoT devices are transient, which means that generated data have a particular lifetime, and after that lifetime, the data become useless and hence discarded [2]. Therefore, it is required to store the data somewhere near the IoT devices [3]. Simultaneously, suppose the data produced by the IoT devices are stored at the cloud server. In that case, it adds communication overhead, as the IoT users need to contact the cloud server whenever they require any data.

Fog computing is a decentralized approach to bring the advantages and intelligence of cloud computing such as storage, applications, and computing services near the end devices somewhere between the cloud and the end devices [4,5]. Fog nodes can be anything such as servers, networking devices (routers and gateways), cloudlets, and base stations. These nodes are aware of their geographical distribution as well as the logical location in the cluster. They can operate in a centralized or in a distributed manner and can also act as stand-alone devices. These nodes receive inputs from the data generators (IoT devices), process them, and provide temporary storage to the data. Fog nodes are intelligent devices that decide what data to store and what to send to the cloud for historical analysis. These devices can be either software or hardware, arranged in a hierarchy, and used to filter data transmitted by the sensors devices. These devices should have less latency, high response time, optimal bandwidth, optimal storage, and decision-making capability. At fog nodes, intelligent algorithms are embedded to store data, computing, and forward data between various layers. The member function of the fog node in the fog-cloud network is depicted in Figure 1.1. In this figure, the *compute* module is responsible for processing data and calculating the desired result. The *storage* module is responsible for storing data reliably so that robustness can be achieved. Further, various *accelerator* units such as digital signal processors and graphics processing units are used in critical tasks to provide additional power. In contrast, the *network* module is responsible for the guaranteed delivery of data.

Fog computing only complements cloud computing by providing short-term analytics, unlike cloud computing which offers long-term analytics. However, it is to be mentioned that fog computing does not replace cloud computing [6]. There are prominent six characteristics that differentiate fog computing from other computing paradigms [7,8].

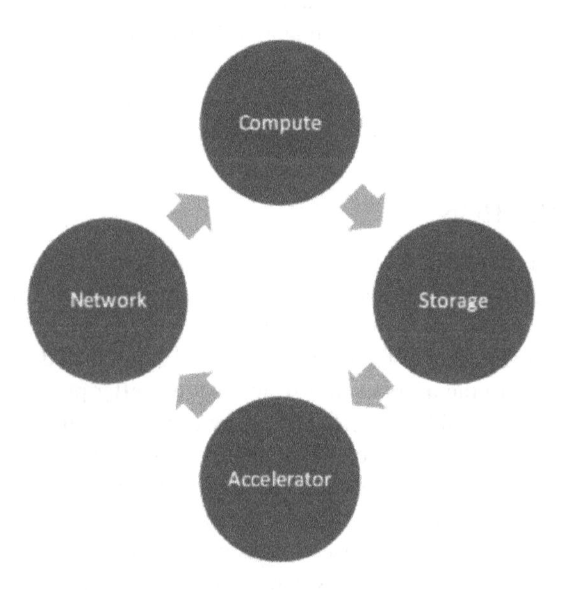

FIGURE 1.1 Functions of fog nodes.

a. **Awareness and Low Latency**: Fog nodes are aware of their logical location in the whole system and offer very low latency and communication costs. Fog nodes are frequently placed near the edge devices, and hence they can return reply and other analysis much faster than the cloud nodes.
b. **Heterogeneity**: Fog nodes generally collect different forms of data and from other types of devices through various kinds of networks.
c. **Adaptive**: In many situations, fog computing deals with uncertain load patterns of various requests submitted by different IoT applications. Adaptive and scaling features of fog computing help it to deal with the scenario mentioned earlier.
d. **Real-Time Interaction**: Unlike cloud computing, which supports batch processing, fog computing supports real-time interaction. The real-time data, which is time -sensitive, is processed and stored at fog nodes and is sent back to the users whenever required. On the contrary, the data which is not time -sensitive and whose life cycle is long is sent to the cloud for processing.
e. **Interoperability**: Because fog computing supports real-time interaction, it requires the cooperation of various providers leading to the interoperable property of fog computing.
f. **Geographically Distributed**: Unlike a centralized cloud, the applications serviced by fog nodes are geographically distributed, like delivering seamless quality videos to the moving vehicles.

Further, the processing time of fog nodes is significantly less (millisecond to subsecond). This technique avoids the need for costly bandwidth and helps the cloud by handling the transient data. To facilitate fog computing, the node should exhibit autonomy (property to take decision independently without the intervention of other nodes), heterogeneity, manageability, and programmability. Figure 1.2 shows fog computing architecture where IoT devices are connected to fog nodes, and then fog nodes are further connected to the cloud nodes [9].

The architecture of fog computing consists of three layers [10]:

a. **Terminal Layer**: This is the lowermost layer and consists of the IoT devices such as mobile phones and sensors, which detect the information from the environment by sensing it and then transmit the detected information to the upper layer. The information is transmitted in the form of data streams. The IoT data streams are the sequence of values emitted by the IoT devices or produced by one application module for another application module and sent to the higher layer for processing.
b. **Fog Layer**: This layer consists of various switches, portals, base stations, and specific servers. This layer lies between the IoT devices and the cloud and is used to process data near the IoT devices. If fog nodes cannot fulfill the terminal layer's request, then the request is forwarded to the cloud layer.
c. **Cloud Layer**: This layer consists of high-performance servers used for the storage of data and performing powerful computing.

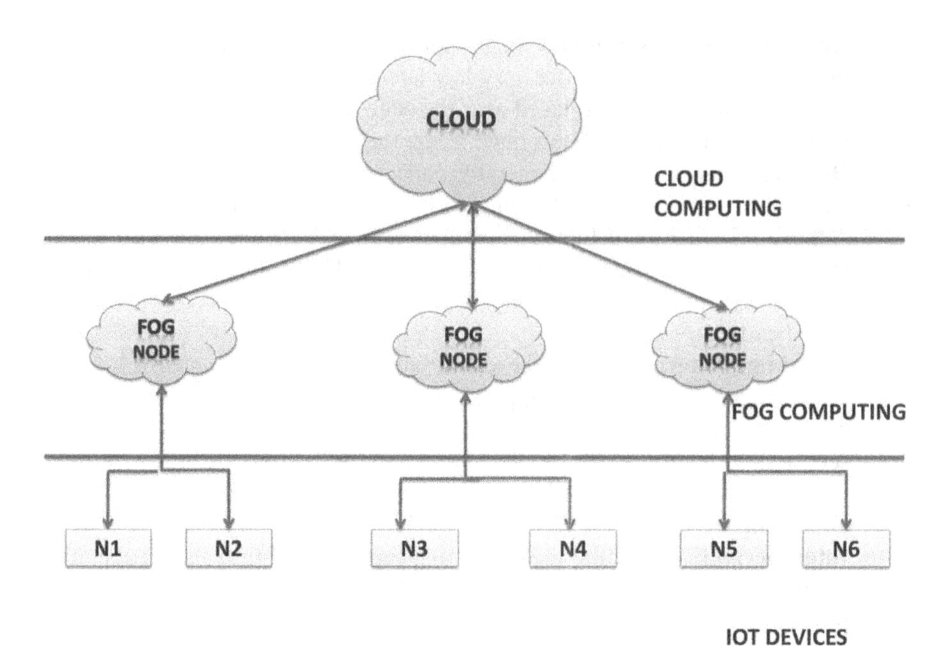

FIGURE 1.2 An architecture of fog computing.

Generally, IoT devices do not have processing power and storage, due to which they suffer from many problems such as performance, reliability, and security [11]. Fog nodes are capable of performing operations that require a large number of resources on behalf of IoT devices which are generally resource-constrained. This makes end devices less complex and also reduces power consumption. Further, fog computing also supports real-time interactions between the IoT devices and fog nodes. The data is available to the IoT devices quickly, unlike cloud computing where batch processing is mostly used. IoT devices are resource-constrained and generally do not have security features for which fog nodes act like proxy servers and provide extra security features. Fog nodes regularly update the software and security credentials and check the safety status of these devices.

Fog computing also offers the implementation of various service models such as Software as a Service (SaaS), Platform as Service (PaaS), and Infrastructure as a Service (IaaS) [12,13]. Due to such advantages, various frameworks such as Google App Engine, Microsoft Azure, and Amazon Web Services using cloud computing have also started supporting fog computing for providing solutions to advanced distributed applications that are geographically dispersed and require low-latency computational resources. They are also using dedicated nodes with low-latency computational power, also called mist nodes (lightweight fog nodes), and are sometimes placed closer to the IoT devices than fog nodes [14,15]. Hence, the integration of IoT with fog computing brings many such advantages.

1.1.1 Importance of Caching at Fog Nodes

The IoT devices do not have to contact the remote server, i.e., cloud, whenever they require some data. The IoT devices first check data in the cache of fog nodes. If required data is present, then the fog nodes return the data to the IoT devices; otherwise, they contact the cloud for the needed data. Hence, caching of data at fog nodes reduces the transactional latency. Moreover, fog computing requires lesser bandwidth to transfer the data [16]. As fog computing supports hierarchical processing, the amount of the data needed to be transferred from the IoT devices to the clouds is more petite. In contrast, the amount of data transmitted per unit of time from the fog node to the IoT devices is more, which leads to improvement in overall throughput. Hence, caching data at fog nodes decreases the general operational expenses. Data is stored in the distributed manner at fog nodes which can be deployed anywhere according to the requirements.

Further, caching of data at fog nodes helps reduce load at the cloud servers as the data whose frequency of interest is more among IoT devices and the probability of reusing the same data is also high is cached at fog nodes. Hence, only selected data is transferred for storage and processing to the cloud, which reduces the latency of contacting the remote server, which is far away from the IoT devices/sensors. Further, storing data at fog nodes ensures continuous services to the IoT devices irrespective of intermittent network connectivity.

Along with the advantages, some challenges need to be addressed to cache data at fog nodes. The biggest challenge of this technique is to decide what to store in the cloud and what to cache at fog nodes. The decision to cache the data at the fog node should be taken so that the hit rate of data at the fog node should be maximized such that the overall throughput is maximized [17,18]. Further, the storage capacity of fog nodes is limited, and they can only store the selected data. Therefore, it is necessary to predict the future demand of the users such that the data frequently required by the users in the future can be stored at the fog node to maximize the hit rate. However, it is difficult to predict the future requirement of the users.

Another challenge that needs to be addressed is maintaining synchronization between the data cached at the fog node or different fog nodes and data at the cloud nodes. Further, the security of data at fog nodes and the selection of ideal fog nodes are also issues of concern [19]. Various machine learning and deep learning techniques can be used to protect the data from attackers. In [20], various deep learning applications for cybersecurity are discussed, which can also be used in fog computing. Further, in [21], the stability of smart grids is predicted using a multidirectional extended short-term memory technique. The cooperation among sensor nodes is exploited in [22] to accelerate the signature verification and provide authentication in a wireless sensor network. The author in [23] also classifies the behavior of various malicious nodes and proposes a detection method of malicious nodes based on similarity in sequences and frequencies of API calls. Further, the authors in [24] also discuss the various trends in malware occurrences in the banking system. Apart from this, the authors in [25,26] also discuss the detection of intruders in the cyber system using machine learning.

Further, the mobility of nodes or virtual machines, which requires maintenance, balancing, and power management, is also a challenge that needs to be addressed. Each fog node may have one or more virtual machines depending upon requests and traffic conditions. The computation and communication required for the process of hand-off and its effect on caching are very much complicated and expensive [27,28].

As discussed above, this chapter focuses on an essential aspect of caching, which is to predict the future demands of the IoT users such that effective caching of data can be done at fog nodes. To address this problem, various machine learning techniques are discussed, which help learn the behavior and pattern of demands of IoT devices and add autoprocessing and auto computing capability to fog nodes. Also, the caching techniques used in wireless networks can be used in fog computing. However, these techniques cannot predict the future demands of the end -users. Hence, directly applying these techniques to fog nodes makes the system less efficient. Therefore, machine learning techniques are used to predict the users' future demand and make caching at the fog more efficient. Before exploring the machine learning techniques, in the next section, various applications of caching at fog nodes and the life cycle of fog data are discussed.

The rest of the chapter is organized as follows. Section 1.2 represents the applications of data caching at fog nodes for IoT devices. Section 1.3 describes the life cycle of fog nodes, and Section 1.4 discusses the machine learning techniques for data caching and replacement. The future research directions are discussed in Section 1.5, followed by the conclusion in the last section.

1.2 APPLICATIONS OF DATA CACHING AT FOG NODES FOR IoT DEVICES

In this section, some real scenarios are discussed where data caching at fog nodes can be very useful [29–36].

a. **Dynamic Content Delivery and Video Streaming**: With the increase in multimedia content, the conventional network suffer from congestion. Further, video traffic acquires half of the traffic, and frames to play the video are required faster, such that there is no interruption. Hence caching of data at fog nodes is a suitable approach for faster delivery of the multimedia contents [37].
b. **Virtual Reality and Online Gaming**: Virtual reality and online gaming require real-time data. In virtual reality, it is required to provide the status of the user as well as the location of the users. Hence, it is required to process the data and provide data to the user as soon as possible, where fog computing seems to be the promising approach for this purpose [38].
c. **Smart Cities**: In smart cities, various IOTs are connected to share data . These IoTs generate a large amount of data that need to be processed near the IoTs. For example, in smart traffic lights, data can be stored and processed at fog nodes and used to send warning signals to the approaching vehicles [39].

d. **Smart Grids**: The data generated by intelligent grids contain complex parameters that are hard to analyze. Fog nodes have the power to investigate and process complex data to perform heavy computations. Hence, fog nodes can be used to store and process the local data generated by smart grids and various IoT devices used in smart cities [40].
e. **Smart Healthcare**: Real-time data processing makes smart healthcare more efficient and faster. Hence fog computing can be used in healthcare in order to make its working more efficient. For example, fog computing may be used to detect falling of the stroke patients [41].
f. **Intensive Computation Systems**: The systems which require intensive computations require low processing and latency time. Hence the data produced by these systems must be processed and stored at fog nodes and provided to the systems whenever needed [42].
g. **Internet of Vehicles**: Fog computing plays a vital role in vehicle-to-vehicle communication and taking safety measures on the road by providing data to the vehicles required to take decisions for traffic control and smart parking. Fog nodes obtain data from the sensors deployed and take decisions for traffic control measures [30].
h. **Wireless Sensors Systems**: The data produced by wireless sensor systems such as oil and gas industries and chemical factories are transient which need to be stored near the users. Hence the data produced by these systems should be cached at fog nodes to improve the performance of the methods [43,44].

In the scenarios mentioned earlier, it is suitable to store the real-time or dynamic contents near the users that are generating data and may also require them in the near future. This requirement can be easily fulfilled by caching the data at fog nodes located near the users or IoT devices.

1.3 LIFE CYCLE OF FOG DATA

As discussed in the introduction section, depending upon the various layers in fog computing, fog data goes through various steps from acquiring data at the terminal layer to processing data and the execution of tasks to constitute a life cycle. The life cycle of the data is shown in Figure 1.3, and various steps involved during the life cycle are explained as follows [45–50]:

a. **Data Acquisition**: The sensors present in the device layer/terminal layer sense the environment and collect data. The acquired data are either sent to the sink node or directly transferred to the fog node for processing.
b. **Lightweight Processing**: This is done at the fog layer and hence includes various tasks such as filtering of data, cleaning of data, eliminating the unwanted data, lightweight manipulation of data, compression/decompression of data, and encryption/decryption of data. Some data are stored at this layer to support real-time processing, and the rest of the data are transferred to the cloud layer for further processing. Further, the feedback and the data are exchanged by the fog layer, as shown in Figure 1.3

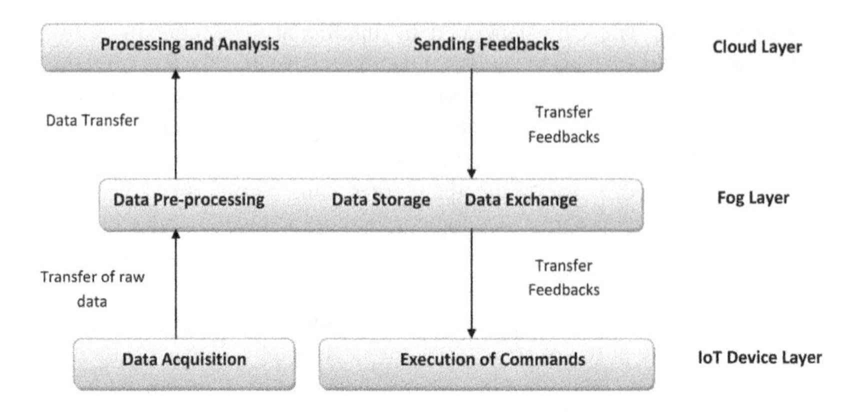

FIGURE 1.3 Life cycle of fog data.

c. **Processing and Analysis**: The data received from the fog layer are processed using different types of analysis to extract the crucial data. The data are permanently stored at the cloud server. According to the processing performed at the data received from the fog layer, reports are generated. Various technologies such as map -reduce are used for data processing in the cloud.
d. **Sending Feedback**: Based on reports generated during data processing, the cloud server sends feedback such as data required by the end devices and proper commands to the device layer to perform the necessary action.
e. **Command Execution**: Based on the feedback received from the cloud server, the actuators perform the respective action, and then required steps are performed on the environment.

It is evident from the above sections that caching played a significant role in fog computing. Efficient caching will help achieve low latency requirements and maintain high QoS and QoE of 5G. Caching is classified as reactive caching, where data caching is done on request, and proactive caching, where prefetching of data is done. To achieve higher spectrum efficiency, proactive caching is better if prediction errors are nearly zero [51]. Therefore, it is crucial to design various techniques to predict the users' future requests, which can be cached at fog nodes such that repetitive requests to the cloud can be avoided.

Various techniques have been used in the literature for data prediction and caching, like fog-to-fog (F2F) caching [52], where multi-agent cooperation is used. The authors in [53] proposed a location customized regression-based caching algorithm to predict the future content demands. According to various activity levels, the authors in [54] distinguished requests on three different popularity levels and then strategically cached data at fog nodes. Apart from caching at fog nodes, device-to-device (D2D) caching has also been done in the fog computing environment where direct communication between the nodes (IoT devices) takes place at a short distance without any infrastructure [55,56]. Whenever data is required by the device, it checks its local cache for the data. If information is not available, it broadcasts the request to the other instruments. The other IoT devices present at the ground tier in hierarchy check

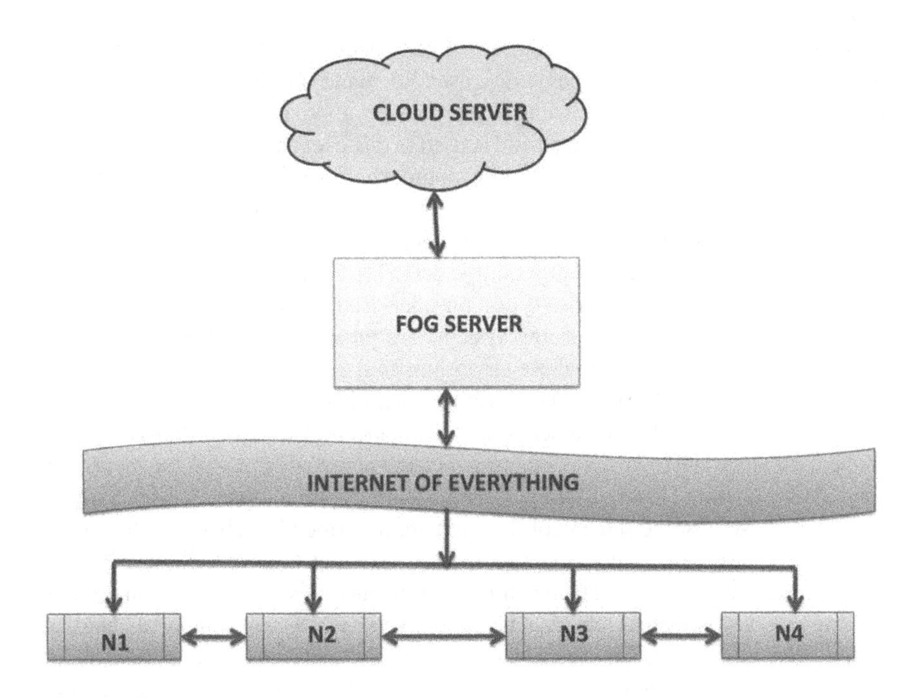

FIGURE 1.4 Caching using D2D technique.

for the data. If the data are present, the respective device replies with the data to the requesting device; otherwise, it replies with the negative acknowledgment. They are then requesting device requests for the data at the fog servers. As stated earlier, if data are not available at the fog server, then it sends the request to the cloud server. Cloud server, in return, sends the data to the fog server, and then the fog server sends data to the requesting nodes. Figure 1.4 shows the interaction between various IoT devices and the interaction between IoT devices and fog nodes.

As mentioned before, the content placement problem relies on the prediction accuracy of the user requirement, the popularity of the content, and caching strategy design. To predict the data content demand , much available data related to similar interests – social and geographic data and history data of users – can be used to predict user demand [57] better. This is effectively implemented using machine learning schemes. In the following section, various machine learning techniques used for data caching at fog nodes are investigated.

1.4 MACHINE LEARNING FOR DATA CACHING AND REPLACEMENT

Caching at fog nodes is influenced by various factors like the varying interest of different IoT users, which changes with different contexts, changed locations, network topology, and so on. Therefore, the future content request is highly unknown before making any caching decision [66]. Machine learning–based algorithms enable each fog node having limited storage to make the right decision in selecting the suitable contents to cache such that the caching performance of the fog node is maximized. Machine learning is used for predicting users' demand and mapping users' input to the output actions. Machine learning is a promising approach for improving network efficiency by predicting users' need and is used to discover early knowledge from large data streams [57]. In the machine learning approach, many data are exploited to determine the content popularity and helpful in filtering the data and knowledge [67–70]. The further processing of these data is valuable for analyzing the correlation between the features and the respective output of the data [71]. Further, machine learning techniques can be categorized into two types: unsupervised learning and supervised learning. In supervised learning, learning systems are provided with learning algorithms with known quantities that help these algorithms make future judgments. On the other hand, in unsupervised learning, the learning system is equipped with unlabeled data, and the algorithm is allowed to act upon it without any guidance. Machine learning can be used at any layer of fog computing, i.e., the terminal layer, the fog layer, or the cloud. At the terminal layer, machine learning is used for data sensing. There are various methods used for the sensing of data and are described in Table 1.1. The complex features of datasets like videos, vibrations, and different modeled readings from the IoT devices can be recognized by machine learning methods [72–77]. Various machine learning algorithms such as recurrent neural network (RNN), convolutional neural network (CNN), and generative adversarial network (GAN) have been used recently [76].

At the fog layer, machine learning is used for data storage and resource management [78]. Using machine learning algorithms, data are sampled from the IoT devices, compressed, and aggregated at fog nodes for further processing. Figure 1.5 shows the data analysis methods for the data produced by the IoT devices and various machine learning techniques that can be used to analyze the data and then decide that what to cache at fog nodes.

TABLE 1.1
Machine Learning Methods Used at Terminal Layer for Data Sensing

Sr. No.	Techniques	Description
1	Residual nets [58]	In this method, in order to reduce the difficulty of training models, shortcut connections are introduced into the convolutional neural networks. Visual inputs are mainly focused on residual nets
2	Long-term recurrent convolutional network [59]	In this method, convolutional neural networks are applied in order to extract the features. In the video, frame sequences are combined with the long short-term memory [60]. Further, the spatial and temporal relationships are exploited between inputs.
3	Restricted Boltzman machine [61]	The performance of human activities is improved by the deep Boltzman machine. In the case of multi-restricted Boltzman machine, performance is improved by multi-nodal DBM [62]
4	IDNET [63]	A convolutional neural network is applied by IDNET for biometric analysis tasks.
5	DeepX [64] and RedEye [65]	In these methods, on the basis of hardware and software, the energy consumption of the deep neural network is reduced

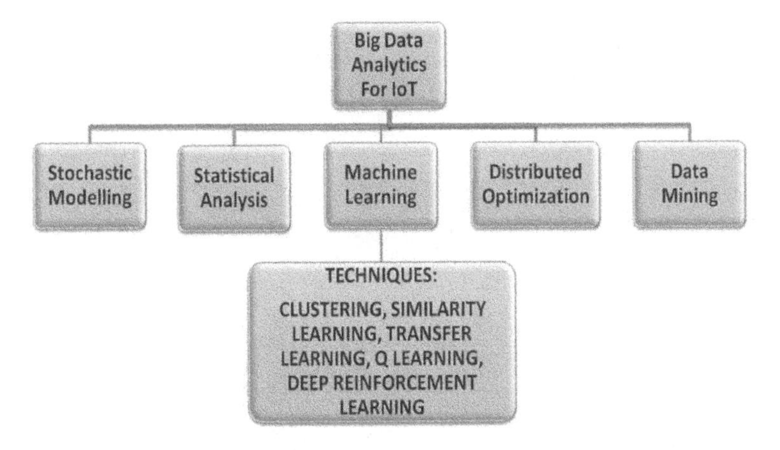

FIGURE 1.5 Big data analytics for the wireless network.

Various machine learning techniques used by the researchers in the literature for data caching at fog nodes are as follows:

1. **Clustering of Fog** Servers: Clustering is a technique used in unsupervised learning in which the information is not guided. In this technique, fog servers are clustered to fulfill the demands of IoT devices [30]. Data are stored at various fog servers after being coded into segments. When the user raises the content request, it is served by the group of fog servers which are clustered based on content stored in them. Suppose the requested content is cached at fog servers . In that case, the IoT device fetches the data from fog servers in ascending order of transmission distance until the obtained segments are sufficient for decoding. Further, if the received components are not enough, the nearest fog server contacts the cloud for the data, fetches the remaining data from the cloud and delivers it to the IoT device. Further, cluster size influences the system's efficiency as the benefit of cooperative caching has vanished if the size of the cluster is immense. However, at the same time, IoT devices can fetch the data from various nodes and increase the cache diversity. Therefore, cluster size should be optimal, which balances the trade-off.
2. **Similarity Learning Approach**: In this approach, fog nodes are given with a pair of similar IoT devices and a pair of less similar devices. From the given set of IoT devices, the intelligent fog node finds the similarity function (or the distance metric function) between the pair of similar devices by learning about their various features [57]. In this technique, two parameters, i.e., common interest and physical relations (link quality), are considered to find the similarity between IoT devices. A one-to-one matching scheme is also used for the pairing of IoT devices. With the help of this function, the intelligent fog node finds whether the new device is similar or not and hence finds the future interest of a new device whose interests are unknown.
3. **Transfer Learning Approach**: In this approach, the intelligent fog node gains knowledge while solving the problem and stores it. It then translates

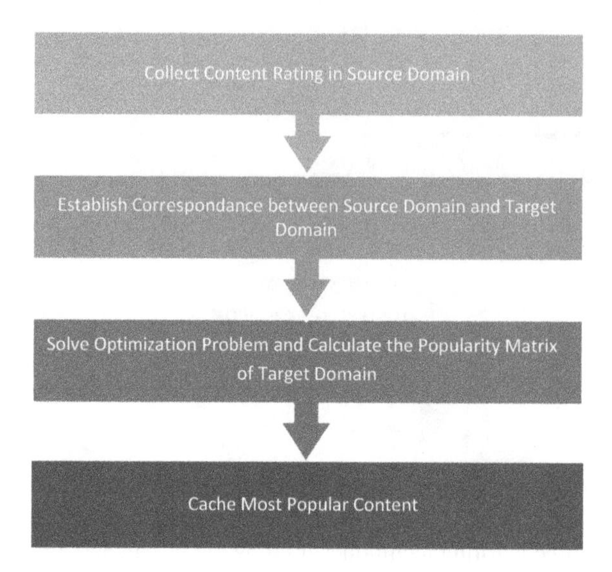

FIGURE 1.6 Training using transfer learning.

the new problem by exploiting the knowledge obtained from the data of existing tasks [79]. The fog node makes use of two domains for training: the source domain and the target domain. The source domain refers to the knowledge obtained from the interactions of IoT devices, while the target domain refers to the pattern of requests made by the IoT devices. The source domain consists of IoT devices and the data requested by machines in the past. Additionally, the source domain contains a popularity matrix (the number of times the particular data is asked by the IoT devices). This technique smartly borrows the user-data relationship from the source domain to understand the target domain by establishing correspondence between both. The correspondence is set to identify the similarly rated content in both disciplines. Further, an optimization problem is formulated by combining both domains to evaluate the popularity matrix of the target domain [80]. According to the values present in the popularity matrix of the target domain, the most popular content is cached at the fog node. Figure 1.6 illustrates the proposed transfer learning caching procedure.

The only problem with this technique is that it cannot give appropriate results if the relation between the source domain and the target domain is not efficient, which means that if the information demanded by the IoT devices is not related to the present information in the system, then transfer learning is not able to take an accurate decision for caching.

4. **Recommendation via Q Learning**: In existing local caching systems, the users do not know about the cached data. Therefore, they cannot send their request even though the requested file is available in the cache, decreasing the system's efficiency considerably. Hence to improve the efficiency

of the system, a recommender algorithm is used [81]. In this system, the fog server broadcasts an abstract to the users to gain knowledge about the presently cached files. An abstract contains one line introduction about the file and the ranking of the file in terms of the number of requests to the file. The value of the abstract also influences the decision of the user to request the file since the request rate of a file and the arrival and departure rate of the IoT devices is unknown in advance. To conquer this problem, Q learning is used, which is a form of deep learning approach used to improve the system's performance by reducing the latency and improving the throughput. It shows perfect accuracy in determining the future demand of the nodes by selecting the Q value. Multiple layers are used in this network, and these layers are used to process data and predict the future demand of the users. Since more data are generated and processed by the lower layers than the higher layers, more layers should be deployed near the users to reduce the network traffic and improve the system's performance.

The requested rate for the *i*th file depends upon the number of IoT devices present in the system and the number of IoT devices that arrived at that particular amount of time. As a result, an unknown number of requests to the *i*th file depends upon the caching action in the previous and present interval. During the learning process, the Q value is selected for each state-action pair which maximizes the reward. Then remaining IoT devices are counted in order to choose the action to the current interval. At the end of the gap, the bonus for the respective action is calculated, and the next state is observed. Then using these values, a new Q value is calculated [81]. This approach increases the long-term reward of the system and hence improves performance.

5. Deep Reinforcement Learning:

According to history, the deep reinforcement learning approach intelligently perceives the environment and automatically learns about the caching policy [82,83]. The emergence of deep neural networks has made it feasible to learn from raw and possibly high-dimensional data automatically. Learning-based caching techniques can be categorized into two approaches: popularity prediction and reinforcement learning approach. In the popularity prediction approach, first content popularity is predicted, and then according to popularity predictions, caching policy is devised. This approach is summarized in Figure 1.7 [66].

Various information like traffic patterns and context information is used to predict the content's popularity. Content popularity is expected in [84] by using user–content correlations and users' social ties through D2D communication. The authors in [85–87] have used various online learning algorithms to predict the content popularity. After the popularity prediction procedure, various caching policies and algorithms can be devised after solving optimization problems by combining estimated popularity with a few network constraints or traditional caching algorithms. However, these caching problems are usually complex and are NP-hard.

In the second approach, which is the reinforcement learning (RL) approach, in place of separating popularity prediction and content placement, the RL-based approach considers both as a single entity and is shown in Figure 1.8.

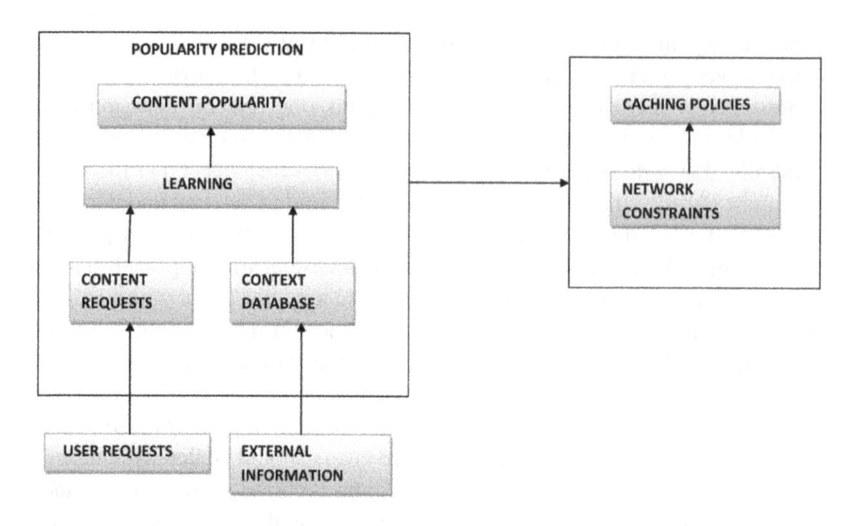

FIGURE 1.7 Popularity prediction approach.

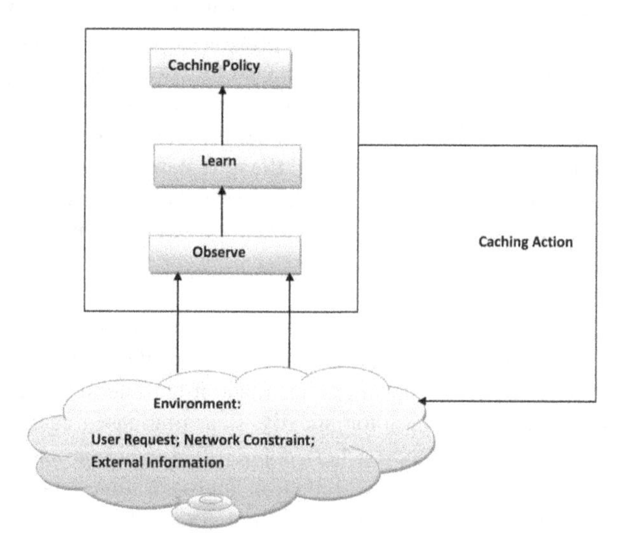

FIGURE 1.8 Reinforcement learning approach.

The RL agent train caching policy with observations based upon its action like QoE or offloaded traffic. This also avoids handling other factors affecting caching performance like node mobility or network topology. Various RL algorithms have been applied for designing fog nodes caching schemes. The advantage of RL-based approaches is that RL agents can be trained directly on raw and high dimensional observations.

One of the RL-based caching schemes is proposed by Zhu et al. [82]. The authors defined the action space where each action indicates the data item to be replaced in the cache. Initially, the fog node observes the environment, and according to the environment, it will obtain the state of the domain. Then according to the caching policy used, it takes the required action and attains the corresponding state. The action vector is represented as follows:

$$A_n = \{a_0, a_1, a_2, \ldots\ldots a_c\}$$

where c is the number of files present in the cache, a0 represents no caching action took place, av represents the new file is cached at the vth position, and the corresponding state vector is defined as follows:

$$S_n = \{s_1, s_2, s_3, \ldots\ldots s_n\}$$

After the action is taken, the system attains the reward, which is fed back to the edge node and the process is repeated. Further, each data item is associated with two data fields: (a) the time-stamp field and (b) the lifetime field. The time-stamp field tgen is used to indicate the time when data is created or generated, while the lifetime field tlife indicates the time up to which the value is valid in the item. The age of the data tage is predicted by finding the difference between the current time and the time generated. If tage < tlife, then the requested data is available at the cache and is fresh, and then data is directly returned to the user from the cache; otherwise, the data available at the cache is not fresh. When data is not fresh and if data is not available, then the node fetches fresh data from the cloud and returns it to the IoT device. Deep reinforcement learning aims at maximizing the reward when the agent takes action at a particular state. Figure 1.9 illustrates the application of deep reinforcement learning at fog nodes and knows the IoT devices' future demands.

6. **Federated Learning**: Conventional machine learning approaches depend upon the data and processing in a central entity. However, this is not always possible as the private data is not sometimes accessible, and also it requires great communication overhead to transmit initial data that is generated by a large number of IoT devices to the central machine learning processors [88–92]. Therefore, federated learning is the decentralized machine learning approach that keeps the data at the generation point itself, and then locally trained models are only transmitted to the central processor. These algorithms also significantly reduce overall energy consumption and network bandwidth by only sharing features rather than the whole data stream. It also responds in real -time to reduce the latency. These machine learning algorithms exploit on-device processing power and efficiently use private data as model training is performed in a distributed manner and keep the data in place, i.e., place of generation. In the content popularity approaches discussed above, direct access to private user data for differentiating content may not be feasible. Till now, federated learning has not been explored for caching in the fog computing environment. The authors in [88] have successfully demonstrated that content popularity prediction can be made with the help of federated learning.

Pros and cons of various machine learning techniques discussed above are summarized in Table 1.2.

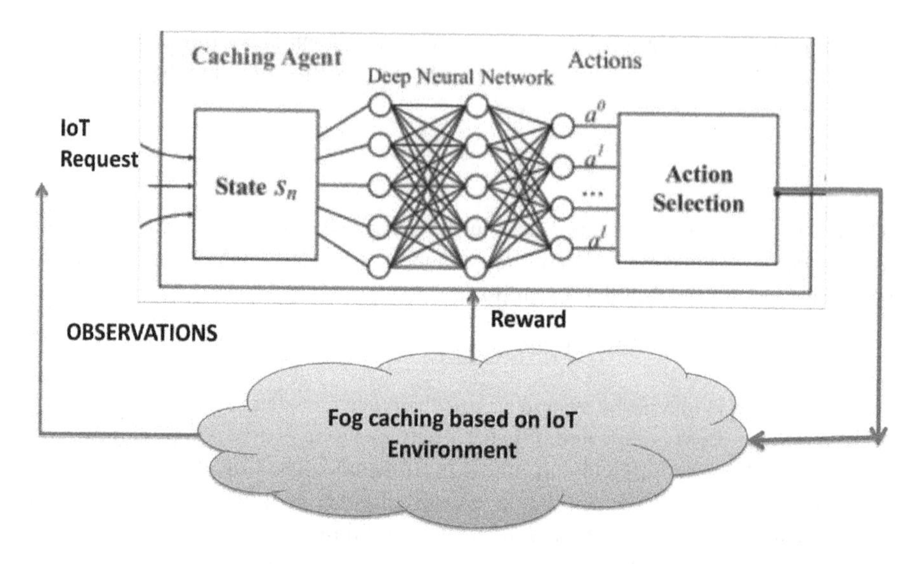

FIGURE 1.9 Applying DRL to fog caching.

TABLE 1.2
Pros and Cons of Machine Learning Algorithms

Techniques	Pros	Cons
Clustering	1. Coded caching is used which improves efficiency 2. Load balancing	1. If the size of the cluster is large, then efficiency is compromised
Similarity learning	1. Use only previous knowledge to find the similarity 2. Efficient if similar device arrives	1. Can't predict demand if a completely new node arrives
Transfer learning	1. Solve the new problem by exploiting existing knowledge	1. Difficult to find the correspondence between the source domain and the target domain
Q Learning	1. End devices know which data is cached and where it is cached and use a priority of data in caching, which improves the efficiency. 2. Extra space is needed for storing priority of the data	1. Delay in forecasting abstract 2. Better accuracy in predicting demands
Deep learning reinforcemen	1. Long-term cost of user fetching data is reduced 2. Overlapping coverage of fog nodes is considered	1. Cooperative caching is not considered 2. Better accuracy in predicting demands
Federated learning	1. Able to use the private data 2. Reduced communication overhead	1. Complex machine learning technique may have to be applied at various levels

1.5 FUTURE RESEARCH DIRECTIONS

In this section, various research issues related to fog nodes caching are discussed. These points may help readers for their future research directions in the area of fog nodes caching.

a. **Lack of Memory Space**: To implement a machine learning -based system, it is necessary to have sufficient data at the learning system for learning purposes. However, fog nodes do not have enough memory space; hence, it is of profound importance to investigate an effective machine learning technique that can learn from limited available data. As discussed before, the reader may explore federated learning, which is not exploited for content prediction in caching.
b. **Heterogeneous IoT Devices**: Most of the time, IoT devices are heterogeneous; e.g., in smart homes, various types of sensors for light and temperature may be installed, which generate a lot of different kinds of traffic. Hitherto, the impact of heterogeneity of IoT devices is not well addressed. In this kind of scenario, network connectivity methods, protocols to handle these devices, and communication methods are not discussed, which increases the latency while communicating with fog nodes.
c. **Synchronization Among Fog Nodes**: In the current research, the synchronization of data present at various fog servers and cloud servers is not discussed. Since the data produced by IoT devices is transient and becomes useless after some time, it is necessary to address the problem of synchronization of data at various fog servers and also with the cloud server.
d. **Game -Theoretic/Auction Models**: In various business models, fog nodes earn by serving the IoT Devices. In this kind of system, fog nodes may not cooperate with each other and may act selfishly. Therefore, various game theory–based or auction-based theories may be applied to solve non-cooperation among fog nodes.

1.6 CONCLUSION

IoT devices generate a lot of data that are stored and processed at cloud servers. To reduce the latency, fog computing has been introduced. However, there is a need for caching data at fog nodes to reduce further communication with cloud nodes. This chapter introduces various advantages of storing data of IoT devices at the fog nodes and subsequently the challenges faced to store data at fog nodes. Further, the life cycle of fog data as well as the architecture of fog computing is discussed. The application of caching data at fog nodes is also discussed in this chapter. This chapter also describes how various machine learning techniques are used to predict the future demand of IoT devices and store the most requested data at fog nodes. The chapter is then concluded with future research directions for readers.

REFERENCES

1. B. Kang, D. Kim, and H. Choo, "Internet of everything: A large-scale autonomic IoT gateway," *IEEE Transactions on Multi-Scale Computing Systems*, vol. 3, no. 3, pp. 206–214, 2017.
2. N. C. Narendra, S. Nayak, and A. Shukla, "Managing large-scale transient data in IoT systems," in *2018 10th International Conference on Communication Systems & Networks (COMSNETS)*. IEEE, 2018, pp. 565–568.
3. S. Vural, N. Wang, P. Navaratnam, and R. Tafazolli, "Caching transient data in internet content routers," *IEEE/ACM Transactions on Networking*, vol. 25, no. 2, pp. 1048–1061, 2016.
4. H. Gupta, A. Vahid Dastjerdi, S. K. Ghosh, and R. Buyya, "iFogSim: A toolkit for modeling and simulation of resource management techniques in the internet of things, edge and fog computing environments," *Software: Practice and Experience*, vol. 47, no. 9, pp. 1275–1296, 2017.
5. R. Mahmud, R. Kotagiri, and R. Buyya, "Fog computing: A taxonomy, survey and future directions," in *Internet of Everything*. Springer, 2018, pp. 103–130.
6. M. Marjani, F. Nasaruddin, A. Gani, A. Karim, I. A. T. Hashem, A. Siddiqa, and I. Yaqoob, "Big IoT data analytics: Architecture, opportunities, and open research challenges," *IEEE Access*, vol. 5, pp. 5247–5261, 2017.
7. S. Kitanov and T. Janevski, "State of the art: Fog computing for 5g networks," in *2016 24th Telecommunications Forum (TELFOR)*. IEEE, 2016, pp. 1–4.
8. L. M. Vaquero and L. Rodero-Merino, "Finding your way in the fog: Towards a comprehensive definition of fog computing," *ACM SIGCOMM Computer Communication Review*, vol. 44, no. 5, pp. 27–32, 2014.
9. Z. Hao, E. Novak, S. Yi, and Q. Li, "Challenges and software architecture for fog computing," *IEEE Internet Computing*, vol. 21, no. 2, pp. 44–53, 2017.
10. P. Hu, S. Dhelim, H. Ning, and T. Qiu, "Survey on fog computing: Architecture, key technologies, applications and open issues," *Journal of Network and Computer Applications*, vol. 98, pp. 27–42, 2017.
11. M. Chiang and T. Zhang, "Fog and IoT: An overview of research opportunities," *IEEE Internet of Things Journal*, vol. 3, no. 6, pp. 854–864, 2016.
12. M. Iorga, L. Feldman, R. Barton, M. Martin, N. Goren, and C. Mahmoudi, "The NIST definition of fog computing," National Institute of Standards and Technology, Tech. Rep., 2017.
13. R. K. Barik, A. Tripathi, H. Dubey, R. K. Lenka, T. Pratik, S. Sharma, K. Mankodiya, V. Kumar, and H. Das, "MistGIS: Optimizing geospatial data analysis using mist computing," in *Progress in Computing, Analytics and Networking*. Springer, 2018, pp. 733–742.
14. J. S. Preden, K. Tammemäe, A. Jantsch, M. Leier, A. Riid, and E. Calis, "The benefits of self-awareness and attention in fog and mist computing," *Computer*, vol. 48, no. 7, pp. 37–45, 2015.
15. H. R. Arkian, A. Diyanat, and A. Pourkhalili, "MIST: Fog-based data analytics scheme with cost-efficient resource provisioning for IoT crowdsensing applications," *Journal of Network and Computer Applications*, vol. 82, pp. 152–165, 2017.
16. M. Aazam, S. Zeadally, and K. A. Harras, "Fog computing architecture, evaluation, and future research directions," *IEEE Communications Magazine*, vol. 56, no. 5, pp. 46–52, 2018.
17. S. Yi, C. Li, and Q. Li, "A survey of fog computing: Concepts, applications and issues," in *Proceedings of the 2015 Workshop on Mobile Big Data*, 2015, pp. 37–42.
18. C. Mouradian, D. Naboulsi, S. Yangui, R. H. Glitho, M. J. Morrow, and P. A. Polakos, "A comprehensive survey on fog computing: State-of-the-art and research challenges," *IEEE Communications Surveys & Tutorials*, vol. 20, no. 1, pp. 416–464, 2017.

19. S. Yi, Z. Qin, and Q. Li, "Security and privacy issues of fog computing: A survey," in *International Conference on Wireless Algorithms, Systems, and Applications*. Springer, 2015, pp. 685–695.
20. M. Alazab and M. Tang, *Deep Learning Applications for Cyber Security*. Springer, 2019.
21. M. Alazab, S. Khan, S. S. R. Krishnan, Q.-V. Pham, M. P. K. Reddy, and T. R. Gadekallu, "A multidirectional LSTM model for predicting the stability of a smart grid," *IEEE Access*, vol. 8, pp. 85454–85463, 2020.
22. C. Benzaid, K. Lounis, A. Al-Nemrat, N. Badache, and M. Alazab, "Fast authentication in wireless sensor networks," *Future Generation Computer Systems*, vol. 55, pp. 362–375, 2016.
23. M. Alazab, "Profiling and classifying the behavior of malicious codes," *Journal of Systems and Software*, vol. 100, pp. 91–102, 2015.
24. A. Azab, M. Alazab, and M. Aiash, "Machine learning based botnet identification traffic," in *2016 IEEE Trustcom/BigDataSE/ISPA*. IEEE, 2016, pp. 1788–1794.
25. A. Azab, R. Layton, M. Alazab, and J. Oliver, "Mining malware to detect variants," in *2014 Fifth Cybercrime and Trustworthy Computing Conference*. IEEE, 2014, pp. 44–53.
26. S. Bhattacharya, P. K. R. Maddikunta, R. Kaluri, S. Singh, T. R. Gadekallu, M. Alazab, and U. Tariq, "A novel PCA-firefly based XGBoost classification model for intrusion detection in networks using GPU," *Electronics*, vol. 9, no. 2, p. 219, 2020.
27. J. Wu, M. Dong, K. Ota, J. Li, W. Yang, and M. Wang, "Fog-computing-enabled cognitive network function virtualization for an information-centric future internet," *IEEE Communications Magazine*, vol. 57, no. 7, pp. 48–54, 2019.
28. A. Strunk, "Costs of virtual machine live migration: A survey," in *2012 IEEE Eighth World Congress on Services*. IEEE, 2012, pp. 323–329.
29. S. Wang, X. Zhang, Y. Zhang, L. Wang, J. Yang, and W. Wang, "A survey on mobile edge networks: Convergence of computing, caching and communications," *IEEE Access*, vol. 5, pp. 6757–6779, 2017.
30. S. Zhang, P. He, K. Suto, P. Yang, L. Zhao, and X. Shen, "Cooperative edge caching in user-centric clustered mobile networks," *IEEE Transactions on Mobile Computing*, vol. 17, no. 8, pp. 1791–1805, 2017.
31. T. M. Fernández-Caramés, P. Fraga-Lamas, M. Suárez-Albela, and M. Vilar-Montesinos, "A fog computing and cloudlet based augmented reality system for the industry 4.0 shipyard," *Sensors*, vol. 18, no. 6, p. 1798, 2018.
32. J. P. Martin, A. Kandasamy, and K. Chandrasekaran, "Unraveling the challenges for the application of fog computing in different realms: A multifaceted study," in *Integrated Intelligent Computing, Communication and Security*. Springer, 2019, pp. 481–492.
33. S. P. Singh, A. Nayyar, R. Kumar, and A. Sharma, "Fog computing: From architecture to edge computing and big data processing," *The Journal of Supercomputing*, vol. 75, no. 4, pp. 2070–2105, 2019.
34. A. Ahmed, H. Arkian, D. Battulga, A. J. Fahs, M. Farhadi, D. Giouroukis, A. Gougeon, F. O. Gutierrez, G. Pierre, P. R. Souza Jr, and M. A. Tamiru, "Fog computing applications: Taxonomy and requirements," *arXiv preprint arXiv:1907.11621*, 2019.
35. P. H. Vilela, J. J. Rodrigues, P. Solic, K. Saleem, and V. Furtado, "Performance evaluation of a fog-assisted iot solution for e-health applications," *Future Generation Computer Systems*, vol. 97, pp. 379–386, 2019.
36. N. M. A. Brahin, H. M. Nasir, A. Z. Jidin, M. F. Zulkifli, and T. Sutikno, "Development of vocabulary learning application by using machine learning technique," *Bulletin of Electrical Engineering and Informatics*, vol. 9, no. 1, pp. 362–369, 2020.
37. S. Zhao, Y. Yang, Z. Shao, X. Yang, H. Qian, and C.-X. Wang, "Femos: Fogenabled multitier operations scheduling in dynamic wireless networks," *IEEE Internet of Things Journal*, vol. 5, no. 2, pp. 1169–1183, 2018.

38. Y. Xiao and C. Zhu, "Vehicular fog computing: Vision and challenges," in 2017 IEEE International Conference on Pervasive Computing and Communications Workshops (PerCom Workshops). IEEE, 2017, pp. 6–9.
39. B. Tang, Z. Chen, G. Hefferman, S. Pei, T. Wei, H. He, and Q. Yang, "Incorporating intelligence in fog computing for big data analysis in smart cities," *IEEE Transactions on Industrial informatics*, vol. 13, no. 5, pp. 2140–2150, 2017.
40. F. Y. Okay and S. Ozdemir, "A fog computing based smart grid model," in *2016 International Symposium on Networks, Computers and Communications (ISNCC)*. IEEE, 2016, pp. 1–6.
41. L. Cerina, S. Notargiacomo, M. G. Paccanit, and M. D. Santambrogio, "A fog-computing architecture for preventive healthcare and assisted living in smart ambients," in *2017 IEEE 3rd International Forum on Research and Technologies for Society and Industry (RTSI)*. IEEE, 2017, pp. 1–6.
42. L. Liu, Z. Chang, and X. Guo, "Socially aware dynamic computation offloading scheme for fog computing system with energy harvesting devices," *IEEE Internet of Things Journal*, vol. 5, no. 3, pp. 1869–1879, 2018.
43. U. Vijay and N. Gupta, "Clustering in WSN based on minimum spanning tree using divide and conquer approach," in *Proceedings of World Academy of Science, Engineering and Technology*, no. 79. World Academy of Science, Engineering and Technolog, 2013, p. 578.
44. W. Lee, K. Nam, H.-G. Roh, and S.-H. Kim, "A gateway based fog computing architecture for wireless sensors and actuator networks," in *2016 18th International Conference on Advanced Communication Technology (ICACT)*. IEEE, 2016, pp. 210–213.
45. R. Buyya, J. Broberg, and A. M. Goscinski, *Cloud Computing: Principles and Paradigms*. John Wiley & Sons, 2010, vol. 87.
46. F. Bonomi, R. Milito, J. Zhu, and S. Addepalli, "Fog computing and its role in the internet of things," in *Proceedings of the First Edition of the MCC Workshop on Mobile Cloud Computing*, 2012, pp. 13–16.
47. M. Asemani, F. Jabbari, F. Abdollahei, and P. Bellavista, "A comprehensive fog-enabled architecture for IoT platforms," in *International Congress on High- Performance Computing and Big Data Analysis*. Springer, 2019, pp. 177–190.
48. M. I. Pramanik, R. Y. Lau, H. Demirkan, and M. A. K. Azad, "Smart health: Big data enabled health paradigm within smart cities," *Expert Systems with Applications*, vol. 87, pp. 370–383, 2017.
49. C. Avasalcai, I. Murturi, and S. Dustdar, "Edge and fog: A survey, use cases, and future challenges," *Fog Computing: Theory and Practice*, pp. 43–65, 2020.
50. L. Andrade, C. Lira, B. de Mello, A. Andrade, A. Coutinho, and C. Prazeres, "Fog of things: Fog computing in internet of things environments," in *Special Topics in Multimedia, IoT and Web Technologies*. Springer, 2020, pp. 23–50.
51. E. K. Markakis, K. Karras, A. Sideris, G. Alexiou, and E. Pallis, "Computing, caching, and communication at the edge: The cornerstone for building a versatile 5g ecosystem," *IEEE Communications Magazine*, vol. 55, no. 11, pp. 152–157, 2017.
52. I. Al Ridhawi, N. Mostafa, Y. Kotb, M. Aloqaily, and I. Abualhaol, "Data caching and selection in 5g networks using f2f communication," in *2017 IEEE 28th Annual International Symposium on Personal, Indoor, and Mobile Radio Communications (PIMRC)*. IEEE, 2017, pp. 1–6.
53. P. Yang, N. Zhang, S. Zhang, L. Yu, J. Zhang, and X. S. Shen, "Content popularity prediction towards location-aware mobile edge caching," *IEEE Transactions on Multimedia*, vol. 21, no. 4, pp. 915–929, 2018.
54. I. Althamary, C.-W. Huang, P. Lin, S.-R. Yang, and C.-W. Cheng, "Popularity- based cache placement for fog networks," in *2018 14th International Wireless Communications & Mobile Computing Conference (IWCMC)*. IEEE, 2018, pp. 800–804.

55. S. Wang, X. Huang, Y. Liu, and R. Yu, "Cachinmobile: An energy-efficient users caching scheme for fog computing," in *2016 IEEE/CIC International Conference on Communications in China (ICCC)*. IEEE, 2016, pp. 1–6.
56. Z. Chen and M. Kountouris, "D2d caching vs. small cell caching: Where to cache content in a wireless network?" in *IEEE 17th International Workshop on Signal Processing Advances in Wireless Communications (SPAWC)*, 2016, pp. 1–6.
57. Z. Chang, L. Lei, Z. Zhou, S. Mao, and T. Ristaniemi, "Learn to cache: Machine learning for network edge caching in the big data era," *IEEE Wireless Communications*, vol. 25, no. 3, pp. 28–35, 2018.
58. K. He, X. Zhang, S. Ren, and J. Sun, "Deep residual learning for image recognition," in *Proceedings of the IEEE Conference on Computer Vision and Pattern Recognition*, 2016, pp. 770–778.
59. J. Donahue, L. Anne Hendricks, S. Guadarrama, M. Rohrbach, S. Venugopalan, K. Saenko, and T. Darrell, "Long-term recurrent convolutional networks for visual recognition and description," in *Proceedings of the IEEE Conference on Computer Vision and Pattern Recognition*, 2015, pp. 2625–2634.
60. K. Greff, R. K. Srivastava, J. Koutník, B. R. Steunebrink, and J. Schmidhuber, "LSTM: A search space odyssey," *IEEE Transactions on Neural Networks and Learning Systems*, vol. 28, no. 10, pp. 2222–2232, 2016.
61. S. Bhattacharya and N. D. Lane, "From smart to deep: Robust activity recognition on smartwatches using deep learning," in *2016 IEEE International Conference on Pervasive Computing and Communication Workshops (PerCom Workshops)*. IEEE, 2016, pp. 1–6.
62. V. Radu, N. D. Lane, S. Bhattacharya, C. Mascolo, M. K. Marina, and F. Kawsar, "Towards multimodal deep learning for activity recognition on mobile devices," in *Proceedings of the 2016 ACM International Joint Conference on Pervasive and Ubiquitous Computing: Adjunct*, 2016, pp. 185–188.
63. D. Figo, P. C. Diniz, D. R. Ferreira, and J. M. Cardoso, "Preprocessing techniques for context recognition from accelerometer data," *Personal and Ubiquitous Computing*, vol. 14, no. 7, pp. 645–662, 2010.
64. C.-Y. Li, C.-H. Yen, K.-C. Wang, C.-W. You, S.-Y. Lau, C. C.-H. Chen, P. Huang, and H.-H. Chu, "Bioscope: An extensible bandage system for facilitating data collection in nursing assessments," in *Proceedings of the 2014 ACM International Joint Conference on Pervasive and Ubiquitous Computing*, 2014, pp. 477–480.
65. E. Miluzzo, A. Varshavsky, S. Balakrishnan, and R. R. Choudhury, "Tapprints: Your finger taps have fingerprints," in *Proceedings of the 10th International Conference on Mobile Systems, Applications, and Services*, 2012, pp. 323–336.
66. H. Zhu, Y. Cao, W. Wang, T. Jiang, and S. Jin, "Deep reinforcement learning for mobile edge caching: Review, new features, and open issues," *IEEE Network*, vol. 32, no. 6, pp. 50–57, 2018.
67. M. Habib ur Rehman, P. P. Jayaraman, S. U. R. Malik, A. U. R. Khan, and M. Medhat Gaber, "Rededge: A novel architecture for big data processing in mobile edge computing environments," *Journal of Sensor and Actuator Networks*, vol. 6, no. 3, p. 17, 2017.
68. C. K.-S. Leung, R. K. MacKinnon, and F. Jiang, "Reducing the search space for big data mining for interesting patterns from uncertain data," in *2014 IEEE International Congress on Big Data*. IEEE, 2014, pp. 315–322.
69. A. Stateczny and M. Wlodarczyk-Sielicka, "Self-organizing artificial neural networks into hydrographic big data reduction process," in *Rough Sets and Intelligent Systems Paradigms*. Springer, 2014, pp. 335–342.
70. A. Rađgyanszki, K. Z. Gerlei, A. Surańyi, A. Kelemen, S. J. K. Jensen, I. G. Csizmadia, and B. Viskolcz, "Big data reduction by fitting mathematical functions: A search for appropriate functions to fit Ramachandran surfaces," *Chemical Physics Letters*, vol. 625, pp. 91–97, 2015.

71. M. Chen, U. Challita, W. Saad, C. Yin, and M. Debbah, "Machine learning for wireless networks with artificial intelligence: A tutorial on neural networks," *arXiv preprint arXiv:1710.02913*, 2017.
72. K. He, X. Zhang, S. Ren, and J. Sun, "Delving deep into rectifiers: Surpassing human-level performance on imagenet classification," in *Proceedings of the IEEE International Conference on Computer Vision*, 2015, pp. 1026–1034.
73. I. Goodfellow, Y. Bengio, and A. Courville, *Deep Learning*. MIT Press, 2016.
74. A. Krizhevsky, I. Sutskever, and G. E. Hinton, "Imagenet classification with deep convolutional neural networks," in *Advances in Neural Information Processing Systems*, 2012, pp. 1097–1105.
75. K. Katevas, I. Leontiadis, M. Pielot, and J. Serrà, "Practical processing of mobile sensor data for continual deep learning predictions," in *Proceedings of the 1st International Workshop on Deep Learning for Mobile Systems and Applications*, 2017, pp. 19–24.
76. S. Yao, S. Hu, Y. Zhao, A. Zhang, and T. Abdelzaher, "Deepsense: A unified deep learning framework for time-series mobile sensing data processing," in *Proceedings of the 26th International Conference on World Wide Web*. International World Wide Web Conferences Steering Committee, 2017, pp. 351–360.
77. A. Graves, "Generating sequences with recurrent neural networks," *arXiv preprint arXiv:1308.0850*, 2013.
78. Y. Wei, F. R. Yu, M. Song, and Z. Han, "Joint optimization of caching, computing, and radio resources for fog-enabled IoT using natural actor–critic deep reinforcement learning," *IEEE Internet of Things Journal*, vol. 6, no. 2, pp. 2061–2073, 2018.
79. B. Bharath, K. G. Nagananda, and H. V. Poor, "A learning-based approach to caching in heterogenous small cell networks," *IEEE Transactions on Communications*, vol. 64, no. 4, pp. 1674–1686, 2016.
80. E. Baffstuffg, M. Bennis, and M. Debbah, "A transfer learning approach for cache-enabled wireless networks," in *13th International Symposium on Modeling and Optimization in Mobile, Ad Hoc, and Wireless Networks (WiOpt)*. IEEE, 2015, pp. 161–166.
81. K. Guo, C. Yang, and T. Liu, "Caching in base station with recommendation via q-learning," in *2017 IEEE Wireless Communications and Networking Conference (WCNC)*. IEEE, 2017, pp. 1–6.
82. H. Zhu, Y. Cao, X. Wei, W. Wang, T. Jiang, and S. Jin, "Caching transient data for internet of things: A deep reinforcement learning approach," *IEEE Internet of Things Journal*, vol. 6, no. 2, pp. 2074–2083, 2018.
83. A. Sadeghi, G. Wang, and G. B. Giannakis, "Deep reinforcement learning for adaptive caching in hierarchical content delivery networks," *IEEE Transactions on Cognitive Communications and Networking*, vol. 5, no. 4, pp. 1024–1033, 2019.
84. E. Bastug, M. Bennis, and M. Debbah, "Living on the edge: The role of proactive caching in 5g wireless networks," *IEEE Communications Magazine*, vol. 52, no. 8, pp. 82–89, 2014.
85. M. Leconte, G. Paschos, L. Gkatzikis, M. Draief, S. Vassilaras, and S. Chouvardas, "Placing dynamic content in caches with small population," in *IEEE INFOCOM 2016-The 35th Annual IEEE International Conference on Computer Communications*. IEEE, 2016, pp. 1–9.
86. S. Li, J. Xu, M. van der Schaar, and W. Li, "Trend-aware video caching through online learning," *IEEE Transactions on Multimedia*, vol. 18, no. 12, pp. 2503– 2516, 2016.
87. X. Zhang, Y. Li, Y. Zhang, J. Zhang, H. Li, S. Wang, and D. Wang, "Information caching strategy for cyber social computing based wireless networks," *IEEE Transactions on Emerging Topics in Computing*, vol. 5, no. 3, pp. 391–402, 2017.

88. S. Niknam, H. S. Dhillon, and J. H. Reed, "Federated learning for wireless communications: Motivation, opportunities and challenges," *arXiv preprint arXiv:1908.06847*, 2019.
89. M. Chen, Z. Yang, W. Saad, C. Yin, H. V. Poor, and S. Cui, "A joint learning and communications framework for federated learning over wireless networks," *arXiv preprint arXiv:1909.07972*, 2019.
90. Z. Yang, M. Chen, W. Saad, C. S. Hong, and M. Shikh-Bahaei, "Energy efficient federated learning over wireless communication networks," *arXiv preprint arXiv:1911.02417*, 2019.
91. W. Y. B. Lim, N. C. Luong, D. T. Hoang, Y. Jiao, Y.-C. Liang, Q. Yang, D. Niyato, and C. Miao, "Federated learning in mobile edge networks: A comprehensive survey," *arXiv preprint arXiv:1909.11875*, 2019.
92. T. Li, A. K. Sahu, A. Talwalkar, and V. Smith, "Federated learning: Challenges, methods, and future directions," *arXiv preprint arXiv:1908.07873*, 2019.

2 ECC-Based Privacy-Preserving Mechanisms Using Deep Learning for Industrial IoT

A State-of-the-Art Approaches

R. Sivaranjani
Anil Neerukonda Institute of Technology & Sciences

P. Muralidhara Rao
Vellore Institute of Technology

P. Saraswathi
Gandhi Institute of Technology and Management

CONTENTS

2.1 INTRODUCTION OF INDUSTRIAL IoT

The IIoT, these days, is showing smartness in production, manufacturing, and packaging by placing inexpensive, intelligent, and small size interconnected components. From the marketing survey and experts' study, the IIoT assures incredible development, and expecting a considerable cost saving, more production rate, and predictable budget may reach up to $124 billion by 2021 [1]. According to Heymann's survey in [2], projected that the revolution cost of transformation traditional industries to start industries might require 267 billion euros by the end of 2025. The IIoT is a network, which connects billions of manufacturing unit devices that could develop businesses by intensifying process efficiency, certainty, security, and safety. However, their breadth is limited to large-scale operations such as railway logistics, power production, power distribution, and aeronautics [3]. According to the records, the IIoT and Industrial Data Analytics (IDA) [4] have raised income to 33.1%, customer gratification to 22.1%, and production excellence up to 11%.

Moreover, the document declares that 68% of industrial firms have their resident IDA plan. Almost 46% have some possible operational frameworks, and approximately 30% have achieved IDA schemes, which represents around 70% of the corporations still not having usual data acquisition and analytics strategies to make the most of their business smarter and lucrative. Although several industries have started using the IIoT extensively, even then the scope is there to upgrade.

According to an investigation conducted by PwC in [5], only 33% of industries in the US are obtaining the smart devices generated data to advance their operational, manufacturing, and integration processes. The connectivity of smart devices allows data gathering, analyzing, and potentially simplifying improvements in yield and proficiency. The significant progress in IoT infrastructures expects to amass 20 billion coupled IoT devices. It integrates the research-oriented technologies such as artificial intelligence, edge and fog computing, machine learning, and data analytics to encourage manufacture and service quality. IIoT enables universally obtain uninterrupted data from locally connected devices through wireless networks, wearable gadgets, and embedded systems. Next, the data collected from the devices send to cloud-based centers for storage, processing, information extraction, and information mining. However, administration and extraction of data effectively and efficiently are challenging tasks to the IIoT administrator. Fortuitously, the development of a study in deep learning affords a new resolution to handle the problems stated previously.

Of late, distributed installation of industrial machinery and generation of the massive amount of data from machinery in various applications started using deep learning. Nowadays, the demand is exponentially growing as it offers an intelligent decision-making on the data present by training a bulk amount of labeled data and continuously updating the related model parameters. On the other hand, we know that the training task should be performed by end-user on a device with limited resources. Due to this, the minor amount of data in the dataset may affect the efficiency of the model, and accuracy sometimes leads to overfitting. Opportunely, federated deep learning has anticipated to be a substitute paradigm [6]. It is an innovative machine learning technique and appropriate for devices with distributed connectivity, to accumulate data, shares the training dataset under the direction of a parameter server. More principally, it is an extension of federated deep learning that enables to modularize, update model parameters and process data in a parallel manner [7]. As a result, it significantly advances model iteration proficiency and convergence rate. Still, the integration of enormous data gathering with deep learning may raise a few tremendous data safety and privacy issues. Jayasri et al. [8] described major thought-provoking problems of IIoT, such as all security services on data, scalability, and interoperability. Some security concerns like data privacy leakage and model's parameters of privacy leakage.

To address scalability, integrity, data privacy leakage, and model parameters leakage, the researcher has started proposing privacy-preserving asynchronous deep learning. Shokri [7] anticipated a technique that permits all participants collectively decide the global model parameters by keeping their training data private in their local data storage and extract the required data from enormous datasets present in the cloud server. Individually participants can able to trains their local dataset using a local neural network (NN) and transfer gradients to a parameter server for additional accumulation. Despite spreading, a slice of gradients could disclose participants' privacy in data. Phong et al. [9] presented a privacy-preserving deep learning outline through additively homomorphic encryption. The authors are the participant's input privacy in the process of training a model that could be improved. However, there still exist a few security flaws, namely all group participants must share a joint homomorphic encryption key. Therefore, to achieve this, each participant's input privacy must know other group participants. In addition, the homomorphic encryption is not safe when the participant's network is exiting from the network. From all these, we can understand that there is a requirement to propose a technique that supports the data security, participants' dynamism to the network, and security to model parameters. To resolve the issues mentioned above, we are proposing two ECC-based privacy-preserving based on proxy re-encryption and group dynamic key management.

2.2 BACKGROUND AND RELATED WORKS

2.2.1 Evolution of Industrial IoT

The evolution of the industrial era is depicted in Figure 2.1. The industrial revolution and advancements in the modern industry have started at the beginning of the 18th century termed as Industry 1.0. Most of the goods, clothes, household items, and clothes were manufactured by human resources or by using animals. This procedure came to an end in the 18th century with an initiation of the manufacturing process. Advancements in industrial research upgrade the industrial machines to work on electrical energy, which initiates Industry 2.0 era from mid-19th century. Later,

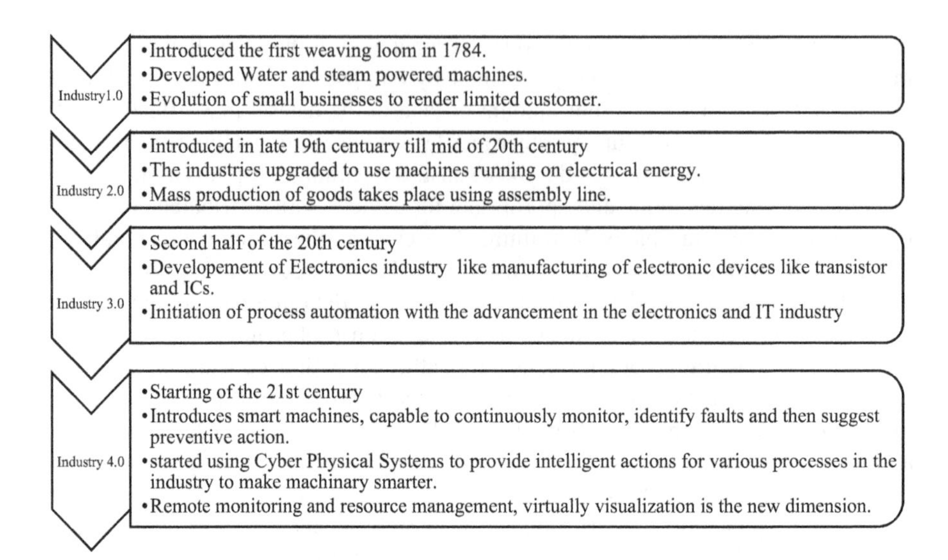

Industry1.0
•Introduced the first weaving loom in 1784.
•Developed Water and steam powered machines.
•Evolution of small businesses to render limited customer.
Industry 2.0
•Introduced in late 19th centuary till mid of 20th century
•The industries upgraded to use machines running on electrical energy.
•Mass production of goods takes place using assembly line.
Industry 3.0
•Second half of the 20th century
•Developement of Electronics industry like manufacturing of electronic devices like transistor and ICs.
•Initiation of process automation with the advancement in the electronics and IT industry
Industry 4.0
•Starting of the 21st century
•Introduces smart machines, capable to continuously monitor, identify faults and then suggest preventive action.
•started using Cyber Physical Systems to provide intelligent actions for various processes in the industry to make machinary smarter.
•Remote monitoring and resource management, virtually visualization is the new dimension.

FIGURE 2.1 Evolution of industrial IoT.

development was extended to the integration of electronics in machines leading to Industry 3.0. Adding smartness to the machines-initiated Industry 4.0 era, where machines became automated and human intervention became limited.

2.2.2 Literature Study on Authentication, Privacy, and Security in IIoT

Today, as we are all in the era of Industry 4.0, it provides the increasing availability, affordability, and proficiency of sensors, processors, and Wireless Sensor Network (WSN) technologies. Moreover, the number of embedded devices used in industrial applications is steadily increasing. It leads to an advancement in the interest of the Industrial Internet of Things (IIoT), where a massive number of devices are connected in a network [10]. These "smart devices" in the Industrial IoT requires lowest or no human intercession to "generate, exchange, and consume data; they often feature connectivity to remote data collection, analysis, and management capabilities".

However, the amount of data extracted from industrial IoT devices usually includes the sensors and aggregators data like machinery working conditions and periodical statistics of the industrial products. Under such a situation, it is understandable that the data proprietors could lose control of their data after "sharing" them. Interested intruders could make irretrievable impairment to users using their sensitive data. Therefore, industrial IoT thus requires wide-ranging security administration techniques to prevent access to interfaces, systems, sensors, (remote) maintenance access points, and equipment by an unauthorized group of participants from authorized processes.

Many privacy-preserving techniques have existed, but we did the literature survey in the direction of privacy mechanisms in machine learning as there is an involvement of continuous data stored on a local database and concentrated on the privacy of input datasets [11–14]. Some of the cryptographic approaches include key management, secret sharing, homomorphic encryption, and a few protocols that have used differential privacy techniques to secure statistics of the data.

As a general system, Mohassel et al. [15] proposed a machine learning model to preserve privacy, using the stochastic gradient descent method. The experimentation performs on two non-colluded servers. Gilad-Bachrach et al. [16] projected a method that used neural networks to predict coded data. Tai et al. [17] proposed a decision tree-based protocol to provide data privacy and also query privacy. Bonawitz et al. [18] anticipated machine learning–based secure data aggregation technique by exploiting secret sharing, which provides dynamism to the participant's network.

As the devices are increasing exponentially, a few studies have concentrated on privacy-preserving deep learning [9, 19–21], with a massive number of participants. This feature has increased the popularity of distributed deep learning nowadays. Deep learning of this kind distributes the training dataset among the participant, afterward based on the query or request the suitable data collects from the selected distributed participants rather than collecting data from a single participant [22]. This model could make more generalized when the deep learning model can train in such a way suitable for more representative data items. However, the distribution, data collection, and aggregation of datasets from the distributed participants increase the concerns related to security issues, which hamper the broader applications of distributed deep learning to some extent [23]. As the distributed participants are sending continuous data to cloud servers, the computational capability of the server should be high along with ample storage capacity to process the collected data. Practically, data sender may concern about their data confidentiality, even when they encrypt their data before transfer their data to the server [24]. Ref. [9] proposed privacy-preserving based on deep learning scheme using additively homomorphic encryption. It is privacy-preserving deep learning. This scheme has used gradients-encrypted asynchronous stochastic gradient descent (ASGD), in combination with learning with errors (LWE)-based encryption and Paillier encryption. This technique has proved that there may be a scope of information leak by the server. It could resolve the extending additive homomorphic encryption scheme, where all the learning participants use the encryption technique. In addition, the gradients are encrypted using the standard public key and then transferred to the server with an assumption by the participant that the server could not know anything about the key used.

The research in deep learning technologies has categorized into three types, termed as federated learning, encryption-based technologies, and differential privacy. Federated learning permits multiple participants collaboratively to learn a unified model and allows the participants to upload the local gradients uploaded and then aggregated through the synchronous aggregation by the server. Recently, Shokri and Shmatikov [25] proposed a technique, which allows multiple participants to collectively and cooperatively train their local dataset in parallel without sharing their local dataset. However, the transfer of a portion of gradients to the server discloses their contribution privacy. Consequently, to overcome these issues, Ref. [9] devised a privacy-preserving deep learning mechanism utilizing additive homomorphic encryption. Whereas assuming a standard encryption key is shared among trainers and kept secret from the server is unreasonable in the real world. Afterward, Phong and Phuong [21] proposed privacy-preserving deep learning to protect each participant's input privacy through weight transmission. Xiaoyu Zhang [26] has proposed a privacy-preserving asynchronous deep learning for industrial IoT, named Deep-PAR and DeepDPA. In both techniques, power function has been used in key derivation processes as well as in key update operation. Results increase in computation cost. However, to meet the requirements of IIoT, computation and communication costs have been taken into consideration. Hence, ECC-based privacy-preserving techniques have been proposed to provide high security with less key length comparing with other asymmetric critical cryptographic techniques [27–31]. In this chapter, we proposed two techniques to provide privacy to data distributed among participants, termed as ECC-based privacy-preserving via re-encryption (ECCRE) and dynamic asynchronous deep learning (ECCAL).

2.3 OBJECTIVES AND MATHEMATICAL BACKGROUND

2.3.1 Objectives of the Proposed Work

- To propose a novel ECC-based privacy-preserving federated learning scheme.
- To propose ECC-based dynamic group key management to derive standard key among the participants to provide privacy-preserving data sharing among the participants.
- To prove the proposed schemes are more secure and lightweight and effectively provide data security to industrial data without any data loss have determined.

2.3.2 Symbols Used and Its Description

Symbol	Description	Symbol	Description
S	Server	Enc	Encryption
O	Proxy system	msg	Message
U_i	ith participant	Dec	Decryption
SK_i	ith participant private key	ω_{global}	Weight vector
PK_i	ith participant public key	η	Learning rate
RK_i	Re-encryption key between U_i and S	g	Gradient vector
P	Group generator of E	Q	A large prime number
param	Common/global parameters	A	Passive Adversary
E	Elliptic curve over Finite field Z_{q^*}	PR_j	jth processing unit

2.3.3 Groundworks

In this section, we describe some prefaces, including ECC mathematical background and the significant modules, were used the proposed techniques.

A. ECC Mathematical Background:

Let F (Z_{q^*}) be a group, q be a prime number. Let E be an Elliptic curve equation: $y = x^3 + ax + b \mod q$, where a, b $\in_R Z_{q^*}$. Let P be a group generator of E.

Definition 2.1 (DLP)

The objective of DLP is to guess an integer $a \in_R Z_{q^*}$ in the attempt to find B = aA for a given two points A and B on E.

Definition 2.2 (ECCDHP)

For the calculated two points A = aP and B = bP on E randomly, its objective is to calculate point G = abP, where a, b $\in_R Z_{q^*}$.

B. ECC Additive Homomorphism:

Ref. [9] introduced "homomorphic encryption", intended to protect the local gradients calculated locally at participants. In general, the homomorphic encryption schemes consist of keygen, encrypt, and decrypt algorithms.

Keygen(param)→(PK, SK): This function takes param as the input, then generates a keypair (PK, SK), initially user choose a random number x, treat it as SK, i.e.,

$$SK = x$$

$$\text{Calculate } PK = SK.P = x.P$$

Enc(PK, msg)→C: For the given message (msg) and the public key(PK), it outputs ciphertext (C) = (R, S). The calculation of the C is as follows, initially choose a random number $r \in Z_{q^*}$ and P, and then calculate

$$R = r.P.msg \tag{2.1}$$

$$S = r.PK \tag{2.2}$$

Dec(SK, C) →msg: For the given C and SK, it outputs plaintext msg. The procedure to extract msg from the C is as follows:

$$\begin{aligned} msg &= SK.(S)^{-1}.R \\ &= msg.(x.r.P)^{-1}.(x.rP) \\ &= msg \end{aligned} \tag{2.3}$$

Definition 2.3 (Additive Homomorphism)

An encryption algorithm is said to be homomorphic over "+", when it satisfies Eq. 2.4:

$$Enc(m_1) + Enc(m_2) = Enc(m_1 + m_2) \cdot \forall m_1, m_2 \in \tag{2.4}$$

Let the participant A wants to send messages msg = $\{m_1, m_2, \dots m_n\}$, to B, participant A choose a random number r_i then calculates its corresponding ciphertext (R_i, S_i), $1 \le i \le n$, then send it to B. After receiving the ciphertext, B calculates the following operations to retrieve the original message:

$$C = \sum_{i=1}^{n} C_i = \sum_{i=1}^{n} enc(m_i) = \sum_{i=1}^{n} (R_i, S_i) \tag{2.5}$$

$$= (R_1, S_1) + (R_2, S_2) + \dots\dots + (R_n, S_n)$$

$$= (R_1 + R_2 + \dots\dots + R_n, S_1 + S_2 + \dots\dots + S_n)$$

$$= \left(\sum_{i=1}^{n} R_i, \sum_{i=1}^{n} S_i\right)$$

$$= \left(\sum_{i=1}^{n} r_i P, \sum_{i=1}^{n} m_i + r_i PK\right)$$

$$= Enc(m_1 + m_2 + \dots\dots + m_{n-1} + m_n)$$

$$Dec(C) = Dec(Enc(m_1 + m_2 + \dots\dots + m_{n-1} + m_n))$$

$$= m_1 + m_2 + \dots\dots m_{n-1} + m_n \tag{2.6}$$

C. Proxy Re-encryption:

This proxy re-encryption phase enables the semi-trusted proxy system to re-encrypt the ciphertext encrypted under the sender's public key to another ciphertext encrypted under the receiver's public key. The re-encryption performs without the involvement of the server being able to decrypt the ciphertext. Figure 2.2 shows its detailed block diagram of data transfer between user A to user B; in the proposed technique, this phase comprises five algorithms (keygen, rekeygen, re-encrypt, encrypt, and decrypt).

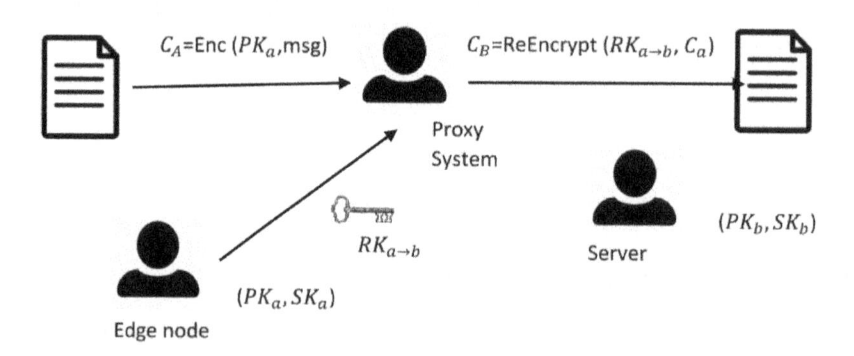

FIGURE 2.2 Block diagram of proxy system re-encryption.

i. Keygen (param) → (PK_a, SK_a): This algorithm takes param as the inputs and outputs $SK_a \leftarrow a \in Z_q^*$ and $PK_a \leftarrow aP$.
ii. Rekeygen (Sk_a, SK_b) → $RK_{a \to b}$: This algorithm takes the two secret keys SK_a and SK_b and then produces the re-encryption key $RK_{a \to b} = SK_a^{-1}.SK_b$.
iii. Enc (PK_a, msg) → C_a: on input the public key PK_a and plaintext (msg), initially choose a random number $r \in Z_q^*$, returns $C_a = (R_a, S_a) = (rP.msg, r.PK_a)$, as described in section B.
iv. Dec(Sk_a, C_a) → msg → msg: on input the secret key (SK_a) and ciphertext (C_a), returns msg using dec () function in section B.
v. Reencrypt ($RK_{a \to b}$, C_a) → C_b: on input the re-encrypt key ($RK_{a \to b}$) and ciphertext (C_a), returns $C_b = (R_b, S_b) = (R_a, RK_{a \to b}.S_a) = (rP.msg, rPK_b)$.

D. Asynchronous Deep Learning [20]:

Let N be the participants connected to the network; each participant contains a local private training dataset. In advance, all participants decide on a common learning objective. A server present in the architecture, whose responsibility is to preserve the up-to-date values of global parameters and make them available to all participants. Each participant trains their local dataset by initializing the parameters. After completing the process of training their local dataset, each participant shares their new gradient parameters to the parameter server and then downloads the updated global parameter values. It permits the participant to converge a set of parameters, critically and independently. After successful training of local datasets, all participants can work independently without interacting with any other participant in the network. The overall task of the participant is to determine the weight variables which minimize a predefined cost function on the given dataset. The cost function can be calculated over subset (t elements) or on the entire training data set. The cost function for the batch of subset elements denoted as $\mathcal{J}_{|Dataset|=t}$.

SGD: Let ω_{global} be the flattened weight vector contains weight values represented as () $\left(\omega_{global}^1,\ldots,\omega_{global}^R\right)$, where R be the number of processing units in the server. Let g be the gradient cost function of $\mathcal{J}_{|Dataset|=t}$ correspond to $\left(\omega_{global}^1,\ldots,\omega_{global}^R\right)$ is denoted as $g^1, g^2, \ldots, g^R$. The weight update rule for the η as

$$\omega_{global} = \omega_{global} - \eta.g \tag{2.7}$$

where $\eta.g = (\eta g^1, \eta g^2, \ldots, \eta g^R)$, in which the η can be altered as described in [32–34].

As long as the gradient is updating, the weight can also be updated accordingly. Therefore, the data used to compute g can be apportioned among N participants. Moreover, the update process could parallelize by considering a separate component vector termed as model parallelism. Specifically, this asynchronous deep learning (ASGD) uses multiple replicas of a NN; in each iteration, each replica could download the updated weights from the server and each replica run over the selected data fragment, which is the subset of the training dataset. To use the parallel computation power of the server, ω_{global} and g are split onto all the processing units PR_j, $1 \leq j \leq R$,

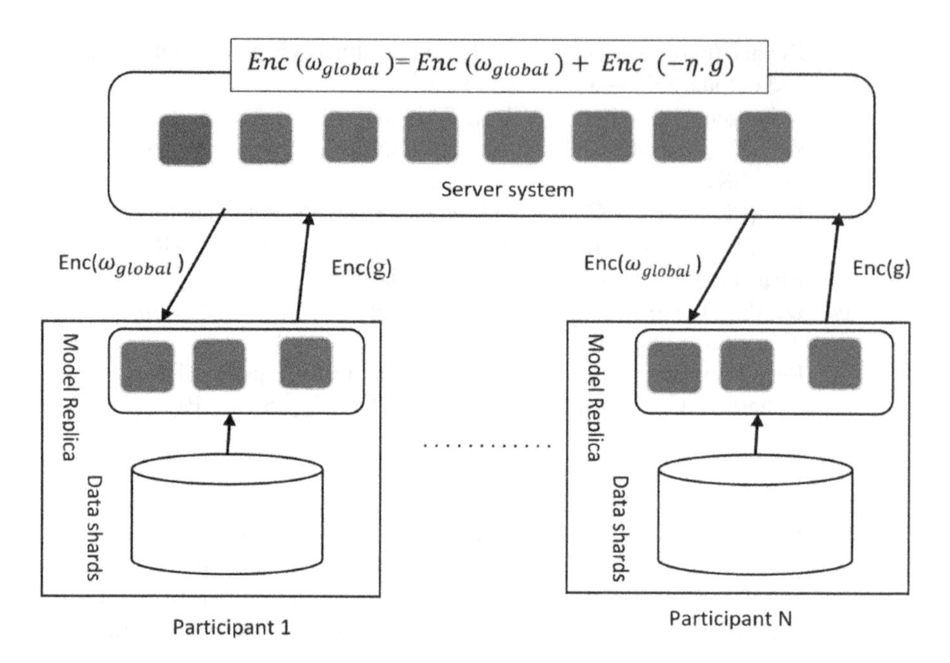

FIGURE 2.3 Block diagram of encrypted ASGD.

namely $\omega_{global} = \left(\omega^{1}_{global}, \ldots, \omega^{R}_{global}\right)$ and $g = (g^1, g^2, \ldots, g^R)$, and then the update rule of the Eq. 2.7 becomes

$$\omega^{j}_{global} = \omega^{j}_{global} - \eta.g^{j}, 1 \le j \le R \tag{2.8}$$

which can compute all the PR_j in parallel; the ASGD can significantly increase deep learning algorithm speed.

The system we proposed has used encrypted gradient ASGD, where the privacy feature added by encrypting the content before doing the update, and then Eq. 2.8 becomes

$$\mathrm{Enc}\left(\omega^{j}_{global}\right) = \mathrm{Enc}\left(\omega^{j}_{global}\right) + \mathrm{Enc}\left(-\eta.g^{j}\right) \tag{2.9}$$

The key used with the decryption process is exclusively known to the participant and is unaware of the server. Thus, the parameter server does not know about each participant's gradient and does not obtain any information from the participant's local dataset. The arrangement of the proposed asynchronous deep learning is depicted in Figure 2.3.

2.4 SECURITY ISSUES IN INDUSTRIAL IoT

The term Industrial IoT includes all the hardware devices that work together using internet connectivity intending to make the industrial processes as smart. IIoT provides intelligent technology insights that can drive significant improvements in efficiency, productivity, and revenue. The most significant Industrial IoT sectors include manufacturing, oil, gas, electricity, water, and transportation. The industrial IoT market nowadays is integrating information technology (IT) and operational technology (OT). These include an industrial control system, data acquisition, and human-computer interface. The motto of integration of all these is to achieve better automation, optimized performance, and considerable cost savings in the entire operational process.

Despite the potentials above, the IIoT is also facing a numeral quantity of challenges like interoperability, device security, privacy, and standardization.

The complicated IIoT system exposes the industrial control system (ICS), process control systems, and other operational technologies to potential cyber-attacks, hacktivism, employment sabotage, and other security risks. The main challenges that the IIoT are facing can be summarized as follows:

1. **Interoperability**: The interoperability operation intended to bridge the gap between the OT and IT could be obstructed by both technologies' challenges and the lack of common software interfaces and standard data formats;
2. **Device Reliability and Durability**: This includes the remote access control, reliability, connectivity, and reliable services provision to the devices present in the manufacturing unit and transportation.
3. **Security and Privacy Issues**: This aims to facilitate ICS security, data safety, and privacy under data protection regulations, also the security of human, industrial assets, and critical infrastructures;

Besides IIoT challenges, ensuring security in the design and manufacturing unit as the number of cybersecurity-related threats are increasing day-to-day life. In contrast, external parties are trying to exploit the weakness in the IT infrastructure and try to attack the devices or the networks to access the sensitive device data and production data. As we are in Industry 4.0 era, the devices, sensors, and sources of data are growing exponentially across the enterprise, and securing data becomes a challenging and essential task. CIOs are the project team often responsible for overseeing strategy development, implementation, maintenance, security, and data management. They are also responsible for implementing new techniques that secure the data from new technologies. Sometimes, these new and exciting technologies force the manufacturers to be more collaborative with their outdoor dealer ecosystem and are causing producers to think differently to manage, share, and secure their sensitive design and manufacturing information. It encourages the industrial environments to evolve from traditional to digital ecosystems and forces companies to share information and work collaboratively in order to quicken product design processes to deliver new products to the market more speedily and confidently ahead of the competition. Based on the detailed analysis of the functionality of industry environment, the principal materials which position a threat to the security include:

- Inclusion of a considerable number of active and passive components
- Cyber-attacks on the industrial plant and the equipment

- Moderately opened networks and systems for customers, suppliers, and partners
- Mistakes made by human error
- Management of cloud components

The effective countermeasures is to overcome the threats in the industrial IoT and to facilitate secure IoT solution communication are as follows:

1. Limiting the access permission to access authorities
2. Hardening ICT components using dedicated software
3. Using encryption processes
4. Strengthening security awareness among employees

2.5 INDUSTRIAL INTERNET OF THINGS SYSTEM ARCHITECTURE

2.5.1 Three Tier Architecture of IIoT

The best features of the smart manufacturing process than the traditional approach include depth analysis of voluminous data, accurate integration, and high correlation. Even though the industries have adapted the smartness, still the manufacturers are facing various technologies in technical data acquisition. So, this section briefs about the Industrial Data Management System framework for a real-time manufacturing process. This framework possesses features like scalable in data collection, transmission, data processing, and preservation in data storage. Figure 2.4 shows the IDMS layers, where the lower layer allows the manufacturers to acquire the raw data from industrial devices and tries to store it partially in their local storage repositories before transferring to the server. The proposed framework has used three-tier architecture that contains edge, platform, and enterprise tiers.

1. **Edge Tier**: This tier comprises all information producers, like the sensors used in manufacturing, production, and packaging wings of the industry, intended to monitor the physical status of machines. Devices, actuators, controls system, servo meters, robots, and PLC's have included. Real-time data collected from the edge nodes can be processed locally or communicated to the next tier through the access network.

   The access network layer has communication protocols such as Bluetooth, Wi-Fi Direct, Z-wave, and ZigBee, which are being used to transfer massive data with guarantee latency and high bandwidth support. These technologies should have the ability to store and process the data at a small level for a considerable amount of time duration. Along with the process capability, it should have data security, provide privacy by protecting the framework from intruders, and make sure safe and secure data transmission in every phase in the data life cycle. A few standard protocols are used to achieve this feature such as IPv6, MQTT, SOAP, REST API, and OPC-UA.

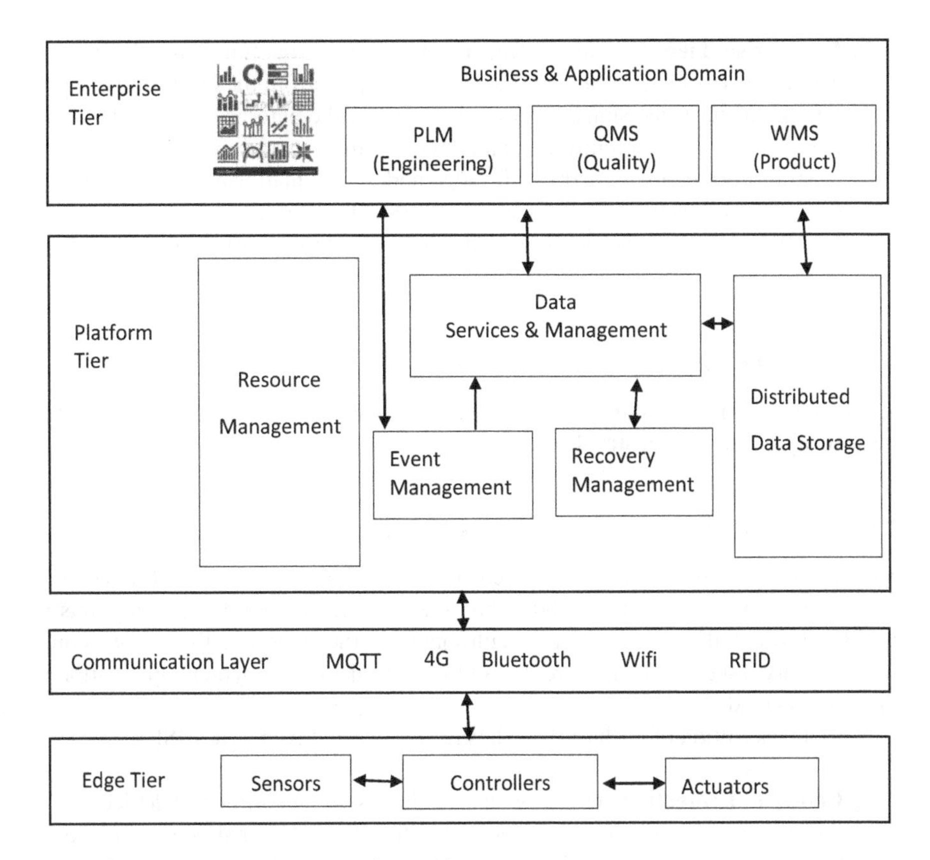

FIGURE 2.4 Industrial IoT three tier architecture.

2. **Platform Tier**: The tier consolidates and analyzes data flow coming from the lower tier. It also forwards management and control management commands from the enterprise to the edge tier. It acts as the intermediator between the lower tier and the upper tier. It is responsible for handling the appeals and replies between these layers. Data managed in this layer is by creating novel methods of acquiring, transferring, and storing at this layer. Various challenges have been resolved that contain indexing, querying transaction, and process handling. In this framework, the middle layer has various components, which convert the raw industrial data into processed data and then sends the data to be circulated to data storage for future use. The data produced by the industrial devices have initially forwarded to the data management component for preprocessing, and later it is delivered for data aggregation component to aggregate into different chunks. Finally, the partially structured and aggregated data sends to the distributed data storage module, where it uses content extraction methods to convert data into a fully structured format.

3. **Enterprise Tier**: It is an abstraction of management functionalities that can use to visualize or analyze the processed data from the platform tier in taking business decisions. Operational users in the enterprise tier can also generate control, configuration, and device management commands, which are transported downstream to the edge nodes. The platform and enterprise tiers have connected over the service network. The service network may use a VPN either over the public internet or a private network equipped with enterprise-grade security.

2.5.2 Security Issues

Apart from the global security threats discussed in Section 2.4, Table 2.1 describes the security issues in Industry 4.0 layers.

2.5.3 Security Framework in Industrial IoT

For the three-tier architecture discussed in the previous section, the IIoT security architecture has to span end-to-end across the three tiers from device endpoints at the edge, through the platform tier, and ultimately to the enterprise tier. To overcome these security issues, a standard real-world industrial IoT deployment application is highly demanded.

A brief description of each of these layers is excerpted from (IIC-IISF):

a. **Endpoint Protection**: This provides the defensive capabilities on devices of the manufacturing unit at the edge and in the cloud. The main concern is to provide physical security functions, cybersecurity [35] procedures, and an authoritative identity. This protection alone is inadequate; as the endpoints must communicate with each other, these communications may be a cause of vulnerability.

   Cryptographic protection of connectivity endpoints depends on endpoint data exchange policies used, type of mutual authentication [36,37] between

TABLE 2.1
Security Issues in IIoT Architecture

Sl. No.	Tier	Devices Present	Security Issues
1	Edge tier	BLE devices GPS and sensors	Unauthorized physical or logical access Confidentiality Data under external manipulation
2	Platform tier	**Service management Databases and service APIs**	**Manipulation of information services Maliciously track information Abuse data services Breach privacy**
3	Enterprise tier	**Smart applications Management interfaces**	Phishing attacks Malware attacks **Track the behavior of their users**

them, and the type of security algorithms used to ensure the security services like confidentiality, integrity, freshness in the data exchanged between the endpoints.

b. **Communications and Connectivity Protection**: The capability of the authoritative identity of the endpoint is used to implement traffic authentication and authorization. Checking the authenticity of the endpoint before allowing them to participate in data communication and authorize the data access permissions(read/write) granted to the endpoints involved in the data exchange. Select the appropriate cryptographic algorithm to ensure the protection of data at the endpoint to data in transmission. Along with the security in connectivity and communication, also preserve the system security throughout by monitoring and control the security configuration management of all the components present in the system.

c. **Security Model and Policy**: The functional layer administrates the policies that assure the security services, availability of the system through its lifecycle, and security implementation techniques. It orchestrates the functionality of elements that works collectively to achieve end-to-end security.

This chapter proposes a technique to provide security in the communication of data in transit by adopting the concepts of homomorphic re-encryption and dynamic group key technique.

2.6 PROPOSED SCHEME

In this section, we describe the complete functionality of ECC-based privacy-preserving deep learning via re-encryption and asynchronous deep learning. The roles identified in both mechanisms include the following:

1. **Server (*S*)**: The server is playing a vital role, responsible for collecting and aggregating the gradients received from the participants. It could equip with N processors, responsible for updating the corresponding ω_{global} parallelly and independently.
2. **Participant or Disk**: The proposed system model has N participants; each holds a partial dataset and train it in a more accurate. Upon downloading a copy of a NN model, each participant chooses a minibatch item from its local storage and estimates updated gradients. Later it encrypts the revised gradients and transfers them onto the server.
3. **Proxy System (*O*)**: It is intended to transfer the ciphertexts of gradient generated using, unlike secret keys into a standard shared key.

Our focus is mainly on training the dataset using accurate NN, on participants in a distributed network, by the usage of their local datasets. To provide more security, asynchronous deep learning is a better choice to achieve data parallelism and model parallelism. Therefore, we proposed three privacy-preserving deep learning techniques on asynchronous stochastic gradient descent.

Firstly, ECC-based deep learning via re-encryption is proposed to achieve privacy for each participant. Each participant holds different public-private key pair.

Therefore, the distributed data set encrypted by using the receiver's public key; the receiver uses his private key to extract the data from the sender. Note that each participant does the encryption locally and independently within the system. Therefore, the distributed dataset encrypted using the participant's secret key cannot be known to other participants throughout the process, observed that all the participants work autonomously in the system.

Then, we have improved our scheme by presenting the thought of group participation in secret/public key derivation, where group participants have a standard public/private key pair. Therefore, the ciphertexts of gradients calculated by the participant on local datasets could aggregate. In addition to that, we have added the dynamism property to the group, where the participants are allowed to join/leave into/from frequently. Hence, privacy through traditional techniques is considered to be unsafe when a participant withdraws from the network. To meet the needs, we are proposing ECC-based privacy-preserving deep learning.

2.6.1 ECC-Based Privacy-Preserving Deep Learning via Re-encryption (ECCRE)

In Industrial IoT data storage management, each participant or disk maintains a NN model by performing the experimentation on local datasets. The primitive participant initializes a random weight ω_{global}, sends $\left(\omega^{1}_{global},\ldots,\omega^{R}_{global}\right)$ to the server, Similarly, the gradient vector $g = (g^1, g^2, \ldots, g^R)$ can be obtained via each iteration of the training. The proposed ECCRE technique mainly consists of five phases namely, keygen(), decrypt(), compute(), re-encrypt(), and aggregate(). Figure 2.5 depicts the detailed description of each phase is as follows:

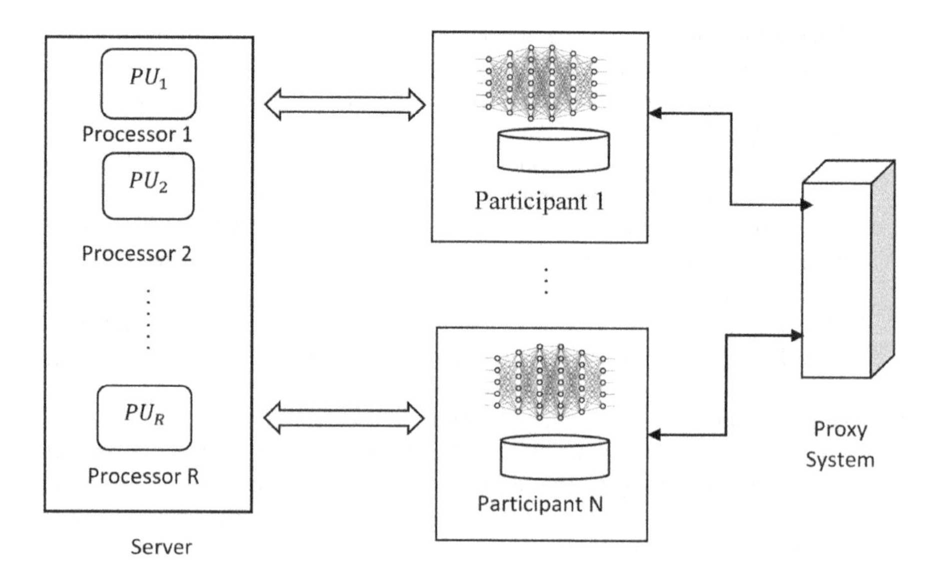

FIGURE 2.5 System model of ECCRE.

a. **Keygen (param)**: This function takes the general param as the input and output re-encryption key as well as the public and private key pairs of all the participants, including the proxy system. Initially, all the N participants decide the param = {E, P, q}, where E: Elliptic Curve over F_q, P: Group generator of $E(F_q)$, and q is the large prime number. Each participant U_i for $1 \leq i \leq N$ chooses a random number $x_i \in Z_q^*$ as its SK_i, computes its PK_i as x_iP, and returns a key pair (PK_i, SK_i). Similarly, the proxy system generates the key pair $(PK_o, SK_o) = (x_oP, x_o)$. Moreover, the re-encryption key (RK_i) used is derived between U_i and ***S*** through ***O*** as follows:

   Step 1: ***S*** chooses a random number $r_i \in Z_q^*$, and sends it to U_i.

   Step 2: U_i calculates $x_i^{-1}.r_i$ and then send it to ***O***.

   Step 3: ***O*** uses its SK_o and computes $x_i^{-1}.r_i.x_o$ and then sends it to ***S***.

   Step 4: ***S*** obtains RK_i by multiplying the received value from ***O*** with r_i^{-1}. $x_i^{-1}.r_i.x_o.r_i^{-1} = x_i^{-1}.x_o$ for ith participant.

b. **Decrypt** $\left(E_{PK_o}(\omega_i), SK_o\right)$: Each participant executes this phase U_i for $1 \leq I \leq N$ and ***O*** obtains a few parts (ω_i) of ω_{global} from ***S***, to be used in the next iteration of learning. The steps of this phase are described as follows:

   Step 1: U_i downloads I_i^{down} blocks of encrypted global weight parameters $E_{PK_o}(\omega_i) = I_i^{down} = \left\{\left(E_{PK_o}\left(\omega_{global}^1\right), \ldots, E_{PK_o}\left(\omega_{global}^{I_i^{down}}\right)\right)\right\}$ from ***S***.

   Step 2: To protect $E_{PK_o}(\omega_i)$, U_i chooses a blinding factor $s_i \in Z_q^*$, then computes $E_{PK_o}(\omega_i + s_i)$ and then sends to ***O*** in order to decrypt the data communicated.

   Step 3: Upon message from U_i, ***O*** uses SK_o to decrypt and then send $\omega_i + s_i$ back to U_i.

   Step 4: U_i obtains the plaintext of the weight parameter $\boldsymbol{\omega_i}$ by subtracting s_i from $(\omega_i + s_i)$.

c. **Compute** $\left(\omega_i, Dataset_B, E_{PK_i}\right)$: This phase uses its weight parameter ω_i, prefetches a minibatch $(Dataset_B)$, conducts the deep learning for one round, and then obtains a new gradient g. Then, U_i encrypts the I_i^{up} blocks of gradients and then forwards them onto the matching processors on the ***S***, denoted as $E_{PK_i}\left(-\eta.g^i\right) = I_i^{up} = \left\{E_{PK_i}\left(-\eta.g^1\right), E_{PK_i}\left(-\eta.g^2\right), \ldots, E_{PK_i}\left(-\eta.g^{I_i^{up}}\right)\right\}$.

d. **Re-encrypt** $\left(E_{PK_i}\left(-\eta.g^i\right), RK_i\right)$: This function is intended to re-encrypt the encrypted gradients using U_i's (PK_i) and using RK_i. ***S*** calculates the ciphertext $E_{PK_o}\left(-\eta.g^i\right)$ using ***O***'s PK_o.

e. **Aggregate** $\left(E_{PK_o}\left(-\eta.g^j\right)\right)$: ***S*** aggregates the global weight parameters and then updates the corresponding blocks $E_{PK_o}\left(\omega_{global}^j\right) = E_{PK_o}\left(\omega_{global}^j + E_{PK_o}\left(-\eta.g^j\right)\right), \left(\forall\ j \in I_i^{up} \subset [1, R]\right)$ simultaneously.

2.6.2 ECC-Based Privacy-Preserving Deep Learning (ECCAL)

This technique uses a common PK and SK, which are derived by all the participants. In Industrial IoT, each operational unit locally holds its datasets and runs the federated learning algorithm. Initially, the participant initializes a random weight ω_{global} and sends $E_{PK}\left(\omega^{1}_{global},\ldots,\omega^{R}_{global}\right)$ to the server, where ω^{i}_{global} is a vector that contains ith part of ω_{global}. The gradient vector g is obtained after splitting it into R parts, namely $g = (g^1, g^2, \ldots, g^R)$, multiply by the η, and then encrypted using the PK, results in the encryption value $E_{PK}(-\eta g^i)(\forall 1 \leq I \leq R)$ from each participant sent to the *S*; the detailed communication arrangement depicted in Figure 2.6.

This technique mainly consists of the keygen(), decrypt(), compute(), aggregate(), and update() phases. The concrete scheme description is as follows:

a. **Keygen(param) → (PK, SK)**: Initially, all the N participants decide the param = {E, P, q}. The selected N participants in the group jointly set up PK and SK used in encryption and decryption operation to perform an additively homomorphic operation. The SK derived in this phase is kept secret against *S*, only known to N participants. Each participant establishes a TLS/SSL channel, independently, to communicate homomorphic ciphertexts.

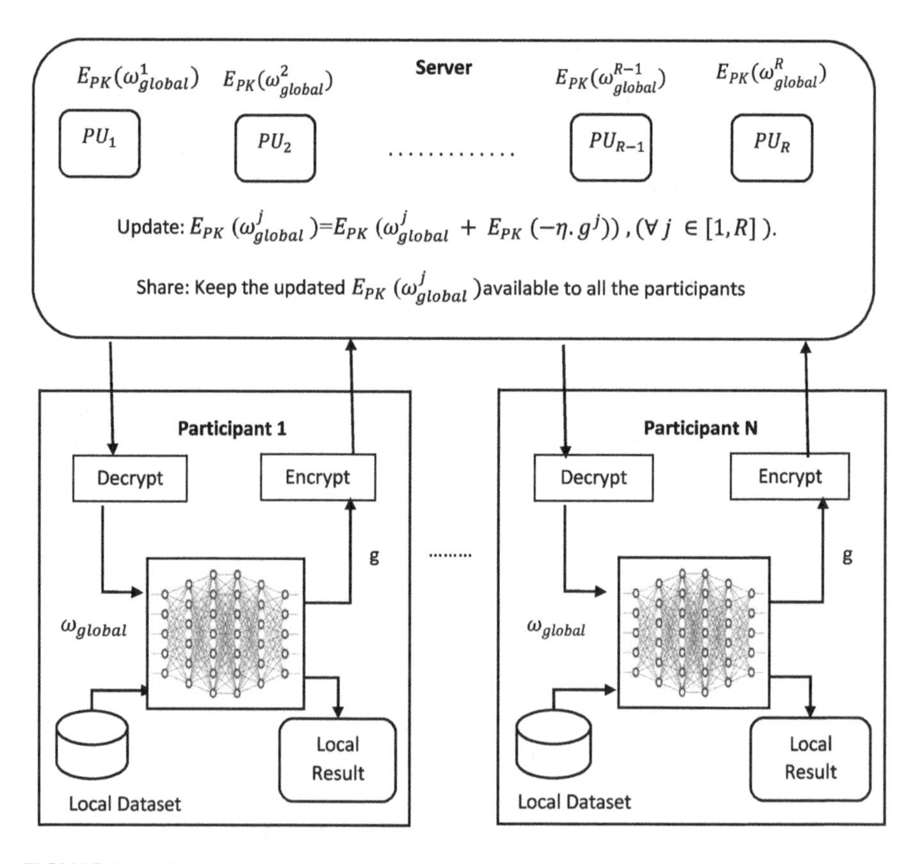

FIGURE 2.6 System model of ECCAL.

Let $U_1, \ldots, U_N$ be the participants contributing to the protocol are arranged in a ring and U_{i-1}, U_{i+1} be the left and right neighbors of U_i for $1 \leq i \leq N - 1$, for participant U_1 the left side neighbor is U_N and the right side neighbor of U_N is U_1.

Procedure keygen (U_i, param)
Input: param, participants U_i for $1 \leq i \leq N - 1$
Output: SK: private key, PK: public key
(Round 1):
for i = 1 to N do in parallel
U_i choose a random number $x_i \in Z_q^*$
compute $X_i = x_i P$ and sends it to his neighbor participants U_{i-1} and U_{i+1}
end for
note that $X_0 = X_N$ and $X_{N+1} = X_1$
(Round 2):
for i = 1 to N do in parallel
U_i computes the left key $K_i^L = x_i X_{i-1} = x_i X_{i-1} P$,
the right key $K_i^R = x_i X_{i+1} = x_i X_{i+1} P$ and $Y_i = K_i^R - K_i^L$
U_i sends Y_i to other group participants.
end for
note that $K_i^R = K_{i+1}^L$ for $1 \leq i \leq (N - 1)$, $K_N^R = K_1^L$ and $K_{i+(N-1)}^R = K_i^L$.
(Key agreement):
for i = 1 to N do in parallel
U_i computes $\tilde{K}_{i+1}^R = Y_{i+1} + K_i^R$
for j = 2 to N − 1 do
U_i computes $\tilde{K}_{i+j}^R = Y_{i+j} + \tilde{K}_{i+(j-1)}^R$
end for
U_i checks if $\tilde{K}_{i+(N-1)}^R = K_{i+(N-1)}^R$
if the condition fails then abort the key agreement protocol
else
U_i computes the session key as $SK = \tilde{K}_1^R + \tilde{K}_2^R + \cdots + \tilde{K}_N^R$
PK = SK*P
end if
end for
end keygen()

The keygen() phase comprises two rounds for intermediate value communication and a pivotal computation phase. During round 1, all the participants choose a random number $x_i \in Z_q^*$ in parallel and compute X_i and send it to their neighbors U_{i-1} and U_{i+1}. In round-2, the user U_i receives the X_{i-1} and X_{i+1} from U_{i-1} and U_{i+1}. Then U_i computes his K_i^L by multiplying X_{i-1} with his contribution x_i, results $x_i x_{i-1} P$, the right key is computed by multiplying X_{i+1} with his contribution x_i, results $K_i^R = x_i x_{i+1} P$, $Y_i = K_i^R - K_i^L$ and sends Y_i to all other participants. Afterward, U_i computes $\tilde{K}_{i+1}^R, \tilde{K}_{i+2}^R, \ldots, \tilde{K}_{i+N-1}^R$ using his K_i^R.

$$\tilde{K}_{i+1}^{R} = Y_{i+1} + K_i^R,\ \tilde{K}_{i+2}^{R} = Y_{i+2} + \tilde{K}_{i+1}^{R},\ldots,\tilde{K}_{i+(N-1)}^{R} = Y_{i+(N-1)} + \tilde{K}_{i+(N-2)}^{R} \quad (2.10)$$

Then U_i verifies if $\tilde{K}_{i+(N-1)}^{R}$ is same as that of the K_i^L (i.e. $K_{i+(N-1)}^{R}$), and the U_i aborts the protocol when the condition verification fails; otherwise, U_i computes $\tilde{K}_1^R + \tilde{K}_2^R + \cdots + \tilde{K}_N^R = x_1x_2P + x_2x_3P + \ldots + x_Nx_1P$ which is equal to $(x_1x_2 + x_2x_3 + \ldots + x_Nx_1)$ P, the x coordinate of the resultant, is considered as the shared private key SK. All the participants compute the public key PK as SK*P.

b. **Decrypt**($E_{PK}(\omega_{global})$, SK) $\rightarrow \omega_i$: Each U_i, $1 \leq i \leq N$ performs these steps:
 Step 1: U_i downloads I_i^{down} blocks of the encrypted global parameters $E_{PK}(\omega_{global}^1, \ldots, \omega_{global}^R)$ stored in the processing units PR_j, $1 \leq \forall j \leq R$ present in the S, represented as $E_{PK}(\omega_i) = I_i^{down} = \{E_{PK}(\omega_{global}^1), \ldots, E_{PK}(\omega_{global}^R)\}$
 Step 2: U_i uses the SK to decrypt the ciphertexts to obtain ω_i and substitutes the values in the local replica.
c. **Compute**($Dataset_B$, ω_i, SK) $\rightarrow E_{PK}(-\eta.g^i)$: This operation selects a minibatch of datasets from its local store, and U_i initiates a new iteration to obtain the gradients g. Then, U_i encrypts the I_i^{up} blocks of gradients and then sends onto the corresponding processors of the S, denoted as $E_{PK_i}(-\eta.g^i)$.
d. **Aggregate**$(E_{PK}(-\eta.g^i) \rightarrow E_{PK}(\omega_{global}))$: Upon receiving encrypted gradients from all the participants, $\mathcal{S}$ aggregates them on the processors in it simultaneously $E_{PK}(\omega_{global}) = E_{PK}(\omega_{global}^j + E_{PK}(-\eta.g^j))$, $(\forall j \in I_i^{up} \subset [1,R])$. For a full upload, $I_i^{up} = [1, R]$, all encrypted gradients have uploaded to $\mathcal{S}$.
e. **Parupdate**(PK, SK) $\rightarrow$ (PK′, SK′): This phase adds dynamism to the participants in the group and supports two operations to join or leave. Join operation is invoked on entering of new participants into the group, and leave operation allows the participants to exit from the group. Both the operations (join or leave) affect the participant's two adjacent participants. Upon join or leave of the participant, the key pair (PK, SK) used currently is updated to a new key pair (PK′, SK′).
i. **Join**(): Let U_1, U_2,..U_N be the existing users, $U_{N+1},\ldots, U_{N+M}$, be a set of new users want to join in the group, then altogether invokes the keygen() algorithm among the users U_1, U_2,..U_N, $U_{N+1},\ldots, U_{N+M}$. We consider a ring of L = M + 3 users $V_1 = U_1$, $V_2 = U_2$, $V_3 = U_N$, $V_i = U_{N+i-3}$, for $4 \leq i \leq L$, each V_i chooses random numbers $y_1 = x_1$, $y_2 = x$(a common seed), $y_3 = x_n$ and $y_i = x_{n+i-3}$. The left and right neighbors of the participant V_i be V_{i-1} and V_{i+1} respectively for $1 \leq i \leq L$ with $V_0 = V_L$ and $V_{L+1} = V_1$. The complete procedure for join operation on N+M users described as:

Procedure join(U_i, param)
Input: param: global parameters and participants U_i for $1 \leq i \leq N + M$
Output: (PK′, SK′): new publickey, new privatekey

Set $L = M + 3$, $V_1 = U_1$, $V_2 = U_2$, $V_3 = U_N$, $V_i = U_{N+i-3}$, for $4 \le i \le L$,
We consider a ring with L users with indexes V[1,..., L]
Call keygen(V[1,..., L], param)
Let for $1 \le i \le L\ \tilde{X}_i = y_i.P, \tilde{X}_0 = \tilde{X}_L, \tilde{X}_{L+1} = \tilde{X}_1$
$\tilde{K}_i^L = y_i\tilde{X}_{i-1} = y_iy_{i-1}P$ and $\tilde{K}_i^R = y_i\tilde{X}_{i+1} = y_iy_{i+1}P$ and $\tilde{Y}_i = \tilde{K}_i^R - \tilde{K}_i^L$
In round 1, V_1 and V_3 additionally send $\tilde{X}_1$ and $\tilde{X}_3$ to all other participants in U[3...N − 1]
In round 2, V_i sends $\tilde{Y}_i$ to all users in U[3...N − 1]
(Key agreement):
for i = 3 to N − 1 do in parallel
U_i computes $\tilde{K}_3^R = \tilde{Y}_3 + K_2^R$
for j = 2 to L − 1 do
U_i computes $\tilde{K}_{2+j}^R = \tilde{Y}_{2+j} + \tilde{K}_{2+(j-1)}^R$
end for
U_i computes the session key as $SK' = \tilde{K}_1^R + \tilde{K}_2^R + \cdots + \tilde{K}_L^R$
$PK' = SK'*P$
end for
end join()

ii. **Leave()**: This procedure can use to generate a new key pair in such a way that participants in the group derive the key by dropping the leaving participants from the rings. In the procedure, both sides of the participant who is willing to leave the group should choose a new secret key and then invoke the keygen() algorithm on new participants in the group. Let U_i for $1 \le i \le N$ be the participants called the keygen() among the participant, let K_i^L and K_i^R be the keys derived with left and right participants by U_i for $1 \le i \le N-1$. Suppose the users $U_{l1}, \ldots, U_{lM}$ want to quit from the participant group U_i for $1 \le i \le N$, then the new participants set is U_i for $1 \le i \le l1 - L \cup U_i$ for $l1 + R \le i \le l2 - L \cup \ldots U_i$ for $lM + R \le i \le N$, where $l1 - L$ and $l1 + R$ are the left and right neighbors of the leaving user U_{li}, for $1 \le li \le M$. Then for any leaving user U_l, $l - L = l - i$, if the users leaving are in consecutive indexes $U_l, U_{l-1}, \ldots, U_{l-(i-1)}$. Afterward, all the N − M participants are renamed with new indexes and denoted as the new participants set by V[1,..., N − M]. Accordingly, both side neighbor key is denoted by $\tilde{K}^L[1, 2, \ldots, N-M]$ and $\tilde{K}^R[1, 2, \ldots, N-M]$ respectively of the participants V_i, $1 \le i \le N-M$. The detailed leave procedure is as follows:

Procedure leave(U_i, param, $\{U_{l1}, \ldots, U_{lM}\}$)
Input: param: global parameters
participants U_i for $1 \le i \le N$ and
$\{U_{l1}, \ldots, U_{lM}\}$: participants intended to leave the group
Output: SK′: new private key, PK′: new public key

(Round 1)

Let K_i^R, K_i^L be the right and left keys of the U_i, $1 \leq i \leq N$, computed and stored in the previous session.

for i = 1 to M do in parallel

let $j_i = l_i - L$ and $j_2 = l_i - R$

participants U_{j1} and U_{j2} randomly choose new secret key values x_{j1}, $x_{j2} \in Z_q^*$, and compute $X_{j1} = x_{j1}P$, $X_{j2} = x_{j2}P$

U_{j1} sends X_{j1} to his neighbor participants U_{j1-1} and U_{j2}, similarly,

U_{j2} sends X_{j2} to his neighbor participants U_{j1} and $U_{j2+1}(U_{N+1} = U_1)$

end for

(Round 2)

for i = 1 to M do in parallel

let $j_1 = l_i - L$ and $j_2 = l_i - R$

set $W = \{j_1 - 1, j_1, j_2, j_2 + 1\}$

U_{j1} modifies his left key as $K_{j1}^L = x_{j1}X_{j1-1} = x_{j1}x_{j1-1}P$ and right key as $K_{j1}^R = x_{j1}X_{j2} = x_{j1}x_{j2}P$

U_{j2} modifies his left key as $K_{j2}^L = x_{j2}X_{j1} = x_{j2}x_{j1}P$ and right key as $K_{j2}^R = x_{j2}X_{j2+1} = x_{j2}x_{j2+1}P$

U_{j1-1} modifies his right key as $K_{j1-1}^R = x_{j1-1}X_{j1} = x_{j1-1}x_{j1}P$

U_{j2+1} modifies his left key as $K_{j2+1}^L = x_{j2+1}X_{j2} = x_{j2+1}x_{j2}P$

end for

Now reindex N – M users U_i for $1 \leq j \leq N \backslash U_{li}$, for $1 \leq i \leq M$, let U_j for $1 \leq j \leq N - M$ be the new participants set and $\tilde{K}^L$ [1, 2, N – M] and $\tilde{K}^R$ [1, 2,…, N – M] be the set of corresponding left and right keys.

for i = 1 to N – M do in parallel

V_i computes $Y_i = \tilde{K}_i^R - \tilde{K}_i^L$ and sends it to rest of the participants in V_j for $1 \leq j \leq N - M$

end for

Note that $\tilde{K}_i^R = \tilde{K}_{i+1}^L$ for $1 \leq i \leq (N - M - 1)$, $\tilde{K}_N^R = \tilde{K}_1^L$ and $\tilde{K}_{i+(N-M-1)}^R = \tilde{K}_i^L$

(Key agreement)

for i = 1 to N – M do in parallel

V_i computes $\hat{K}_{i+1}^R = Y_{i+1} + \hat{K}_i^R$

for j = 2 to N – M – 1 do

V_i computes $\hat{K}_{i+j}^R = Y_{i+j} + \hat{K}_{i+j-1}^R$

end for

V_i checks if $\tilde{K}_{i+(N-M-1)}^R = \hat{K}_{i+N-M-1}^R$ (i.e. $\tilde{K}_i^L = \hat{K}_{i+N-M-1}^R$)

if the condition fails then abort the key agreement protocol

else

V_i computes the session key as $SK' = \hat{K}_1^R + \hat{K}_2^R + \cdots + \hat{K}_{N-M}^R$

$PK' = SK'{*}P$

end if

end for

end Leave()

2.7 SECURITY ANALYSIS

This section describes the security efficiency of the proposed technique.

Computational Diffie–Hellman (CDH) problem: Let G be an additive group of some large prime order q. Then CDH is defined as follows:

Instance: (P; aP; bP) for some a, b $\in Z_q^*$

Output: abP

The probability of getting success using the polynomial-time algorithm in solving CDH problem in G is defined to be

$$\text{Succ}_{G,\mathcal{A}}^{CDH} = \text{prob}\left[\mathcal{A}(P, aP, bP) = abP;\ a, b \in Z_q^*\right] \tag{2.11}$$

CDH assumption: For every probabilistic, polynomial-time algorithm $\mathcal{A}$ is negligible.

2.7.1 Security Analysis of ECC Based Re-encryption

Theorem 2.1

Each participant guarantees input and model privacy when the homomorphic encryption is safe since all the participants $\mathcal{O}$ and $\mathcal{S}$ are non-collusion.

Proof: Let us assume that the algorithm is running in the semi-honest model, the process used in the generation of RK_i never discloses SK_i of U_i to neither $\mathcal{O}$ nor $\mathcal{S}$. Furthermore, each participant then calculates the encrypted gradients using its PK_i, so the outsiders cannot obtain the real values of the local gradients; this achieves the privacy of the participants. Along with that, PK_0 is used to encrypt weight parameters and store them in $\mathcal{S}$. Considering that both $\mathcal{S}$ and $\mathcal{O}$ do not collide with one another, the weight parameters do not know to $\mathcal{S}$ in both re-encrypt() and aggregate() phases. During the decrypt() phase, to obtain some parts of ω_i, U_i randomly selecting r_i and adds it to the ciphertext of $E_{PK_0}(\omega_i)$. Therefore, one-time-pad and homomorphic encryption can promise that the model parameters can't be revealed to $\mathcal{O}$. Hence, model privacy is guaranteed.

2.7.2 Security Analysis of ECC Base Encryption

Theorem 2.2

The UP keygen() presented in (a) of section VI could secure against passive challenger underneath ECC-DDH assumption, accomplish forward secrecy (FS), and achieve the following:

$\text{Adv}_{UP}^{BCGKA}(t,\ q_e) \leq \text{Adv}_{G}^{ECDDH}(t') + q_e/|G|$, where $t' = t + O(|P|.q_{e.}t_{sm})$,

where t_{sm} is the time required to compute scalar multiplication (sm) over $G = E(F_p)$, and q_{ex} is the amount of ExecuteQuery(XQ) that a challenger may ask.

Proof: Given $\mathcal{A}$, construct an algorithm $\mathcal{D}$ that undertakes EC-DDH with a negligible advantage.

$\mathcal{A}$ has access to the XQ and Reveal (RQ) Queries. Since the long-secret key term does not utilize in the UP, corrupt query (CQ) may just omit, and the scheme achieves

FS naturally. Let us regard $\mathcal{A}$ pretense a single EQ. Let us draw Real and Fake distributions for (transcript, session key) pairs (T, SK) as under:

$$\text{Real} := \left\{ \begin{array}{c} x_1, x_2, \ldots\ldots x_N \leftarrow Z_q^*; \\ X_1 = x_1P, X_2 = x_2P, \ldots\ldots, X_N = x_NP, \\ K_1^R = K_2^L = x_1x_2P, K_2^R = K_3^L = x_2x_3P, \ldots\ldots, \\ K_N^R = K_1^L = x_1x_NP : \{T,\ SK\} \\ Y_1 = K_1^R - K_1^L, Y_2 = K_2^R - K_2^L, \ldots\ldots\ldots, Y_N = K_N^R - K_N^L, \\ T = \left(X_1, X_2, \ldots., X_N; Y_1, Y_2, \ldots, Y_N\right); \\ SK = \sum_{i=1}^{N} K_i^R \end{array} \right\}$$

$$\text{fake}' = \left\{ \begin{array}{c} x_1, x_2, \ldots\ldots x_N \leftarrow Z_q^*; \\ X_1 = x_1P, X_2 = x_2P, \ldots\ldots, X_N = x_NP, \\ K_1^R = K_2^L = x_1x_2P, K_2^R = K_3^L = x_2x_3P, \ldots\ldots, \\ K_{N-1}^R = K_N^L = x_2x_3P, K_N^R = K_1^L = x_1x_NP : \{T,\ SK\} \\ Y_1 = K_1^R - K_1^L, Y_2 = K_2^R - K_2^L, \ldots\ldots\ldots, Y_N = K_N^R - K_N^L, \\ T = \left(X_1, X_2, \ldots., X_N; Y_1, Y_2, \ldots, Y_N\right); \\ SK = \sum_{i=1}^{N} K_i^R \end{array} \right\}$$

First, we consider the following lemmas, which deal with EQ and RQ.

Lemma 2.1

For every challenger running in time t, we ascertain as under:

$$\left|\text{Prob}\left[(T, SK) \leftarrow \text{Real} : \mathcal{A}(T, SK) = 1\right] - \text{Prob}\left[(T, SK) \leftarrow \text{fake}' : \mathcal{A}(T, SK) = 1\right]\right|$$
$$\le \text{Adv}_G^{ECC-DDH}(t'') + 1/|G|, \text{ where } t'' \le t + 2.t_{sm}$$

Proof. Let (A, B, C) $\in G^3$ be any incidence dealing with ECC-DDH. Using $\mathcal{A}$, we construct a $\mathcal{D}$ for dealing with ECC-DDH that captures (A, B, C) as input and anything $\mathcal{A}$ outcomes. Let dist′ be drafted as below utilizing (A, B, C):

$$\text{dist}' := \left\{ \begin{array}{c} x_1, x_2, \ldots\ldots x_N \leftarrow Z_q^*; \\ X_1 = x_1 A, X_2 = x_2 P, \ldots\ldots, X_N = x_N B, \\ K_1^R = K_2^L = x_1 x_2 A, K_2^R = K_3^L = x_2 x_3 P, \ldots\ldots, \\ K_{N-1}^R = K_N^L = x_2 x_3 B, K_N^R = K_1^L = x_1 x_N C : \{T,\ SK\} \\ Y_1 = K_1^R - K_1^L, Y_2 = K_2^R - K_2^L, \ldots\ldots\ldots, Y_N = K_N^R - K_N^L, \\ T = \left(X_1, X_2, \ldots, X_N; Y_1, Y_2, \ldots, Y_N\right); \\ SK = \sum_{i=1}^{N} K_i^R \end{array} \right\}$$

Now we examine the output of $\mathcal{D}$. The real distribution and the distribution:

$\{a, b \leftarrow Z_q^*, A = aP, B = bP, C = abP; (T, SK) \leftarrow \text{Dist: } (T, SK)\}$ are equal statistically every time the multiples x_1, x_2 utilized are arbitrary in dist′. On the different half, the distribution: $\{a, b \leftarrow Z_q^*, c \leftarrow Z_q^* - \{a, b\}, A = aP, B = bP, C = cP; (T, SK) \leftarrow \text{dist}': (T, SK)\}$ and the distribution fake′ are statistically the same within a factor of $1/|G|$. The distinctive variety is in fake′ the SK value is picked at random from G, but in dist′ this value is picked arbitrarily from $G - \{abP\}$, or the random self-reducibility property of EC-DDH, these two distributions are equal statistically. Consequently, we set up as below:

$$\left|\text{Prob}\left[(T, SK) \leftarrow \text{Real} : \mathcal{A}(T, SK) = 1\right] - \text{Prob}\left[(T, SK) \leftarrow \text{fake}' : \mathcal{A}(T, SK) = 1\right]\right|$$

$$\begin{aligned} &\leq \left|\text{Prob}\left[a, b \leftarrow Z_q^* : \mathcal{D}(aP, bP, abP) = 1\right]\right. \\ &\quad \left.-\text{Prob}\left[a, b \leftarrow Z_q^*, c \leftarrow Z_q^* - \{a, b\} : \mathcal{D}(aP, bP, cP) = 1\right]\right| \\ &\quad + 1/|G| \leq \text{Adv}_G^{\text{ECDDH}}(t'') + 1/|G| \end{aligned}$$

Here, t goes beyond by the time t'' of $\mathcal{A}$, i.e., $t'' = t + 2t_{sm}$, where t_{sm} is the time needed to carry out sm in G.

Subsequently, we provide the Proof of final distribution Fake as follows:

$$\text{Fake} = \left\{ \begin{array}{c} x_1, x_2, \ldots\ldots x_N \leftarrow Z_q^*; \\ X_1 = x_1 P, X_2 = x_2 P, \ldots\ldots, X_N = x_N P, \\ K_1^R = K_2^L, K_2^R = K_3^L, \ldots\ldots, K_{N-1}^R = K_N^L, K_N^R = K_1^L \leftarrow G : \{T,\ SK\} \\ Y_1 = K_1^R - K_1^L, Y_2 = K_2^R - K_2^L, \ldots\ldots\ldots, Y_N = K_N^R - K_N^L, \\ T = \left(X_1, X_2, \ldots, X_N; Y_1, Y_2, \ldots, Y_N\right); \\ SK = \sum_{i=1}^{N} K_i^R \end{array} \right\}$$

Lemma 2.2

For every $\mathcal{A}$ running in time t, we ascertain as under:

$$\left|\text{Prob}\left[(\text{T, SK}) \leftarrow \text{fake}' : \mathcal{A}(\text{T, SK}) = 1\right] - \text{Prob}\left[(\text{T, SK}) \leftarrow \text{Fake: } \mathcal{A}(\text{T, SK}) = 1\right]\right|$$

$$\leq \text{Adv}_{\text{G}}^{\text{ECC-DDH}}(\text{t}'') + 1/|\text{G}|, \text{ where } \text{t}'' \leq \text{t} + 2.\text{t}_{\text{sm}}$$

Proof: Let (A, B, C) $\in G^3$ be any incidence for dealing with ECC-DDH. Utilizing $\mathcal{A}$, we construct a distinguisher $\mathcal{D}$ for dealing with ECC-DDH that captures (A, B, C) as input, pair (T, SK) generated from Dist, $\mathcal{A}$ runs on (T, SK) generates anything $\mathcal{A}$ outcomes.

$$\text{Dist} = \left\{\begin{array}{c}
x_1, x_2, \ldots\ldots x_N \leftarrow Z_q^*; \\
\text{if } N \,\%\, 2 == 0 \text{ then} \\
\text{for } i = 1 \ldots N \text{ do} \\
X_i = x_i A,\ X_{i+1} = x_{i+1} B \\
\text{end for} \\
\text{else} \\
\text{for } i = 1 \ldots N-2 \text{ do} \\
X_i = x_i A,\ X_{i+1} = x_{i+1} B \\
\text{end for} \\
X_N = x_N A \\
\text{end if} \\
\text{for } i = 1 .. N-2 \text{ do } : (\text{T, SK}) \\
K_i^R = K_{i+1}^L = x_i x_{i+1} C, \\
K_{i+1}^R = K_{i+2}^L = x_{i+1} x_{i+2} C \\
\text{end for} \\
K_{N-1}^R = K_N^L = x_{N-1} x_1 C, \\
K_N^R = K_1^L \leftarrow G \\
Y_1 = K_1^R - K_1^L, Y_2 = K_2^R - K_2^L, \ldots\ldots\ldots, Y_N = K_N^R - K_N^L, \\
T = (X_1, X_2, \ldots., X_N; Y_1, Y_2, \ldots, Y_N); \\
SK = \sum_{i=1}^{N} K_i^R
\end{array}\right\}$$

Let dist′ be sketched as below utilizing (A, B, C): Now we examine the output of $\mathcal{D}$. The fake′ distribution and the distribution: {a, b $\leftarrow Z_q^*$, A = aP, B = bP, C = abP; (T, SK) ← Dist: (T, SK)} are equal statistically every time the multiples x_1, x_2 utilized are arbitrary in Dist. On the different half, the distribution: {a, b $\leftarrow Z_q^*$, c $\leftarrow Z_q^* - \{a, b\}$, A = aP, B = bP, C = cP; (T, SK) ← Dist: (T, SK)} and the distribution Fake are statistically the same within a factor of 1/|G|. The distinctive variety is in Fake, and the SK value is picked at random from G. However, in dist′, this value is picked arbitrarily from G − {abP}, or else, then by the random self-reducibility property of EC-DDH, these two distributions are equal statistically. Consequently, we set up as below:

$$\left|\mathrm{Prob}\left[(\mathrm{T},\mathrm{SK}) \leftarrow \mathrm{fake}' : \mathcal{A}(\mathrm{T},\mathrm{SK}) = 1\right] - \mathrm{Prob}\left[(\mathrm{T},\mathrm{SK}) \leftarrow \mathrm{Fake}: \mathcal{A}(\mathrm{T},\mathrm{SK}) = 1\right]\right|$$

$$\leq \left|\mathrm{Prob}\left[\mathrm{a},\mathrm{b} \leftarrow Z_q^*: \mathcal{D}(\mathrm{aP},\mathrm{bP},\mathrm{abP}) = 1\right]\right.$$

$$\left. - \mathrm{Prob}\left[\mathrm{a},\mathrm{b} \leftarrow Z_q^*, \mathrm{c} \leftarrow Z_q^* - \{\mathrm{a},\mathrm{b}\}: \mathcal{D}(\mathrm{aP},\mathrm{bP},\mathrm{cP}) = 1\right]\right|$$

$$+ 1/|G| \leq \mathrm{Adv}_G^{\mathrm{ECDDH}}(t'') + 1/|G|$$

Here, t goes beyond by the time t″ of $\mathcal{A}$ i.e., $t'' = t + 2t_{sm}$, where t_{sm} is the time required to carry out sm in G.

Security of the parupdate() Protocol: In Dynamic Agreement Protocol (DAP), it assumes DSig is secure, we can convert any adversary attacking the protocol into an unauthorized protocol.

Theorem 2.3: The dynamic parupdate() protocol described section (e) of chapter VI satisfies the following:

$\mathrm{Adv}_{\mathrm{DAP}}^{\mathrm{BDACGKA}}(t, q_e, q_j, q_l) \leq \mathrm{Adv}_{\mathrm{UP}}^{\mathrm{BCGKA}}(t', (q_l + q_j) + q_e/2) + |P|\,\mathrm{Adv}_{\mathrm{DSig}}(t')$ where $t' = t + t_{DAP}.O(q_j + q_l + |P|\, q_e)$ is the time needed for implementing DAP by any of the participants.

Proof: Let $\mathcal{A}'$ be an adversary targets to attack authorized protocols (AP); using this, we need to construct $\mathcal{A}$, who attacks the parupdate() method, which is an unauthorized protocol.

Lemma: Suppose an event Forge occurs. Then $\mathcal{A}'$ makes a send queries and constructs a Forge algorithm ($\mathcal{F}$), which tries to forges the digital signature as follows: For the given PK, $\mathcal{F}$ chooses U ∈ P and set PK_U = PK. The other critical pairs for the system are created genuinely by $\mathcal{F}$. The $\mathcal{F}$ tries to simulate the oracle models of $\mathcal{A}'$ by implementing the protocol itself and tries to obtain required signatures concerning PK_U, as needed from its signing oracle. Thus $\mathcal{F}$ facilitates the faultless simulation to $\mathcal{A}'$, for any valid signature w.r.t PK_U = PK output by $\mathcal{A}'$, then $\mathcal{F}$ outputs keypair as its forgery. The probability of succeeding by $\mathcal{F}$ is Prob [Forge] $\leq |P|\,\mathrm{Adv}_{\mathrm{DSig}}(t')$.

Now we can describe the construction of the $\mathcal{A}$, who attacks UP that uses the $\mathcal{A}'$, who attacks AP. $\mathcal{A}$ uses a list Tlist and contains the transcripts in Tlist. C simulates the oracle queries of $\mathcal{A}'$ using his own queries to the Execute oracle. The motto is to obtain the transcript T of UP for each executes a query of $\mathcal{A}'$.

Execute the query: $\mathcal{A}$ defines the U and then sends the execute the query to it, receives T as the output on the execution of UP. It appends (U, T) to Tlist.

Send a query: $\mathcal{A}$ sets the users list U and makes an execute query to its own execute oracle. It receives the T in return and stores (U, T) in the Tlist. Apart from the general queries, there exist two particular queries, Send_{lv} and Send_{jn}

If the set $U_l = \{U_{N+1},...,U_{N+M}\}$ are the participants willing to join in the group $U = \{U_1,...,U_N\}$, then $\mathcal{A}'$ could send queries which initiate Join (U, U_l), and participants in U might have completed either keygen() or join() or leave() algorithm previously. $\mathcal{A}$ initially finds any unique entry in the form (1). (U, T) in Tlist, or (2). (U', U'', T) in Jlist with $U = U' \cup U''$, or (3). (U', U'', T) in List with $U = U' \setminus U''$, it checks for the availability of the entries in the list by making an XQ to its random oracle on U, obtains the T, and stores (U, T) in Tlist. If (U, T) $\in$ Tlist, $\mathcal{A}$ generates a reveal query to any instance in U to obtain secret key SK corresponding to T, computes the public key PK = SK.P, and simulates the algorithm for join by querying execute and RQ. Then $\mathcal{A}$ obtains T′ and stores (U, U_l, T′) in Jlist. $\mathcal{A}$ thus simulates the T′ of join using execute and reveal oracles. In cases (2) and (3), T is generated by itself, and so $\mathcal{A}$ can simulate T′ by calling join from T.

Similarly, when $U_2 = \{U_{l1},.., U_{lM}\}$ are the participants willing to leave from the group $U = \{U_1,...,U_{N+M}\}$, $\mathcal{A}'$ sends a query which initiates leave (U, U_2). Accordingly, $\mathcal{A}$ tries to finds any one of the unique entries in the form (1). (U, T) in Tlist, or (2). (U', U'', T) in Jlist with $U = U' \cup U''$, or (3). (U', U'', T) in List with $U = U' \setminus U''$, it checks for the availability of the entries in the list by making an execute the query to its random oracle on U, gets the T, and stores (U, T) in Tlist.

$\mathcal{A}$ simulates the leave() by himself and extracts a T′ from T as follows: Initially, $\mathcal{A}$ trace identifies the positions in T where new messages overwrite the old messages. $\mathcal{A}$ does the necessary modification in T according to the leave () algorithm mentioned in Section 2.6 and gets the updated T′. Thus $\mathcal{A}$ expands T into T′ for leave() by storing (U, U_2, T′) in the list.

Suppose $\mathcal{A}'$ places a SendQuery to an instance Π^i_V. After completion of appropriate checking, $\mathcal{A}$ identifies a unique entry (U, T) in Tlist such that (V, i) $\in$U. If $\mathcal{A}$ is unable to find the (U, T), then $\mathcal{A}$ finds a unique entry in (U, U_l, T′) in Jlist such that (V, i) $\in$U, which means join() has already initiated in that session. Next, $\mathcal{A}$ obtains subsequent public information for T′ to be output by Π^i_V, and then send it to $\mathcal{A}'$. When the $\mathcal{A}$ finds an entry (U, U_2, T′) in Llist such that (V, i) $\in$U, then we can conclude that answer to the inquiry is found from T′.

Join Query: Suppose $\mathcal{A}'$ will sends an inquiry which initiates Join(U, U_l), where $U_l = \{U_{N+1},..., U_{N+M}\}$ are the participants willing to add in the group U = {U_1,..., U_N}. Let the instances $\Pi^i_{U_{N+1}}, \Pi^i_{U_{N+2}}, ..., \Pi^i_{U_{N+M}}$, wants to join the group, then, $\mathcal{A}$ discoveries an entry in the form (U, U_l, T′) in Jlist. If no entry is present, then $\mathcal{A}$ yield nothing to $\mathcal{A}'$, otherwise $\mathcal{A}$ sends T′ to $\mathcal{A}'$.

Leave Query: Let $\mathcal{A}'$ sends queries which initiate Leave(U, U_2), where U_2 = $\{U_{l1},.., U_{lM}\}$ are the participants willing to leave from the group $U = \{U_1,...,U_{N+M}\}$. Let the instances $\prod^1_{U_{l1}}, \prod^i_{U_{l2}}, ..., \prod^i_{U_{lM}}$ wants to leave from the group, then, $\mathcal{A}$

discoveries an entry (U, U_2, T′) in Jlist. If no such entry is identified, then *A* return nothing to $\mathcal{A}'$, otherwise $\mathcal{A}$ output T′ to $\mathcal{A}'$.

Reveal query: Assume $\mathcal{A}'$ makes a Reveal(V, i) for an instance $\prod_V^i$ for which $acc_V^i = 1$. At this movement T′ in which $\prod_V^i$ participants has finalized. If T′ corresponds to the transcript of the AP, then $\mathcal{A}$ finds a unique pair (U, T) in the Tlist such (V, i) ∈U. Assuming that forge has not occurred, T is the exclusive unauthenticated transcript corresponds to T′. Then $\mathcal{A}$ makes the appropriate reveal query to one of the instances in T and returns the result to $\mathcal{A}'$; otherwise, T′could be the transcript for join or leave. Since T′ has been replicated by $\mathcal{A}$, it can be able to compute the modifies key in session and send a suitable answer to $\mathcal{A}'$.

Now $\mathcal{A}$ finds (U, T) in Tlist such that (V, i) ∈U, guessing that the forge has not occurred, then simulation for $\mathcal{A}'$ is flawless. Whenever the case of forge occurs, $\mathcal{A}$ aborts and outputs a random bit. So $Prob_{\mathcal{A}', AP}[succ|Forge] = \frac{1}{2}$, then

$$Adv_{\mathcal{A}, UP} := 2\left|Prob_{\mathcal{A}, DAP}[succ] - \frac{1}{2}\right|$$

$$\mathcal{A} = 2\left|Prob_{\mathcal{A}', DAP}\left[succ \wedge \overline{Forge}\right] + Prob_{\mathcal{A}', DAP}[succ \wedge Forge] - \frac{1}{2}\right|$$

$$= 2\left|Prob_{\mathcal{A}', DAP}\left[succ \wedge \overline{Forge}\right] + Prob_{\mathcal{A}', AP}[succ \mid Forge]Prob_{\mathcal{A}', DAP}[Forge] - \frac{1}{2}\right|$$

$$= 2\left|Prob_{\mathcal{A}', DAP}\left[succ \wedge \overline{Forge}\right] + \frac{1}{2}Prob_{\mathcal{A}', DAP}[succ \wedge Forge] - \frac{1}{2}\right|$$

$$= 2\left|Prob_{\mathcal{A}', DAP}[succ] - Prob_{\mathcal{A}', DAP}[succ \wedge Forge] + \frac{1}{2}Prob_{\mathcal{A}', DAP}[Forge] - \frac{1}{2}\right|$$

$$\geq \left|2Prob_{\mathcal{A}', DAP}[succ] - 1\right| - \left|Prob_{\mathcal{A}', DAP}[succ \wedge Forge] - 2Prob_{\mathcal{A}', DAP}[succ \wedge Forge]\right|$$

$$\geq Prob_{\mathcal{A}', DAP} - Prob[Forge]$$

So, $Adv_{\mathcal{A}, UP} \geq Adv_{\mathcal{A}', DAP} - Prob[forge]$

The $\mathcal{A}$ makes an XQ for each XQ generated by $\mathcal{A}'$, in general $\mathcal{A}'$ makes q_J and q_L are join and leave queries, initialized by $Send_J$ and $Send_L$ of $\mathcal{A}'$. Now, each $Send_J$ and $Send_L$ of $\mathcal{A}'$ makes at most one XQ. Thus, the maximum $q_J + q_L$ $\mathcal{A}$ can make XQ to respond to all the $Send_J$ and $Send_L$ inquiries made by $\mathcal{A}'$.

Also, $\mathcal{A}$ makes XQ for each change in session SK initiated by $\mathcal{A}'$ using Send queries. Since a session includes at least two instances, such as XQ is made after at least two Send queries of $\mathcal{A}'$, thus a total of $(q_s + q_J - q_L)/2$ XQ of $\mathcal{A}$ to respond to all other Send queries generated by $\mathcal{A}'$, where q_s be the number of Send queries made by $\mathcal{A}'$. Altogether, a total of XQs made by $\mathcal{A}$ is at a maximum of $(q_s + q_J - q_L)/2 + (q_E + q_J + q_L) = q_E + (q_s + q_J + q_L)/2$. Also, since

$$Adv_{DAP}^{KA}(t, q_E, q_s, q_J, q_L) \leq Adv_{UP}^{KA}(t', q_E + (q_s + q_J + q_L)/2) + Prob[Forge]$$

2.8 EXPERIMENTATION AND RESULTS

The efficiency of the proposed technique tested to sing PySyft and PyTorch Python libraries for secure and private deep learning on the SECON dataset, which contains the semiconductor manufacturing process data. It consists of 591 attributes, and the signals have recorded throughout different trails. This dataset aims to analyze the data and finding correlations between failing the test and the attributes of the trial to enable engineers to firstly gain insight into the cause of the failures, and secondly, the attribute which has the highest impact on the failure. The experimentation concentrated on distributing available datasets among the three participants. The performance analysis is done on computation and communication costs. Table 2.2 shows the basic operations and its computation cost, Table 2.3 shows the privacy features

TABLE 2.2
Execution Time of Basic Operations

Symbol	Description	Execution Time (s)
T_{mp}	Multiplication operation with pairings	1.27
T_a	Addition operation with pairings	0.148
T_{sa}	Scalar point addition	0.038
T_{sm}	Scalar point multiplication	0.326
T_h	Hash function operation	0.010
T_{ep}	Exponentiation operation	3.88

TABLE 2.3
Comparison of Privacy-Preserving Parameters

Scheme	Cryptographic Technique	Security against Server	Security against Participant	Trainer Transmission	Dynamic Secrecy	Number of Roles
[9]	-	No	No	Gradients	-	2
[25]	LWE-based paillier	Yes	No	Encrypted gradients	No	2
[26]-DeepPAR	Proxy re-encryption	Yes	Yes	Encrypted gradients	Yes	3
[26]-DeePDPA	Key management-Paillier	Yes	No	Encrypted gradients	Backward secrecy	2
ECCRE	ECC re-encryption	Yes	Yes	Encrypted gradients	Yes	3
ECCAL	ECC key management	Yes	No	Encrypted gradients	Backward secrecy	2

comparative analysis with the other techniques, the proposed technique is providing same privacy as same as in DeepPAR and DeePDPA. Tables 2.4 and 2.5 show the computation and communication cost of all the phases in proposed techniques ECCRE and ECCCAL. Tables 2.6 and 2.7 describe the performance analysis of the proposed techniques with few other existing techniques, as the proposed algorithm is ECC addition and subtraction operations, the execution time is less compared to other techniques, where they have used the exponentiation operations.

TABLE 2.4
Computation and Communication Cost of ECC-RE

Keygen()		Decrypt()	Compute()	Reencrypt()	Aggregate()
Computation cost					
U_i	$2T_{sm} + I$	Enc() + A	Enc() + DL + A	-	-
O	$2T_{sm}$	Dec()	-	-	-
S	$T_{sm} + I$	-	-	Enc()	$\lvert g^i \rvert A$
Communication cost					
U_i	O (m)	$O(2\lvert\omega_i\rvert m)$	$O(\lvert g^i\rvert m)$	-	-
O	O (Nm)	$O(\lvert\omega_i\rvert m)$	-	-	-
S	O (Nm)	-	-	-	-

A, Addition in homomorphic ciphertext; Dec, Decryption; Enc, Encryption; m, message length.

TABLE 2.5
Computation and Communication Cost of ECCAL

Keygen()		Decrypt()	Compute()	Aggregate()	Join()	Leave
Computation cost						
U_i^*	$3T_{sm}$ + sub + (N − 1)Add	$\lvert\omega_i\rvert$ Dec	$DL + \lvert g^i\rvert(Enc() + A)$	-	$3T_{sm}$ + sub + (M − 1)Add	-
U_j				-	(M + 1)Add	4M + 1sub
S	–	-	-	$\lvert g^i \rvert A$	-	-
Communication cost						
U_i^*	O (N − 1)m	$O(\lvert\omega_i\rvert m)$	$O(\lvert g^i\rvert m)$	-	O ((M + 1)m)	-
U_j				-	O ((M − 1)m)	O ((M + 1) m)
S	-	-	-	-	-	-

Sub, subtraction; Add, Addition.

TABLE 2.6
Comparative Analysis of Computation Cost

	Keygen()	Decrypt()	Compute()	Reencrypt()	Aggregate()
[25]	-	-	$DL+\|g^i\|+\|\omega_i\|$	-	$\|g^i\|$
[9]	-	$(Add + Mul)\|\omega_i\|$	$DL+\|g^i\|Enc$		
DeepPAR	Exp + Inv + Mul	Enc + Add	DL + Enc + Mul	Exp	$\|g^i\|A$
DeepDPA	Inv + hash + Div + (N + 1)Exp + NMul + PKG	$\|\omega_i\|Dec$	$DL+\|g^i\|(Enc+Mul)$	-	$\|g^i\|A$
ECCRE	$5T_{sm}+2I$	Enc + Add + Dec	Enc() + DL + A	Enc()	$\|g^i\|A$
ECCAL	$3T_{sm}+sub+(N-1)Add$	$\|\omega_i\|Dec$	$DL+\|g^i\|(Enc()+A)$	-	$\|g^i\|A$

DL, Deep Learning computation cost.

TABLE 2.7
Comparison of Communication Cost

	Keygen()	Decrypt()	Compute()	Join()	Leave()
[25]	-	$O(2\|\omega_i\|m)$	$O(\|g^i\|m)$	-	-
[9]	-	$O(2\|\omega_i\|m)$	$O(\|g^i\|m)$	-	-
DeepPAR	O (Nm)	$O(2\|\omega_i\|m)$	$O(\|g^i\|m)$	-	-
DeepDPA	O (N + 1)m	$O(\|\omega_i\|m)$	$O(\|g^i\|m)$	O ((M)m)	O ((M-1)m)
ECCRE	O (N − 1)m	$O(2\|\omega_i\|m)$	$O(\|g^i\|m)$	-	-
ECCAL	O (Nm)	$O(\|\omega_i\|m)$	$O(\|g^i\|m)$	O ((M-1)m)	O ((M+1)m)

2.9 CONCLUSION

In this chapter, we proposed two ECC-based privacy-preserving protocols; it is a shared model, which was learned asynchronously in parallel by maintaining the input privacy as well as model privacy. Based on the re-encryption through the proxy system, ECCRE enables a technique, which prevents each participant's information secured from revealing to other participants. It supports the participant's dynamic update in a lightweight manner. Moreover, the other technique, ECCAL, is a technique working on the principle of group key management. It offers more dynamism as well as backward secrecy in dynamic privacy-preserving using deep learning. From the results, both the mechanisms' performances are more compared with other techniques.

REFERENCES

1. SCOOP The Industrial Internet of Things (IIoT): The Business Guide to Industrial IoT. i-SCOOP. Available online: https://www.i-scoop.eu/internet-of-things-guide/industrial-internet-things-iiot-savingcosts-innovation (accessed on 21 April 2018).
2. E. Heymann, German Auto Industry: WLTP Followed by Lacklustre Demand. Talking point, Deutche Bank. Available online: https://www.dbresearch.com/PROD/RPS_EN-PROD/PROD0000000000489273.pdf (accessed on 19 December 2018).
3. K. Moskvitch, Industrial Internet of Things: Data Mining for Competitive Advantage. Available online: https://eandt.theiet.org/content/articles/2017/02 (accessed on 9 April 2018).
4. Industrial Analytics 2016/2017. Available online: https://digital-analytics-association.de/wp-content/uploads/2016/03/Industrial-Analytics-Report-2016-2017-vp-singlepage.pdf (accessed on 5 January 2019).
5. PwC 20 years inside the mind of the CEO: What's next? 20th CEO Surv. Rep. 2017.
6. J. Konen, H. B. McMahan, F. X. Yu, P. Richtrik, A. T. Suresh, and D. Bacon, "Federated learning: Strategies for improving communication efficiency," 2016, *arXiv:1610.05492*.
7. R. Shokri and V. Shmatikov, "Privacy-preserving deep learning," in *Proc. 22nd ACMSIGSACConf.Comput.Commun. Secur.*, 2015, pp. 1310–1321.
8. R. Jayasri, R. Vidya, and J. A. D. Rex. A survey on industrial automation based on IoT with Arduino microcontroller. *International Journal of Contemporary Research in Computer Science and Technology (IJCRCST)*, vol. 4, 24–27, 2018.

9. L. T. Phong, Y. Aono, T. Hayashi, L. Wang, and S. Moriai, "Privacy-preserving deep learning via additively homomorphic encryption," *IEEE Transactions on Information Forensics and Security*, vol. 13, no. 5, pp. 1333–1345, May 2018.
10. P. Daugherty and B. Berthon. *Winning with the Industrial Internet of Things: How to Accelerate the Journey to Productivity and Growth.* Technical Report. Dublín: Accenture, 2015.
11. T. Graepel, K. Lauter, and M. Naehrig, "ML confidential: Machine learning on encrypted data," in *Proc. Int. Con. Inf. Secur. Cryptographic*, 2012, pp. 1–21.
12. A. P. Singh et al., "A novel patient-centric architectural framework for blockchain-enabled healthcare applications," in *IEEE Transactions on Industrial Informatics*, 2020.
13. P. Mohassel and P. Rindal, "ABY3: A mixed protocol framework for machine learning," in *Proc. ACMSIGSACConf.Comput.Commun. Secur.*, 2018, pp. 35–52.
14. Z. Liao et al., "Distributed probabilistic offloading in edge computing for 6G-enabled massive Internet of Things," in *IEEE Internet of Things Journal*, 2020.
15. P. Mohassel and Y. Zhang, "SecureML: A system for scalable privacy-preserving machine learning," *in Proc. IEEE Symp. Secur. Privacy*, San Jose, CA, USA, 2017, pp. 19–38.
16. R. Gilad-Bachrach, N. Dowlin, K. Laine, M. Naehrig, and J. Wernsing, "Cryptonets: Applying neural networks to encrypted data with hig throughput and accuracy," *in Proc. Int. Conf. Mach. Learn.*, 2016, pp. 201–210.
17. R. K. H. Tai, J. P. K. Ma, Y. Zhao, and S. S M. Chow, "Privacy-preserving decision trees evaluation via linear functions," in *Proc. Eur. Symp. Res. Comput. Secur.*, 2017, pp. 494–512.
18. K. Bonawitz et al., "Practical secure aggregation for privacy-preserving machine learning," in *Proc. ACMSIGSAC Conf. Comput. Commun. Secur.*, 2017, pp. 1175–1191.
19. P. Paillier, "Public-key cryptosystems based on composite degree residuosity classes," in *Proc. Int. Conf. Theory Appl. Cryptographic Techn.*, 1999, vol. 99, pp. 223–238.
20. L. T. Phong and T. T. Phuong, "Privacy-preserving deep learning via weight transmission," *IEEE Transactions on Information Forensics and Security*, vol. 14, no. 11, pp. 3003–3015, Nov. 2019.
21. J. M. Pollard, "Monte Carlo methods for index computation (mod p)," *Mathematics of Computation*, vol. 32, no. 143, pp. 918–924, 1978.
22. J. Dean et al., "Large scale distributed deep networks," in *Proceedings of the Advances in Neural Information Processing Systems*, Lake Tahoe, ND, USA, 3–6 December 2012, pp. 1223–1231.
23. Y. Liang, Z. Cai, J. Yu, Q. Han, and Y. Li, "Deep Learning-based inference of private information using embedded sensors in smart devices," *IEEE Network*, vol. 32, pp. 8–14, 2018. [CrossRef].
24. W. Wu, U. Parampalli, J. Liu, and M. Xian, "Privacy-preserving k-nearest neighbor classification over the encrypted database in outsourced cloud environments," *World Wide Web*, vol. 22, pp. 101–123, 2019. [CrossRef].
25. R. Shokri and V. Shmatikov, "Privacy-preserving deep learning" in *Proc. 22nd ACMSIGSAC Conf.Comput.Commun. Secur.*, 2015, pp. 1310–1321.
26. X. Zhang, X. Chen, J. K. Liu, and Y. Xiang, "DeepPAR and DeepDPA: Privacy-preserving and asynchronous deep learning for industrial IoT," *IEEE Transactions on Industrial Informatics*, vol. 16, no. 3, pp. 2081–2090, March 2020, DOI:10.1109/TII.2019.2941244.
27. M. Alazab, "Profiling and classifying the behaviour of malicious codes," *Journal of Systems and Software*, vo. 100, pp. 91–102, February 2015.
28. "From zeus to zitmo: Trends in banking malware," in *2015 IEEE Trustcom/BigDataSE/ISPA*, vol. 1, pp. 1386–1391. IEEE, 2015.

29. "Machine learning based botnet identification traffic," in *2016 IEEE Trustcom/BigDataSE/ISPA*, pp. 1788–1794. IEEE, 2016.
30. "Mining malware to detect variants," in *2014 Fifth Cybercrime and Trustworthy Computing Conference*, pp. 44–53. IEEE, 2014.
31. S. Bhattacharya et al., "A novel PCA-firefly based XGBoost classification model for intrusion detection in networks using GPU," *Electronics*, vol. 9, no. 2, p. 219, 2020.
32. Stanford Deep Learning Tutorial. Available online: http://deeplearning.stanford.edu.
33. J. Duchi, E. Hazan, and Y. Singer. "Adaptive subgradient methods for online learning and stochastic optimization," *Journal of Machine Learning Research*, vol. 12, pp. 2121–2159, July 2011.
34. S. K. Dasari, K. R. Chintada, and M. Patruni, "Flue-cured tobacco leaves classification: a generalized approach using deep convolutional neural networks," in *Cognitive Science and Artificial Intelligence*, Springer, Singapore, 2018, pp. 13–21.
35. M. Alazab and M. Tang, (Eds.). *Deep Learning Applications for Cyber Security. (Advanced Sciences and Technologies for Security Applications).* Cham: Springer. DOI: 10.1007/978-3-030-13057-2, 2019.
36. S. More et al., "Security assured CNN-based model for reconstruction of medical images on the internet of healthcare things," *IEEE Access*, vol. 8, pp. 126333–126346, 2020.
37. A Makkar et al., "FedLearnSP: Preserving privacy and Security using federated learning and edge computing," IEEE Magazine of Consumer Electronics, 2020.

3 Contemporary Developments and Technologies in Deep Learning–Based IoT

Prishita Ray, Rajesh Kaluri, Thippa Reddy G., Praveen Kumar Reddy M., and Kuruva Lakshmanna
Vellore Institute of Technology

CONTENTS

3.1 INTRODUCTION

The human central nervous system consists of two essential components: the brain and the spinal cord. This is responsible for controlling all functions of the body, reactions to stimuli, and the making of important decisions. Yet the basic building blocks of this highly complex control architecture are neurons, also known as nerve cells. Each of these neurons has a cyton or cell body with extending dendrites, which is connected to the terminal nodes through a link called the axon. The nodes of one neuron are responsible for releasing chemical signals that are received by the dendrites of another neuron, also known as nerve impulses, which is how messages are passed across the body from the brain.

With the advent of artificial intelligence (AI), there has been an attempt to model a similar behavior among computers, so that they can make decisions autonomously, just like humans. A significant subcategory of AI is machine learning (ML) [1–3], where given some input, a machine can learn on its own from the available data and make accurate decisions or predictions on unseen data with the help of algorithms. When these learning algorithms are based on the neuron architectures of the CNS, this further categorizes ML into deep learning (DL) [4–6]. Inspired by the natural neurons, in computer science terminology they are called artificial neural networks (ANNs) or just neural networks in general. DL [7] is a highly popular area of research, particularly for its versatility to deal with various kinds of data, and for a wide range of applications.

Many different forms of neural network architectures are in use today. The most basic form is the single-layer perceptron which can further be extended to the multilayer perceptron model. For image-based input data, Convolutional Neural Networks (CNNs) are used for learning [8]. A popular application, facial recognition, uses two such networks, which form General-Adversarial Networks, when the amount of input data available is limited. Another framework [9], Recurrent Neural Networks (RNNs) is helpful when working with time-series data. Autoencoder (AE) models are used for feature selection and dimensionality reduction. Some other architectures such as Restricted Boltzmann Machines (RBMs) and emerging ones like Ladder Networks and Siamese Networks have further extended the usability of neural networks in almost all domains. Thus, it has now become possible to imitate the working of the brain among computers, greatly reducing the need for human intervention and extending its capabilities for complex decision-making.

3.2 DL ARCHITECTURE

The artificial rendition of the neuron is called a node. Each node receives some inputs like the dendrites which are then multiplied with weights. These weighted inputs are then summed and passed through an activation function to produce an output [8–10]. A visualization of a neuron and a node can be observed in Figures 3.1 and 3.2.

3.2.1 Neural Network

A neural network is a collection of these nodes in multiple layers. In its most basic form, it contains an input layer, a hidden layer, and an output layer, each having a specific number of nodes [11]. The input layer consists of as many nodes as features in each example of a training dataset, and the output layer for predicting a value for that training instance [12]. The learning process happens within the hidden layers where an activation function such as sigmoid and ReLU adjusts the weight connections between the layers to give an output as accurate as possible, based on the loss between the actual and predicted outputs, i.e., di-Oi. This is taken care of using a learning rate. Each node in the hidden layer receives inputs from multiple other nodes, and its calculated output is propagated to many other nodes in further layers as shown in Eq. 3.2. In addition, optimizers are used to ensure convergence of the

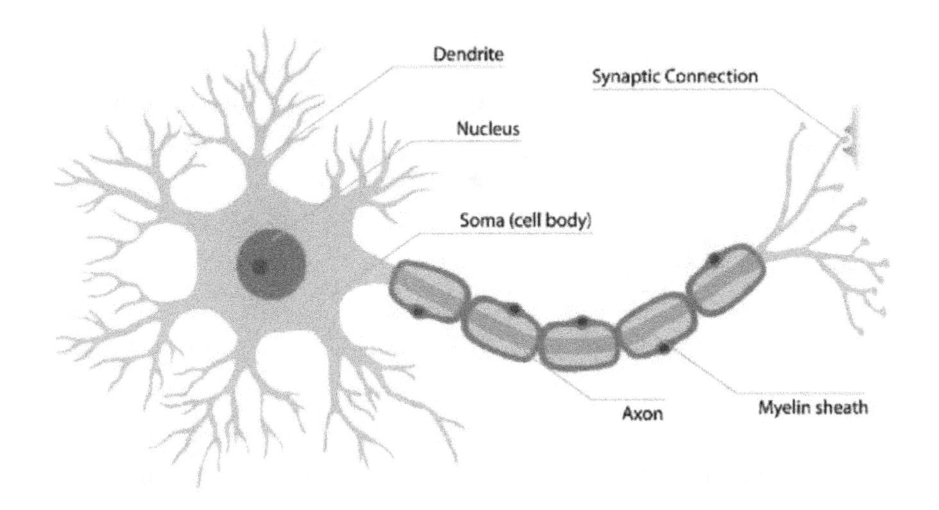

FIGURE 3.1 Anatomy of a typical neuron or nerve cell.

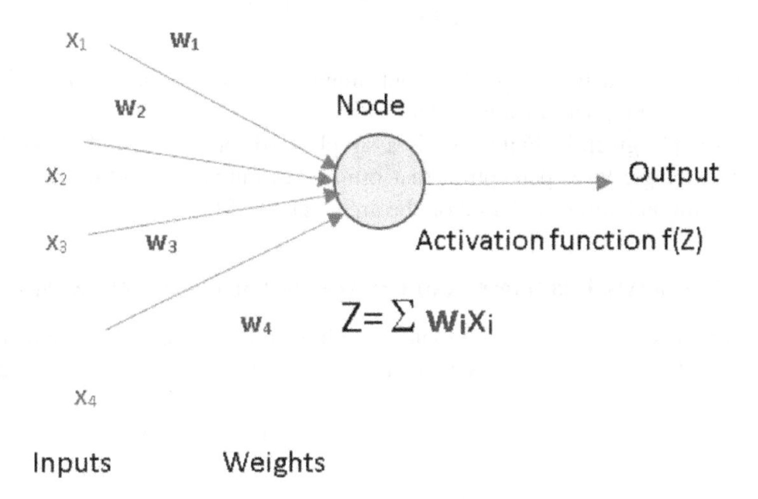

FIGURE 3.2 A node in a neural network.

model to a global minimum of obtained loss by changing the weights, biases, and learning rate accordingly. This is also known as gradient descent.

$$a_j^l = \sigma \sum_k w_{jk}^l a_k^{l-1} + b_j^l \tag{3.1}$$

where a is the output of node j in layer l, σ is the activation function, w is the weight matrix between layer l − 1 having k number of nodes and layer l, and b is the bias of node j. There are many different loss functions. Some of the most commonly used ones are mean-squared error, cross-entropy loss (C), maximum likelihood

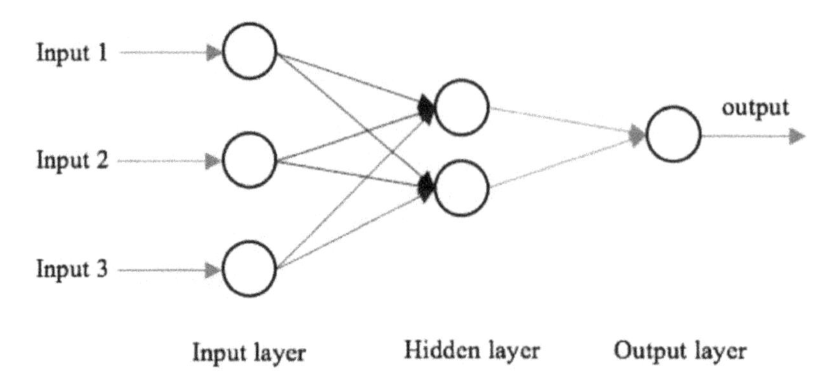

FIGURE 3.3 A basic neural network architecture.

estimation, and Kullback-Leibler (KL) divergence shown in Eq. 3.2. The final loss of the network is calculated at the output layer using the MSE formula as follows:

$$c = \frac{1}{2n}\sum_{x} \| y(x) = a^{L}(x) \| \tag{3.2}$$

Here y denotes the actual output for input sample x, a^L is the calculated output, and n gives the total number of training data.

The example given in Figure 3.3 is a simple feed-forward neural network also known as a single-layer perceptron that minimizes output loss. Many versions of shallow neural networks are based on the architecture of Figure 3.3.

3.2.2 Multilayer Perceptron and Convolutional Neural Networks

DL involves neural networks with many such hidden layers. Multilayer perceptrons are the most basic example of a deep neural network (DNN) [13,14] that can be used for representing XOR functions, as observed from Figure 3.4. The purpose behind using many layers is to find hidden abstractions from the input data with increasing hierarchy within the network. Each node in the first hidden layer receives inputs from each node in the input layer and uses a set of weights and associated bias to calculate its output with the help of an activation function, which is then passed as an input to every node in the next hidden layer [7,15,16]. This process is continued till the output layer is reached, and the loss is calculated, which the optimizer then uses to change the learning rate, that adjusts the weight and bias values of each node in every layer thereafter using backpropagation of the calculated loss. Backpropagation algorithms, such as EBPTA, change the hyperparameters (weights and biases) for each layer, based on a loss calculated using its own outputs, and the loss from its next immediate layer, propagated backward, using a chain-rule formula.

This process of training all the layers is run on the entire data, or on a mini-batch of the data, in one training cycle known as an epoch. Multiple epochs are run, so as to train the network over time. To improve the rate of convergence, some techniques such as adding momentum and learning rate decay can be introduced. Additionally,

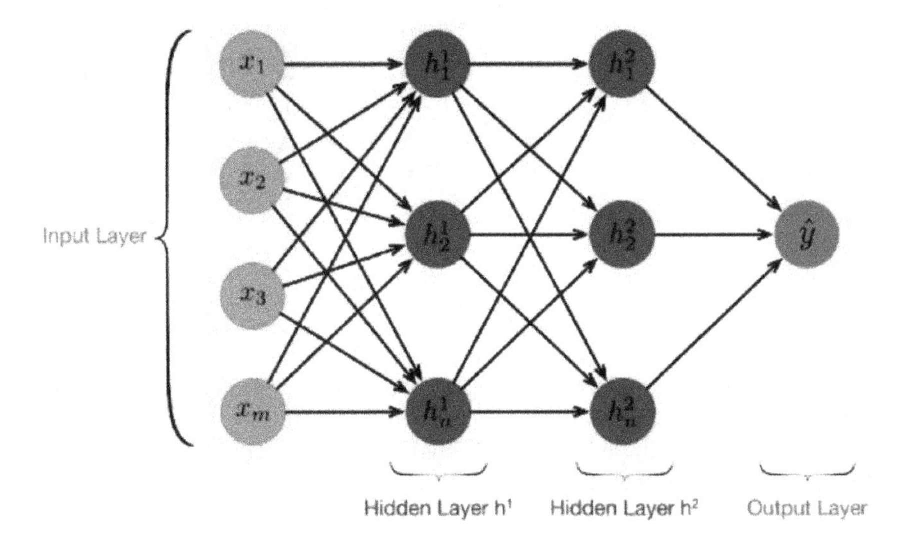

FIGURE 3.4 Multi-layer perceptron.

to prevent model overfitting, imposing regularization such as lasso (L1), ridge (L2), or dropout is a common practice. Occasionally normalization of input data is helpful when each feature has a huge range in values that can significantly affect error calculation.

A CNN such as in Figure 3.5 contains convolution and pooling layers, to reduce the dimensionality required to represent the input images. The images are first represented using the RGB values of the pixels, and then a representative value is calculated for each square subset of this matrix at the convolution layer. This value is then collected for each subset, in the pooling layer [17–20]. The procedure continues till a final fully connected layer classifies the training example. An RNN Figure 3.6 uses just one hidden layer that trains on its output along with the provided input value repeatedly for multiple time steps. This hidden layer is known as the cell, and its cell state denotes the value at the current training epoch as shown in Figure 3.6. Some

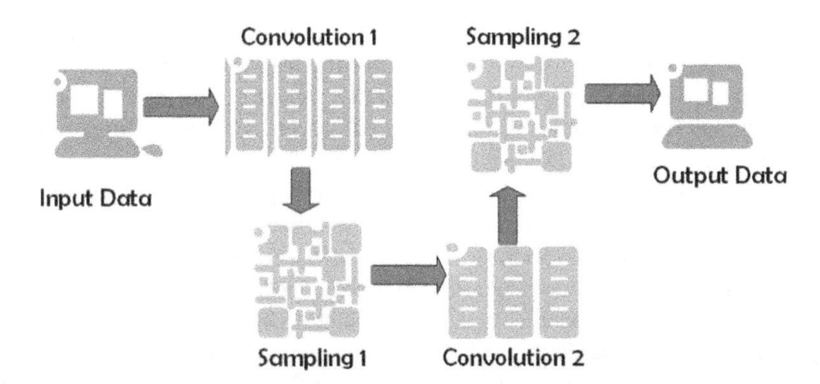

FIGURE 3.5 Convolutional neural networks.

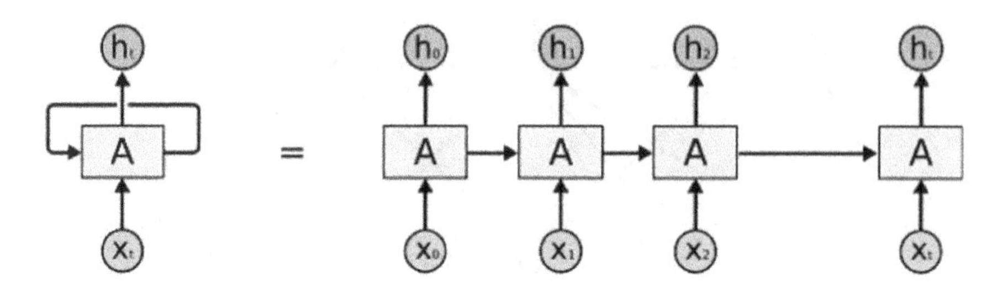

FIGURE 3.6 Unfolded architecture of a RNN.

important variants such as LSTM (Long Short-Term Memory) contain specific gates along with the basic RNN structure [17,21]. LSTMs have input-output and forget gates to retain previously learned representations, whereas Gated Recurrent Units (GRUs) contain just two gates – update and reset.

3.2.3 Learning Methods

The loss calculation for network training can be done in multiple ways, based on the type of input data that is available. If the data contains examples with their known targets or outputs, the training is supervised, with the difference between the actual and predicted values serving as the loss [8,22]. However, if the targets are unknown, unsupervised learning is used. This involves clustering algorithms and is primarily used for dimensionality reduction and feature selection such as in AEs. In such cases, the loss is calculated based on feature values and their correlations.

Another type of training is semi-supervised learning, where only some labeled samples are available, and the majority of the input data is unlabeled. The total training loss is the sum of the labeled and unlabeled costs obtained through supervised and unsupervised learning respectively. Ladder network architectures make use of this learning mechanism.

3.3 INTERNET OF THINGS (IoT)

Internet of Things also known as IoT is an emerging field where devices are able to connect and communicate with each other over the network, without human intervention. Over the past decade, it has particularly gained a growing interest as it is ecofriendly and requires lesser installation and transmission costs. Moreover, it is an easy-to-use technology. Devices that are a part of the IoT [23,24] can send and receive data over the cloud [25]. This can be used to deploy applications that run on the cloud such as Amazon Web Services. An important aspect of IoT is that every device is uniquely identifiable across the network, so as to prevent overload.

The connected devices can range from mobile phones, laptops, etc. to much more large-scale and complex machines such as airplanes. Data collection is done using sensors attached to these devices which are then passed onto a gateway for transmission. This can then be analyzed or processed on the cloud or on the device [26]. A significant advantage of IoT is that real-time data is used making it much more applicable. Owing to the large amounts of data that is obtained, it becomes infeasible to store them physically. Therefore, cloud computing is slowly gaining popularity, so that data is processed or analyzed only when required, preventing the requirement for long-term storage. This also leads to more speed and efficiency.

Today, almost every industry is utilizing IoT to simplify its tasks. Starting from healthcare, automobile, manufacturing, any data-driven task or decision is now being implemented with IoT. However, a major problem that needs to be kept in mind is to ensure security of this data [27], as businesses invest huge sums of money and any compromise can cause massive losses [25]. As data is sent and received across networks, new technologies such as 5G can further enhance the capability of IoT as now transfers can be faster, more secure, and can cover much larger areas with many more devices. Smart cities, the infrastructural prototypes of which are currently under heavy development, exploit the utility of IoT to the fullest. Every aspect, from vehicular movement control, power saving, communication within houses, security, renewable energy generation, etc., can be achieved with this technology if given sufficient effort and investment [24].

3.3.1 At the Intersection of DL and IoT

IoT has become a highly popular technology today, as it allows multiple devices to communicate with each other through networking via the cloud. Any kind of data can be transferred using IoT, be it bimodal, multimodal, or any kind of multimedia. Combining the powers of DL and IoT [16,22], the speed and efficiency with which complex tasks are accomplished across many connected computers are unmatched.

DL can help dynamize the data collection process of IoT devices and make it easier, especially in noisy and complex environments. Additionally, it can help generate predictions on the data with very low inference time, which is a great advantage especially for large-scale tasks such as predicting future energy consumption. Even though traditional and shallow machine learning algorithms such as Support Vector Machines (SVMs) [28], K-Nearest Neighbors (KNNs), and Principal Component Analysis (PCA) [29] are available, they have several drawbacks in view of being integrated with IoT, thus rendering them to be unsuitable and giving DL an upper hand. These algorithms work well for relatively simpler classification tasks, with smaller datasets. Also parameter selection, feature extraction, and data preprocessing are highly essential to ensure proper performance. DL models, on the other hand, can handle these problems quite well, along with ensuring better classification accuracy using high dimensional data. Unlike normal data, IoT data has specific characteristics such as large-scale streaming data that is continuous, high in volume, and heterogeneous; has time and space correlation; and contains a lot of noise formed during acquisition and transmission. Streaming data is generated within short intervals of time; therefore, it becomes imperative to quickly analyze insights from it. Any algorithms should be designed to work with such kinds of data. Some common problems that need to be addressed in IoT applications include collaboration among multiple sensors that includes acquisition and fusion of their readings as well as encoding features that are essential to the desired task, working with IoT devices having low-end and limited resources, reliability assurances, and time-consuming data labeling processes. In order to integrate DL with IoT, one needs to keep in mind the resource requirements such as memory and computation power, execution bottlenecks, and performance characteristics for running such models on target IoT devices [8]. The algorithms must be compatible with the hardware that they must be run on to ensure efficient inference execution. Moreover, most of this computation is distributed and done using cloud infrastructure, which brings with it another set of challenges, in particular, privacy issues and network fluctuations that occur due to a large number of connected heterogeneous devices and their mobility. Energy efficiency is also a significant factor that needs to be taken into consideration, when deploying such models using the hardware architecture.

Embedded systems are commonly used as IoT devices [8], such as the ARM System on a Chip (SoC), and Zuloco that contains four ARM v7 cores. The Zuloco can runs at 1 GHz, with a storage of 512 Mbytes of RAM, peak power consumption of 3 W, and is highly affordable. Other commonly available SoCs include the Qualcomm Snapdragon 800, which is widely used in phones and tablets; Intel Edison in wearables and form-factor sensitive IoT; and Nvidia Tegra K1, which has great GPU performance. Along with software, and DL libraries like TensorFlow integrated with it, neural networks can be employed to train on real-time information and make predictions or decisions. Using the sensor data, estimation and classification can be performed based on whether it is continuous or categorical in nature. Estimation-oriented problems typically include tracking and localization using physical values that are recorded but contain a lot of noise which can lead to errors and biases. Classification-oriented problems are mainly used for activity and context recognition, using carefully selected features by encoding local, global, and temporal information. These devices additionally support other services such as WiFi and sensors for recording values to name a few. Channel measurement from Channel State Information (CSI) can also be obtained from the WiFi NIC cards. Sensors, such as temperature measurers, cameras, proximity sensors, accelerometers, gyroscopes, and touchscreens, are often used to capture real-time data. However, unlike traditional data used for training DL models [29,30], very small amounts of real-time data are labeled. Therefore, these networks have to make inferences taking into account the scarcity of available output, in order to minimize human intervention. For this purpose, semi-supervised learning algorithms have received more attention lately. To address security issues, new DL models such as software-designed networks (SDNs) are becoming popular [19].

3.3.2 Recent Trends in DL-Based IoT

Lack of sufficient resources is a major drawback in IoT devices, due to energy constraints and low computation capability when trying to deploy DL architectures for various applications [26,31]. DL-based IoT is especially becoming popular in industries such as infrastructure, social networks, and content delivery. Deep CNNs have become the most popular architecture, due to their high performance and reliability as well as low-energy consumption. Dimensionality reduction, parallelism, distributed computing, filtering, etc. are some factors that need to be considered while designing the hardware for such systems, rendering the traditional Von Neumann

Machine unsuitable. A lot of memory storage, memory accesses, and computations have to be dealt with. Therefore, Graphical Processing Units (GPUs) can provide the necessary infrastructure to reduce the computation load on these devices, high parallelism, and multiple convolutional kernel support. Moreover, to cater to the needs of multiple industries, custom systems are being created, such as neuromorphic and bio-inspired approaches, digital and analog CMOS circuits, and memristors.

NVIDIA's Jetson architecture provides full support for DL models that can be run on low-power embedded devices. To defer computational load toward the edge, weight quantization, network pruning, and AE networks are useful for near-sensor deployment. Performance is of utmost importance on safety critical applications such as autonomous driving. Two key points to ensure the reliability of such systems are the identification of sensitive neurons in an architecture, where errors can potentially affect the accuracy of the system, and the identification of sensitive hardware modules used for simulation [23]. Some security vulnerabilities like confidence reduction, targeted misclassification, random misclassification, source misclassification, etc. can compromise the privacy of user data during training and inference stages. The Tensor Processing Unit (TPU), an ASIC from Google, is another recent framework that can handle complex computations using a SIMD hardware machine. Another ASIC, Field Programmable Gate Arrays (FPGAs) along with the reVision stack from Xilinx are also very popular today.

3.4 POPULAR FRAMEWORKS AND MODELS

3.4.1 Models

The type of DL architecture that would be suitable for an IoT application depends largely on its requirements and type of data. When semi-supervised training needs to be performed, a Ladder Networks model is suitable. In this architecture, two encoders are used, a clean encoder that processes the input normally and a corrupted encoder, that processes the input with added Gaussian noise. The decoder network denoises the output to reconstruct the original input data. They follow two different pathways: one for supervised learning on labeled samples, and another for unsupervised learning on unlabeled samples. Vision-based IoT analytics can exploit this model.

Deep Belief Networks (DBNs) are used for prediction tasks. They contain a single visible input layer and multiple hidden layers, which are used to represent more hierarchically important latent variables from the input. Every hidden layer is a RBM that is trained over the previous hidden layer [32]. So the architecture includes a set of stacked RBMs, such that a bipartite graph is formed between any two layers and for training, the product of the probabilities of the nodes in the visible layer should be maximized. Some common areas of application include intrusion detection systems, emotion prediction, feature extraction, etc.

In cases where only unlabeled data is available for training, AEs are helpful. They are commonly used for unsupervised learning and consist of two parts: an encoder and a decoder. The input and output layers have the same number of nodes, with the encoder network forming a latent variable representation of the input, and the decoder network trying to reconstruct the original input from this representation.

Thus, the loss is calculated as the difference between the reconstructed and original values. AEs can be of various types such as denoising, sparse, variational, and undercomplete. Variational autoencoders (VAEs) further use backpropagation for semi-supervised learning over scarcely labeled data. The innermost layer in this AE represents the latent variable using a normal distribution. Diagnosis and fault detection IoT systems are two areas where such networks can be helpful.

Long Short-Term Memory (LSTM) is a variant of RNNs that are capable of retaining learned network parameters for a longer time period from real-time data [15]. Applications with time-series or sequential data can use such networks as here as the input is dependent on the output of the previous time step. This is accomplished by using a feedback loop. Each node in such a network has three gates: an input, output, and forget gate that contain binary values. Based on the value of the forget gate information is retained in the node or forgotten. IoT applications such as human activity recognition and weather prediction can be accomplished using LSTMs, as shown in Figure 3.7. Such data is usually observed in user authentication, WiFi, RFID, VWB, and denial of service (DoS) attacks. Therefore, models trained on these data features can monitor bad data injection, intrusion detection in networks, or any other malicious activity that may have the potential of compromising the privacy of user information. Similarly, GRUs with an update and reset gate are an alternative RNN architecture suitable for similar tasks.

When new data needs to be created from existing data, General Adversarial Networks (GANs) are used. They contain two competing networks, i.e., a generative network that generates new data based on the training data distribution learned by the model, and a discriminator that classifies whether this new data is fake or real. Such networks can be used in localization and path-finding IoT. As described in the architecture section, CNNs can handle image-based data, which is very essential in many applications. They have convolution layers that find representational values for subsets of the RGB matrix of the pixels. These values are then collected using pooling layers, and finally, a fully connected network with a ReLU activation function classifies the data. Any IoT device with a camera for obtaining real-time images and

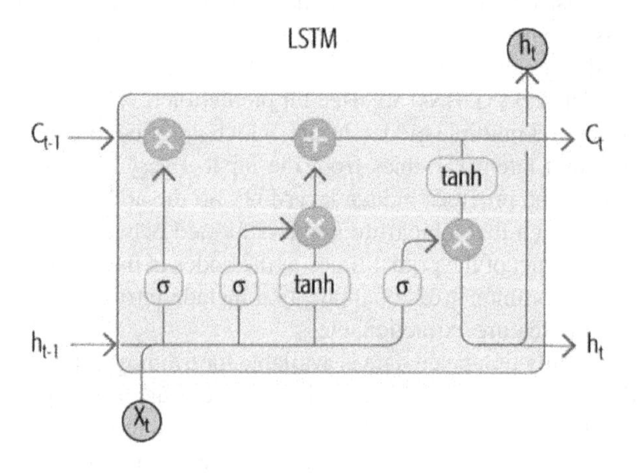

FIGURE 3.7 Architecture of LSTMs.

video, such as smartphones, airplanes, and drones, will be able to make predictions and detect with trained CNN models.

Other learning approaches such as Deep Reinforcement Learning (DRL) and Transfer Learning have become popular recently. DRL uses DNNs to help an agent learn an optimal policy to maximize its total obtained reward accumulated over time, when trying to achieve a goal, with the knowledge of its initial state and possible state and action spaces for exploration as part of its learning environment. Transfer learning utilizes the knowledge gathered from one problem to solve another related problem. This can include pretrained models, or custom-developed models, trained on language or image data.

In addition to the above models, combinations of these architectures can be experimented in different ways based on the type of application. For example, CNNs can be followed by RNN layers to perform human activity recognition.

3.4.2 Frameworks

Several DL frameworks are available to effectively train the above models on IoT devices and infrastructure. Some of the most widely used frameworks are Theano, H_2O, Torch, Tensorflow, Keras, Caffe, Neon, etc. Neon is very versatile for working with cross-platform hardware backends and can run several models such as CNNs, GANs, and AEs to name a few, and is open source. Theano is also an open-source framework that works with the CUDA library to optimally run codes on GPUs. Another major advantage that it offers is to parallelize training models when run using CPU. The training on hyperparameters with the help of graph representations. The H_2O ML framework provides [33] interfaces for multiple programming languages such as R, MATLAB, Javascript, and Python. Moreover, it can be run in multiple modes and allows complete training of AE models, instead of layer-by-layer execution. TensorFlow is a highly popular ML framework, as all possible DL models can be implemented with it ranging from feed-forward neural networks to multiple stacked architectures. Like Theano, it also uses graph representations to train models. Keras is a more simplified alternative to Tensorflow but has limited functionality.

Caffe is a C++-based framework that is open source and has a collection of reference models. Using the CUDA library, it performs computations on the GPU. A positive aspect of this framework is its ability to separate representation from implementation. It also provides interfaces for MATLAB and Python programming languages. Torch supports the effective development and fast implementation of many DL models on both CPUs and GPUs along with parallelization, and is written using the Lua language.

3.4.3 Applications

By taking advantage of the available IoT infrastructure and integrated DL frameworks that can run complex neural network models, the implementation of multiple real-time applications becomes possible.

RF Sensing: Typically, an IoT application consists of three layers: the sensing layer, which records values using sensors; the gateway layer, which is responsible for connecting the IoT device to a network and helping it communicate; and the cloud layer, which performs computation tasks with data stored on the cloud. RF (Radio Frequency) sensing is one such area [23], where signals that are obtained by any means from ordinary objects such as through reflection, scattering etc. can record physical values such as speed, activity, position, and direction. These will then be monitored and analyzed using DL algorithms, such as CNNs, LSTMs, and AEs, to generate predictions or perform other tasks. This is particularly helpful, as RF sensing can cover large areas with multiple subjects, and is cost effective as well. IoT devices used for this purpose should additionally provide support for WiFi, RFID sensors, acoustics, and ultra-wideband along with a GPU for running models using TensorFlow or Keras frameworks.

Since such real-time signals contain a lot of noise, it is essential that a sufficient amount of preprocessing is done to remove errors from this calibrated data. This includes randomness errors like Packet Boundary Detection (PBD), Sampling Frequency Offset (SFO), and Central Frequency Offset (CFO). The data is then converted into different formats for each type of DL model, such as image-based for CNNs, time-series for LSTMs [15], and the original format for AEs. Finally, the entire prediction process involves an offline training stage and an online prediction stage. Applications can be used for recognition and detection tasks, some of which are indoor localization with the help of WiFi NIC CSI and AEs, activity recognition like security surveillance through anomaly detection, and health sensing by monitoring vital signs. If the environment in which the predictions are to be made changes, transfer learning can be exploited.

When dealing with multimodal data, GANs and DRL proves to be helpful when there is a dearth of training data, and fusion from multiple data sources needs to be performed. In such cases, different DL channels to deal with each type of data will be helpful.

Brain-Computer Interface (BCI): Brain signals serving as input to an IoT device can be analyzed to automatically convert them to commands for controlling actions performed by these devices [34]. The BCI can therefore provide a direct interface for transferring information from the brain to the computer, avoiding the requirement of any external media to do so. In order to extract the most essential features from this input data, a selective-attention mechanism based on reinforcement learning (RL) is formulated. Moreover, LSTMs process the interdimensional information obtained from the SAM. Two advantages have led to BCI becoming so popular today, due to its privacy factor owing to the nature of brain activity [35,36], and for minimal physical exertion as only thinking needs to be performed in real time for the interaction to take place.

However, there are many issues that occur while working with brain signals acquired using EEG, MEG, fNIR, etc. Due to their low signal-to-noise ratio, it becomes very difficult to model the actual activity as spatial and temporal resolutions are low. Such signals are highly susceptible to environmental activity and sentiment statuses. Moreover, the quality of data preprocessing is human-dependent. Therefore, DL models can take care of these problems as they can work with any type of data, irrespective of dimensionality and improve the data collection process.

The obtained raw brain signals are forwarded to the cloud where a pretrained DL model analyzes it and generates new signals for controlling the connected devices. Three neural network architectures form the training model: Replicate and Shuffle processing (RS), Selective Attention Learning, and the LSTM network for classification [37]. In order to preprocess the data, the input is replicated and shuffled to increase latent dependency among all feature dimensions. RL-based SAM then tries to fine tune the dependency to depend on only the most relevant subset of features, which is modeled as a focal zone. The state of the RL agent is described by the position of the focal zone. Finally, the LSTM classifier uses this information to generate commands. Rewards obtained by the RL agent are based on the classification performance of the LSTM [15]. BCI can be employed in applications such as brain typing systems and cognitive smart-controlled home service robots [18].

Edge Computing: DL models have a layered structure, thus making it suitable for edge computing. A major disadvantage of cloud computing is that huge amounts of data cannot be processed with limited network connectivity [38], which makes it imperative to defer such tasks to edge computers near the IoT devices [39]. This is especially helpful when the intermediate data size is smaller than the input data size. Some IoT networks defer the more computationally heavy tasks to the edge devices from the central server. In addition, DL models can maintain data privacy. These networks consist of two layers: (i) an edge layer that consists of IoT devices and the IoT gateway [40] and controls network access in LANs, and (ii) a cloud layer that consists of IoT devices connected to cloud services using internet connections and cloud servers. Such a system is highly useful for performing automatic object detection.

The edge computing structure consists of two parts: the edge layer collects the data and processes it using the DL models represented in Figure 3.8. The lower layers of the model are closer to the IoT gateway, whereas the higher layers are nearer to the cloud. In order to solve the scheduling problem, an online and offline learning algorithm is proposed. Tasks with the maximum bandwidth are removed. Tasks to be run on online devices are decided by the online learning algorithm. These edge devices are highly efficient and capable of performing complex processing in a short time. They can handle much more computation than the centralized cloud servers and can work with huge volumes of data. Moreover, edge computing ensures lesser response time and energy consumption, making it easier to run DL models and improving their performance. Based on existing IoT infrastructure, the gateways are the edge devices that forward the collected data to the cloud using the network.

A trained CNN model on the Caffe Framework can be run on an Nvidia GPU using this technology for video-sensing applications. With the help of a camera, image-like data can be obtained from the IoT gateways using WiFi connections. This data is then transferred to the cloud for further analysis. As only the intermediate processed information is sent for cloud computing, it reduces the pressure on the network by a great margin. Therefore, more layers in the DL model are dedicated to the edge, and lesser layers are dedicated to the cloud to bring down network traffic. With the Zuloco IoT device and trained CNN models that are run using Tensorflow on the ARM Computation Library (ACL), a perception-based system can also be modeled for object labeling and tracking device positions, using data from the cloud.

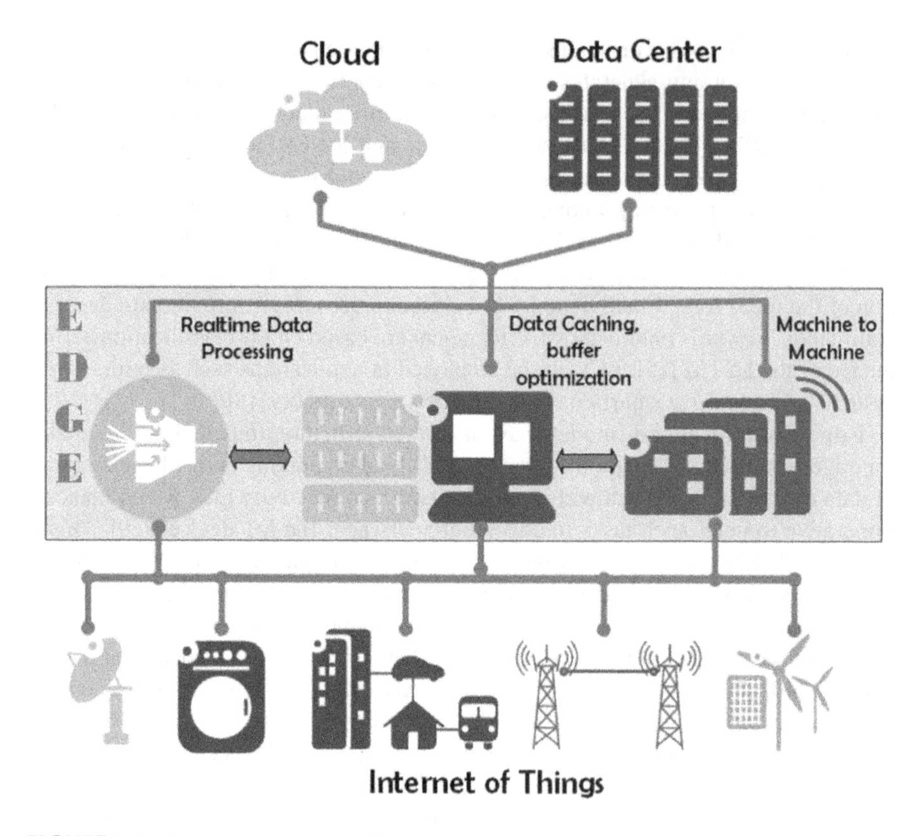

FIGURE 3.8 Edge computing in IoT.

Activity Recognition and Motion Analysis: One major challenge that arises while designing DL models for IoT applications is the dearth of labeled data available for training. Therefore, semi-supervised and supervised algorithms are becoming more popular today. A recent architecture called DeepSense uses a combination of various neural network models to deal with different kinds of sensory inputs, using the Intel Edison IoT device. It has a dual-core Intel Atom CPU, with 1GB memory for running the neural networks. Individual CNN models connected to each sensor encode their local recorded features. This information is then passed on to a global CNN model that fuses the individual encoded sensor outputs. An RNN then extracts temporal patterns from this data that can then be used for estimation or classification tasks. Moreover, to ensure that the DL architecture can run on available IoT devices, dropout layers are also introduced using a compression algorithm so that redundant inputs are discarded. This helps to speed up computation and reduce resource consumption while ensuring good prediction accuracy.

This architecture can be used in applications such as heterogeneous human activity recognition (HHAR) and user identification with biometric motion analysis (UserID). HHAR requires input from motion sensors such as accelerometers and gyroscopes of unseen users for estimation, whereas UserID trains on the same kind of input with a fixed set of users for classification. To address model reliability issues, an uncertainty estimation algorithm is utilized. This performs principled uncertainty estimation with the help of supervised and semi-supervised training schemes and deep IoT compression algorithms. Also, the loss function that is used for error calculation influences the estimation of uncertainties and must be chosen carefully. Other applications where such architectures could be helpful include real-time detection systems such as earthquake monitoring and weather data.

GANs can also be used to deal with the problem of labeled data using a semi-supervised approach. By formulating an adversarial game, three components of the network can be trained simultaneously. A classifier is responsible for predicting labels for the input data, which a generator then uses to recreate the original input from the learned representations which will surely include perturbations, and a discriminator tries to identify the real samples from the fake ones, given the labels. However, exploiting GANs for such applications would require support for high network complexity, must be able to eliminate training instability, and would require suitable preprocessing of multimodal data. Therefore, an activation function like the Wasserstein metric can prevent such issues.

In addition to HHAR and UserID, another area of application is WiFi signal-based gesture recognition that recognizes hand gestures using received signal strength indicators from WiFi signals [33]. The input consists of hand gestures made near a smartphone in different spatial and traffic scenarios. HHAR is trained using the cross-validation approach. Some improvements such as attention-based structures for multi-sensor data fusion, extending the models to perform regression tasks and reducing computation intensity, can help to make the architecture more scalable.

Information-Centric IoT: Unlike the current centralized host-centric cloud computing servers, Information-Centric Network (ICN) is a better alternative. Functionalities in such networks are based on the content name instead of the IP address of the content producer. Here, the content itself is secure, instead of taking extra measures for channel security. Moreover, the network layer supports caching, which can increase communication delay, improve content access, and improve dissemination, thus leading to better performance by preventing constant communication with the signal producer. This also resolves the issue of limited storage in edge devices. In order to deal with heterogeneity and mobility that are common problems in IoT applications, ICN is a good solution as it uses content names and prevents the need for middlewares. As it can also be integrated with edge computing, deploying applications becomes easier. To further speed up the execution process, some modifications like distributed DL and batch-orchestration algorithms can be introduced.

Local communication and networks may be disrupted by natural disasters and power failures, therefore faster processing can be achieved by implementing ICN over IoT infrastructure. Collaborative on-path caching can enhance data distribution and diversity in the entire IoT network. Since inputs in information-centric IoT are multimodal in nature, the Graph Convolutional Network (GCN) integrates graph topology and local vertex features in the convolutional layers of a typical CNN model. This helps to create subgraphs for representing such input data. For online processing of real-time data, RNNs and LSTMs can help. Some applications like handover controllers (HO) learn policies using RL in a large-scale wireless environment. It is framed as a collaborative model between a DNN and HO controller. Smart city applications with IoT also use a semi-supervised deep RL model.

Secure Software Defined Networks (SDNs): Privacy of data is a major challenge while trying to deploy IoT applications. A new architecture, SDNs, has recently become popular for its ability to enhance the security and resilience of IoT. This can prove to be useful in massive IoT deployments which have to deal with a lot of network traffic and security [41] is critical. The SDN model has a control plane containing a network controller for the logic and forward planes to switch the data flow. Hardware devices such as switches, routers, firewalls, and intrusion detection systems (IDS) form the forward plane. Simulation and programming for these devices are taken care of by the network controller, also known as the Network Operating System (NOS). It provides network state and topology information, as well as a northbound API to communicate with applications and a southbound API serving as an interface between itself and the forwarding devices with the help of OpenFlow protocol.

(IDS in networks can be implemented using the RBMs DL architecture for network anomaly detection [20,27]. Key advantages of using SDNs include ease of management, fault tolerance, interoperability, and network resilience. IDS are hardware or software systems dedicated toward identifying security threats in network traffic. Collection of data, analysis, and response launch are the steps to be followed in the process. The analysis can be performed using either a signature-based approach that compares the network signatures with an existing database of violation signatures to identify possible known threats to security, or an outlier analysis that detects abnormal patterns or a stateful protocol analysis, where network activity is compared with the expected behavior, and an alert is made in case of profile violation. A hybrid of the above approaches can improve the accuracy of detection that is made.

The controller layer will run the intrusion detection DL model consisting of RBMs. Those input samples that have the highest reconstruction error are identified as anomalies. This type of training could be either semi-supervised or unsupervised. Using the KDD99 dataset, five types of attacks, namely DoS, Probe, User to Root, Remote to Local, and Reconnaissance attacks, were tracked.

Smart Buildings: Automatic occupant activity recognition is important to control individual components that form the IoT network of smart buildings [31,42,43]. This can be achieved using CSI input from WiFi-enabled devices, e.g., thermostats, soundbars, and televisions. Occupancy detection, counting, and localization are necessary for overall occupant activity recognition [26]. A combined DL model using AEs to eliminate noise, CNNs to perform feature extraction of salient features, and LSTMs to identify temporal dependencies among features is used. The basic principle behind this is the interference of signal reception during human movement.

WiFi is a favorable input medium owing to its low cost and privacy preservation properties. Open-wrt is a lightweight Linux-based OS that is suitable for this task. Two WiFi routers, a transmitter and receiver of signals, will identify any interference [44] as a potential activity and then CSI frames are generated from this input.

Handheld Devices and Wearables: Tracking of user and surrounding conditions performed by handheld devices [45] and wearables provide support for a vast range of applications. Moreover, CNNs and other DL architectures are helpful while working with audio and image-based sensor data. Some representative deep learning models such as DeepEar, AlexNet, SVHN, and Deep KWS are possible implementations. Three hardware platforms such as the Qualcomm Snapdragon 800, Nvidia Tegra K1, and Intel Edison provide the IoT infrastructure. The Torch and Caffe frameworks run the DL models on these platforms. However, two key aspects – memory bottlenecks and partitioning of inference – across processors must be ensured for efficient performance.

Big Data and Streaming Analytics: Speech and gesture recognition applications involve working with a great amount of big data [45,46], primarily in the healthcare industry followed by transportation, agriculture, and urban infrastructure. IoT is one of the main producers of big data, whereas the big data [47] industry sees IoT as an important target for improvement. Some applications like autonomous driving, fire prediction, and health condition recognition deal with big data analytics in real-time obtained from sensors. High-performance computing systems and cloud platforms are essential to process such huge volumes of information. Streaming data analytics includes concepts of data parallelism where a large dataset is divided into several smaller datasets on which analytics run parallel, and incremental processing performs small tasks quickly in a pipeline of computation tasks. However, they must be accomplished within a time-bound setting.

Some characteristics of IoT big data are volume, the velocity of data production and processing in real-time, variety of data including structured, semistructured and unstructured data as well as text, audio, video, image etc., veracity for ensuring quality, consistency and trustworthiness of the data for accuracy, and variability to deal with differences in data flow and value that transforms big data to useful information. DL models [9] such as CNNs can enhance the time complexity of the training process as it performs local connectivity over a subset of the inputs obtained via cameras for an image recognition task. Applications such as landslide prediction and traffic signal detection exploit such networks. RNNs and LSTMs work with time-series data to predict movement patterns, estimate energy consumption levels, etc. AEs are used for fault detection and diagnosis-related tasks due to their ability to denoise data and extract important features.

Variations of these traditional models, such as Faster R-CNN, are able to detect objects in real-time. To deal with the incremental volume of data, online ML algorithms [32] are helpful. Approaches such as Deep RL ensure the autonomy of IoT platforms and transfer learning can compensate for a lack of training datasets, while stream analysis of big data can be performed using online learning. For automatic speech or voice recognition, energy-intensiveness and resource consumption are two major concerns that need to be addressed. The recognition can be performed in three levels. A simple neural network identifies voice activity through environmental noise, the more complex network checks whether the voice can be classified as speech, and finally the network with the highest complexity is used to detect individual words.

Indoor localizations such as smart homes and hospitals provide location-aware services. Input can be acquired through WiFi, RFID, Bluetooth, etc. CNNs architectures are able to fuse both magnetic and visual sensing data. Data from Inertial Navigation Systems (INS) processed using LSTMs are analyzed to predict robot positions. Combined CNN and RNN models can perform gesture recognition. An

Encoder-Recurrent-Decoder architecture for human body pose recognition and motion prediction in videos has been proposed recently. It consists of an encoder network, followed by an RNN and then by a decoder network.

To ensure security and privacy in IoT networks [48], especially from False Data Injection (FDI), a conditional Deep Belief Network (DBN) identifies attack features and uses them for detection in real time. Moreover, smartphones are susceptible to hacking, thus inducing a sense of vulnerability to user data privacy. Energy load forecasting using CNNs, LSTMs, and LSTM sequence-to-sequence models is also becoming a popular application today. Smart cities incorporate agriculture, transportation, energy, and many such IoT domains. To ensure more efficient working of DL models on IoT infrastructure, several measures such as increasing machine availability, lowering maintenance expenses, and optimizing monitoring sensors can be taken.

Another application of a real-time crowd density prediction system with user's caller data record (CDR) from their mobile phones, which includes user ID, time, location, and communication details, along with an RNN model, has been explored. Vision-based classifications using CNN models perform waste management and garbage classification. Also, air quality monitoring and pollution control can be accomplished through a similar method. Intelligent Transport Systems (ITS) employ RNN and RBM models in a parallel computing environment with the help of GPS data. This can effectively prevent traffic congestion.

Object Tracking: Multiple object visual tracking [49,50] can be performed using CNNs. With an NVIDIA Jetson architecture, along with a camera and wireless connection, results obtained through this combination can be applied for multiple real-time tasks. Computer vision is an emerging technology using ML that teaches objects to identify relevant features from image and video input in real time. Such applications can also be deployed through edge computing. By interpolation of identified features, using algorithms such as YOLO and Mask RCNN, real-time tracking of objects can be achieved.

Automatic Medical Diagnosis: With electrocardiogram (ECG) data and a hierarchical edge-based DL model [31] of CNNs, automatic heart rate monitoring of patients can be performed. Moreover, for such health monitoring applications, cloud–fog-based architectures are also becoming popular. This also integrates the latest technology such as Infrastructure-as–a-Service (IaaS) and virtual machines in the final deployment to identify health problems based on monitored heart rate. In addition to the above applications, the resourcefulness of DL in IoT can be exploited for multiple other applications such as manufacturing, supply chain management, precision agriculture, and animal tracking.

3.5 CONCLUSION

Fast and efficient processing of real-time data is being widely utilized in various industries today. Though IoT devices mainly consist of embedded systems with attached sensors, they have limited computation capability and reliability to run complex DL architectures. Therefore, modifications such as parallelism and cloud-based data handling are being introduced to compensate for these constraints.

Every architecture is suited for dealing with specific kinds of input. CNNs can deal with image and video data very well, whereas RNN architectures like LSTMs perform well with time-bound data. Moreover, multiple DL frameworks with CPU and GPU support are available today to run these complex models. Due to the versatile nature of DL-based IoT, its capabilities can be exploited for multiple implementations like RF sensing, BCI, and edge computing. These are then helpful for applications such as smart homes, health monitoring, and object tracking to name a few. However, there are several challenges that need to be addressed to ensure proper performance and reliability of such a technology. Nevertheless, DL-based IoT is one domain that is gaining momentum and is sure to receive massive investments in the near future.

REFERENCES

1. Ismaeel Al Ridhawi, Yehia Kotb, Moayad Aloqaily, Yaser Jararweh, and Thar Baker. A profitable and energy-efficient cooperative fog solution for IoT services. *IEEE Transactions on Industrial Informatics*, 16(5):3578–3586, 2019.
2. Mohammad Saeid Mahdavinejad, Mohammadreza Rezvan, Mohammadamin Barekatain, Peyman Adibi, Payam Barnaghi, and Amit P Sheth. Machine learning for internet of things data analysis: A survey. *Digital Communications and Networks*, 4(3):161–175, 2018.
3. G Thippa Reddy, S Bhattacharya, S Siva Ramakrishnan, Chiranji Lal Chowdhary, Saqib Hakak, Rajesh Kaluri, and M Praveen Kumar Reddy. An ensemble based machine learning model for diabetic retinopathy classification. In *2020 International Conference on Emerging Trends in Information Technology and Engineering (IC- ETITE)*, pages 1–6, 2020.
4. Ahmad Azab, Robert Layton, Mamoun Alazab, and Jonathan Oliver. Mining malware to detect variants. In *2014 Fifth Cybercrime and Trustworthy Computing Conference*, pages 44–53. IEEE, 2014.
5. Thippa Reddy, Swarna Priya RM, M Parimala, Chiranji Lal Chowdhary, Saqib Hakak, and Wazir Zada Khan. A deep neural networks based model for uninterrupted marine environment monitoring. *Computer Communications*, 157:64–75, 2020.
6. Ahmad Azab, Mamoun Alazab, and Mahdi Aiash. Machine learning based bot- net identification traffic. In *2016 IEEE Trustcom/BigDataSE/ISPA*, pages 1788–1794. IEEE, 2016.
7. Mehdi Mohammadi, Ala Al-Fuqaha, Sameh Sorour, and Mohsen Guizani. Deep learning for IoT big data and streaming analytics: A survey. *IEEE Communications Surveys & Tutorials*, 20(4):2923–2960, 2018.
8. Hakima Khelifi, Senlin Luo, Boubakr Nour, Akrem Sellami, Hassine Moungla, Syed Hassan Ahmed, and Mohsen Guizani. Bringing deep learning at the edge of information-centric internet of things. *IEEE Communications Letters*, 23(1):52–55, 2018.
9. Shuochao Yao, Yiran Zhao, Huajie Shao, Chao Zhang, Aston Zhang, Shaohan Hu, Dongxin Liu, Shengzhong Liu, Lu Su, and Tarek Abdelzaher. Sensegan: Enabling deep learning for internet of things with a semi-supervised framework. *Proceedings of the ACM on Interactive, Mobile, Wearable and Ubiquitous Technologies*, 2(3):1–21, 2018.
10. Mamoun Alazab and MingJian Tang. *Deep Learning Applications for Cyber Security*. Cham: Springer, 2019.
11. Thierry Bouwmans, Sajid Javed, Maryam Sultana, and Soon Ki Jung. Deep neural network concepts for background subtraction: A systematic review and comparative evaluation. *Neural Networks*, 117:8–66, 2019.

12. A Galusha, J Dale, JM Keller, and A Zare. Deep convolutional neural network target classification for underwater synthetic aperture sonar imagery. In *Detection and Sensing of Mines, Explosive Objects, and Obscured Targets XXIV*, volume 11012, page 1101205. International Society for Optics and Photonics, 2019.
13. Najla Etaher, George RS Weir, and Mamoun Alazab. From zeus to zitmo: Trends in banking malware. In *2015 IEEE Trustcom/BigDataSE/ISPA*, volume 1, pages 1386–1391. IEEE, 2015.
14. He Li, Kaoru Ota, and Mianxiong Dong. Learning IoT in edge: Deep learning for the internet of things with edge computing. *IEEE Network*, 32(1):96–101, 2018.
15. Mamoun Alazab, Suleman Khan, Somayaji Siva Rama Krishnan, Quoc-Viet Pham, M Praveen Kumar Reddy, and Thippa Reddy Gadekallu. A multidirectional LSTM model for predicting the stability of a smart grid. *IEEE Access*, 8:85454–85463, 2020.
16. Jie Tang, Dawei Sun, Shaoshan Liu, and Jean-Luc Gaudiot. Enabling deep learning on IoT devices. *Computer*, 50(10):92–96, 2017.
17. G Glenn Henry, Terry Parks, and Kyle T O'brien. Neural network unit with output buffer feedback for performing recurrent neural network computations, February 4 2020. US Patent 10,552,370.
18. Celestine Iwendi, Mohammed A Alqarni, Joseph Henry Anajemba, Ahmed S Alfakeeh, Zhiyong Zhang, and Ali Kashif Bashir. Robust navigational control of a two-wheeled self-balancing robot in a sensed environment. *IEEE Access*, 7:82337–82348, 2019.
19. Celestine Iwendi, Zunera Jalil, Abdul Rehman Javed, Thippa Reddy, Rajesh Kaluri, Gautam Srivastava, and Ohyun Jo. Keysplitwatermark: Zero water- marking algorithm for software protection against cyber-attacks. *IEEE Access*, 8:72650–72660, 2020.
20. Celestine Iwendi, Mueen Uddin, James A Ansere, Pascal Nkurunziza, Joseph Henry Anajemba, and Ali Kashif Bashir. On detection of sybil attack in large-scale vanets using spider-monkey technique. *IEEE Access*, 6:47258–47267, 2018.
21. Chafika Benzaid, Karim Lounis, Ameer Al-Nemrat, Nadjib Badache, and Mamoun Alazab. Fast authentication in wireless sensor networks. *Future Generation Computer Systems*, 55:362–375, 2016.
22. Ahmed Dawoud, Seyed Shahristani, and Chun Raun. Deep learning and software-defined networks: Towards secure IoT architecture. *Internet of Things*, 3:82–89, 2018.
23. Xuyu Wang, Xiangyu Wang, and Shiwen Mao. Rf sensing in the internet of things: A general deep learning framework. *IEEE Communications Magazine*, 56(9):62–67, 2018.
24. Xiang Zhang, Lina Yao, Shuai Zhang, Salil Kanhere, Michael Sheng, and Yunhao Liu. Internet of things meets brain–computer interface: A unified deep learning framework for enabling human-thing cognitive interactivity. *IEEE Internet of Things Journal*, 6(2):2084–2092, 2018.
25. Swarna Priya RM, Sweta Bhattacharya, Praveen Kumar Reddy Maddikunta, Siva Rama Krishnan Somayaji, Kuruva Lakshmanna, Rajesh Kaluri, Aseel Hussien, and Thippa Reddy Gadekallu. Load balancing of energy cloud using wind driven and firefly algorithms in internet of everything. *Journal of Parallel and Distributed Computing*, 142:16–26, 2020.
26. Nicholas D Lane, Sourav Bhattacharya, Petko Georgiev, Claudio Forlivesi, and Fahim Kawsar. An early resource characterization of deep learning on wearables, smartphones and internet-of-things devices. In *Proceedings of the 2015 International Workshop on Internet of Things Towards Applications*, pages 7–12, 2015.
27. Sweta Bhattacharya, Praveen Kumar Reddy Maddikunta, Rajesh Kaluri, Saurabh Singh, Thippa Reddy Gadekallu, Mamoun Alazab, and Usman Tariq. A novel PCA-firefly based XGBoost classification model for intrusion detection in networks using GPU. *Electronics*, 9(2):219, 2020.

28. Dinesh Valluru and I Jasmine Selvakumari Jeya. IoT with cloud based lung cancer diagnosis model using optimal support vector machine. *Health Care Management Science*, pages 1–10, 2019.
29. Thippa Reddy Gadekallu, Neelu Khare, Sweta Bhattacharya, Saurabh Singh, Praveen Kumar Reddy Maddikunta, In-Ho Ra, and Mamoun Alazab. Early detection of diabetic retinopathy using PCA-firefly based deep learning model. *Electronics*, 9(2):274, 2020.
30. Shreshth Tuli, Nipam Basumatary, Sukhpal Singh Gill, Mohsen Kahani, Rajesh Chand Arya, Gurpreet Singh Wander, and Rajkumar Buyya. Healthfog: An ensemble deep learning based smart healthcare system for automatic diagnosis of heart diseases in integrated IoT and fog computing environments. *Future Generation Computer Systems*, 104:187–200, 2020.
31. Han Zou, Yuxun Zhou, Jianfei Yang, and Costas J Spanos. Towards occupant activity driven smart buildings via WiFi-enabled IoT devices and deep learning. *Energy and Buildings*, 177:12–22, 2018.
32. Muhammad Shafique, Theocharis Theocharides, Christos-Savvas Bouganis, Muhammad Abdullah Hanif, Faiq Khalid, Rehan Hafız, and Semeen Rehman. An overview of next-generation architectures for machine learning: Roadmap, opportunities and challenges in the IoT era. In *2018 Design, Automation & Test in Europe Conference & Exhibition (DATE)*, pages 827–832. IEEE, 2018.
33. Zahoor Uddin, Muhammad Altaf, Muhammad Bilal, Lewis Nkenyereye, and Ali Kashif Bashir. Amateur drones detection: A machine learning approach utilizing the acoustic signals in the presence of strong interference. *Computer Communications*, 154:236–245, 2020.
34. Rajesh Kaluri and Ch Pradeep Reddy. Optimized feature extraction for precise sign gesture recognition using self-improved genetic algorithm. *International Journal of Engineering and Technology*, 8(1):25–37, 2018.
35. Rajesh Kaluri and Ch Pradeep Reddy. An overview on human gesture recognition. *International Journal of Pharmacy and Technology*, 8(04):12037–12045, 2016.
36. Rajesh Kaluri and Ch Pradeep Reddy. A framework for sign gesture recognition using improved genetic algorithm and adaptive filter. *Cogent Engineering*, 3(1):1251730, 2016.
37. Muhammad Numan, Fazli Subhan, Wazir Zada Khan, Saqib Hakak, Sajjad Haider, G Thippa Reddy, Alireza Jolfaei, and Mamoun Alazab. A systematic review on clone node detection in static wireless sensor networks. *IEEE Access*, 8:65450–65461, 2020.
38. Quoc-Viet Pham, Seyedali Mirjalili, Neeraj Kumar, Mamoun Alazab, and Won-Joo Hwang. Whale optimization algorithm with applications to resource allocation in wireless networks. *IEEE Transactions on Vehicular Technology*, 69(4):4285–4297, 2020.
39. Xiaoying Jia, Debiao He, Neeraj Kumar, and Kim-Kwang Raymond Choo. Authenticated key agreement scheme for fog-driven IoT healthcare system. *Wireless Networks*, 25(8):4737–4750, 2019.
40. Noshina Tariq, Muhammad Asim, Feras Al-Obeidat, Muhammad Zubair Farooqi, Thar Baker, Mohammad Hammoudeh, and Ibrahim Ghafir. The security of big data in fog-enabled IoT applications including blockchain: A survey. *Sensors*, 19(8):1788, 2019.
41. Dharmendra Singh Rajput, Rajesh Kaluri, and Harshita Patel. Security threat assessment of aircraft system using FSS. In *Proceedings of the 2019 7th International Conference on Information Technology: IoT and Smart City*, pages 453–457, 2019.
42. R Vinayakumar, Mamoun Alazab, Sriram Srinivasan, Quoc-Viet Pham, Soman Kotti Padannayil, and K Simran. A visualized botnet detection system based deep learning for the internet of things networks of smart cities. *IEEE Transactions on Industry Applications*, 56:4436–4456, 2020.

43. Kai Lin, Min Chen, Jing Deng, Mohammad Mehedi Hassan, and Giancarlo Fortino. Enhanced fingerprinting and trajectory prediction for IoT localization in smart buildings. *IEEE Transactions on Automation Science and Engineering*, 13(3):1294–1307, 2016.
44. G Thippa Reddy, Rajesh Kaluri, Praveen Kumar Reddy, Kuruva Lakshmanna, Srinivas Koppu, and Dharmendra Singh Rajput. A novel approach for home surveillance system using IoT adaptive security. Available at SSRN 3356525, 2019.
45. A Priyanka, M Parimala, K Sudheer, Rajesh Kaluri, Kuruva Lakshmanna, and M Reddy. Big data based on healthcare analysis using IoT devices. In *Materials Science and Engineering Conference Series*, volume 263, page 042059, 2017.
46. J Archenaa and EA Mary Anita. A survey of big data analytics in healthcare and government. *Procedia Computer Science*, 50:408–413, 2015.
47. G Thippa Reddy, M Praveen Kumar Reddy, Kuruva Lakshmanna, Rajesh Kaluri, Dharmendra Singh Rajput, Gautam Srivastava, and Thar Baker. Analysis of dimensionality reduction techniques on big data. *IEEE Access*, 8:54776–54788, 2020.
48. M Eswar Kumar, G Thippa Reddy, K Sudheer, M Reddy, Rajesh Kaluri, Dharmendra Singh Rajput, and Kuruva Lakshmanna. Vehicle theft identification and intimation using GSM & IoT. In *Materials Science and Engineering Conference Series*, volume 263, page 042062, 2017.
49. Heng Fan, Liting Lin, Fan Yang, Peng Chu, Ge Deng, Sijia Yu, Hexin Bai, Yong Xu, Chunyuan Liao, and Haibin Ling. Lasot: A high-quality benchmark for large-scale single object tracking. In *Proceedings of the IEEE Conference on Computer Vision and Pattern Recognition*, pages 5374–5383, 2019.
50. Seong-Wook Joo and Rama Chellappa. A multiple-hypothesis approach for multiobject visual tracking. *IEEE Transactions on Image Processing*, 16(11):2849–2854, 2007.

4 Deep Learning–Assisted Vehicle Counting for Intersection and Traffic Management in Smart Cities

Guy M. Lingani, Danda B. Rawat, and Moses Garuba
Howard University

CONTENTS

4.1 INTRODUCTION

A smart city is expected to use information and communication technologies to increase operational efficiency, share information with the public, and improve both the quality of government services and citizen welfare [1]. Intelligent transportation systems, one of many components of smart cities, are expected to improve traffic efficiency and road safety by reducing traffic accidents through an informed decision. Traffic data collection is a key input ingredient to understanding and improving traffic management and safety. There are different types of mechanisms to collect traffic data such as magnetic loops, pressure tubes, radar guns, microwave sensors, and cameras [2]. Statistics regarding street and highway accidents are so vital to any comprehensive understanding and treatment of the safety problem that their collection and analysis are essential [3]. The goal of implementing traffic monitoring systems is to enable better decision-making through the use of collected data for all stakeholders such as government, business, pedestrians, drivers, and residents [1].

The focus of any smart city should be its people, providing benefits such as a better quality of life for residents and visitors; economic competitiveness to attract industry and talent; and an environmentally conscious focus on sustainability [1]. Accurate traffic data collection helps to make an informed decision.

In this chapter, we propose an automated method for counting vehicles using deep learning–based video processing for identifying traffic patterns and offering an adaptive traffic management system. Specifically, the proposed approach counts vehicles in the intersection which helps to make an informed decision about making streets one-way or two-way adaptive traffic light system depending not only on the number of vehicles in the intersection but also based on the time of the day. The proposed approach detects and counts the number of vehicles going straight and turning left and right, and pedestrians at a traffic intersection in a real-time manner (as well as in recorded videos in case of offline processing) using deep learning. Then the collected statistics is used to make an informed decision. The proposed approach gives better results (in terms of time and accuracy) compared to state-of-the-art work.

There have been recent related researches such as counting number of vehicles on the road using recorded videos offline [4], video-based counting in a straight road [5], and vehicle detection based on networking [6]. Furthermore, vehicle counting in the intersection of Washington DC streets is done manually based on recorded videos [7], which is not real-time and tedious since in general humans should take, on average, a 17 minutes break after working for 52 minutes to be efficient [8]. However, none of the existing approaches considers monitoring and counting the vehicles/traffic in the intersection where vehicles could go straight, take right, take left, or take a U-turn.

To address these issues, we develop and evaluate a real-time vehicle counting algorithm in the intersection using deep learning [9] so that the collected data can be used to make adaptive traffic lights and make streets one-way or two-way depending on the time of the day and traffic patterns and when big trucks are allowed/disallowed in the city streets.

The remainder of the chapter is organized as follows. Section 4.2 presents a system model used for deep learning for monitoring and counting the vehicles in the intersection that go straight, take right, take left, or take a U-turn. Section 4.3 presents the mathematics behind how we count vehicles that go straight, take right turns, and left turns. Section 4.4 presents results that show the accuracy, precision, and recall of the proposed vehicle counting approach using deep learning. Finally, Section 4.5 presents the conclusions.

4.2 SYSTEM MODEL

The concept of the system model is based on a traffic intersection with four branches as shown in Figure 4.1, where each branch is crossed by a virtual line materialized by a yellow line and define by the line equation line $y = ax + b$ and labeled based upon the location of the line (North, South, East, or West). Deep learning with real-time YOLO [10] Convolutional Neural Network (CNN) is capable to mimic the functions of the human brain by processing data in order to detect objects. The CNN will then generate precious data such as vehicle coordinates allowing it to be tracked [9] and count when the centroid comes across a given virtual line. Since each virtual line

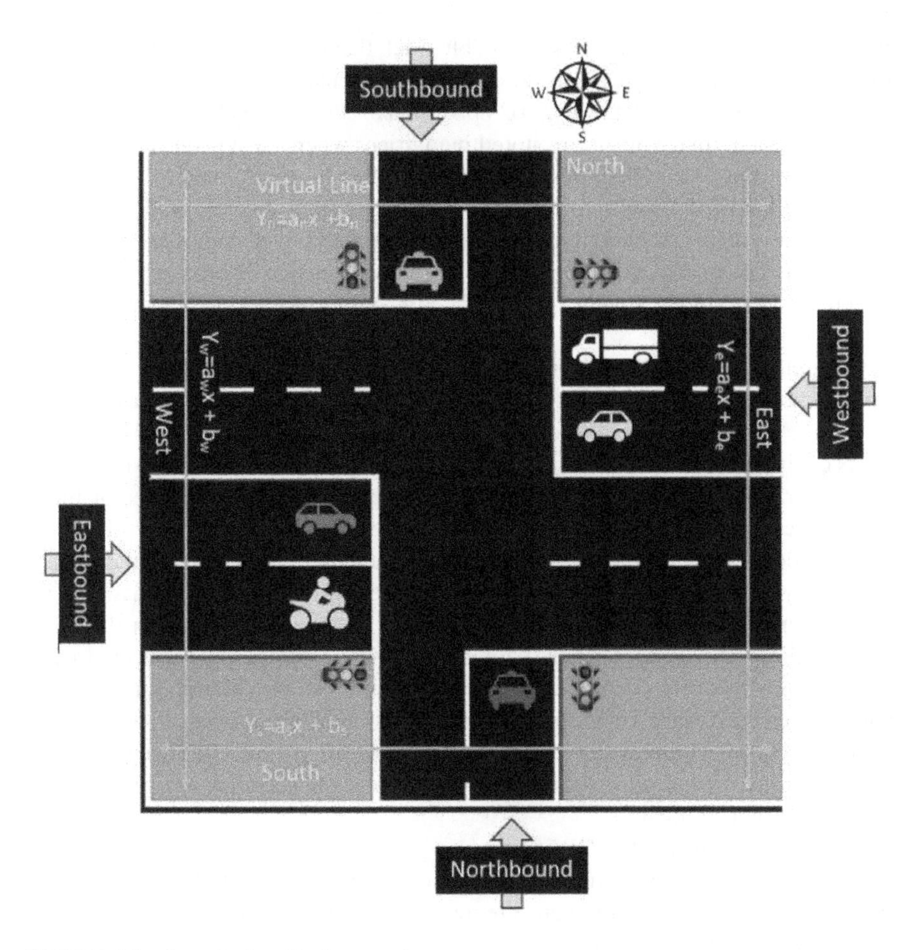

FIGURE 4.1 A typical traffic intersection.

is known and the detected objects centroids are know over the time relatively to the virtual lines there for it possible to precisely determine if an object has crossed a virtual line. Once an object is detected, image classification will determine the types of detected objects, such as cars, trucks, and motorcycles. By tracking the moving classified objects' positions, it becomes possible to count how many went straight, took a left turn, right turn, or U-turn. Specifically, we use deep learning to detect vehicles and their types (cars, trucks, motorcycles) and then we assign unique IDs to count them uniquely in the intersection and use the proposed vehicle counting approach to count vehicles that go straight, take right turns, and take left turns.

4.3 THE PROPOSED APPROACH

One of the major challenges in tracking objects is to accurately identify the same object's centroid coordinates across consecutive video frames. To assess the challenge, it is important to keep in mind the process in which computer perceives images, how it detects them, and how it can possibly track them. An image is represented by a set of picture elements or pixels. Each pixel represents one color. For instance, an image with a resolution of 1280 by 1024 pixels has 1280 × 1024 (1,310,720 pixels). In computer vision, image pixels are stored in matrices where each pixel is represented by a binary value. Gray image has an intensity value ranging from 0 to 255 with 0 being black and 255 being white. Color image RGB (Red, Green, Blue) is a three-dimensional matrix (channel) image. It contains the Red Green and Blue color image in three separate matrices. Each pixel is represented from 8 to 32 bit. Each pixel color is represented as an RGB value in Figure 4.2 (a combination of red, green, and blue ranging from 0 to 255).

Computer Vision problem is image classification problem also called image recognition problem where, from a given input image the algorithm try to determine if the picture is representing a specific object. Using Convolutional Neural Networks (CNN), it is possible to detect object edges in a given picture with successive convolution operations by applying a filter to detect objects. In the case of video pictures, images are processed frame by frame to detect objects. Multiple Object Tracking (MOT) allows several objects to be tracked concomitantly across all the frames where they are detected. The real problem resides in the accuracy of assigning a unique identifier to each object. The uniqueness of objects is critical in the sense where it allows the coordinates (x, y) of each object's centroid to be collected without ambiguity and then count them. In a two-dimensional coordinate system, object consecutive coordinates (x, y) indicate the object's movements over time. The initial goal is to detect and count vehicles at an intersection. Therefore, keeping each object identifier unique across the frames determines the accuracy in counting each detected object (Figure 4.3).

Binary Large OBject (BLOB) refers to a group of connected pixels in a binary image such as a car or pedestrian in the frame [11]. The notion of "Large" is used to distinguish sizes of objects of interest represented by groups of connected pixels of a certain size as opposed to "small" group of binary connected pixels considered to be noise. In computer vision, there exist various libraries with algorithms designed to detect BLOBs connected regions in digital images with interesting features such as

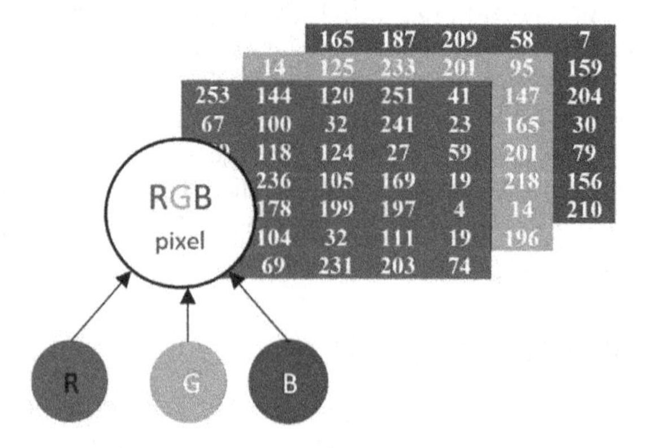

FIGURE 4.2 One pixel consists of three channels/layers.

FIGURE 4.3 Object (car, pedestrian, etc.) detection with category, ID/names (just to count), and centroid (faded dot).

blob contour, orientation, centroid, color, area, mean, and standard deviation of the pixel values in a targeted area. BLOBs can be detected using Laplacian of Gaussian (LoG) algorithm, Difference of Gaussian (DoG), or Determinant of Hessian (DoH) [11]. Centroid is considered to be the center of gravity of a BLOB. The BLOB centroid represents the average x-position and y-position of the BLOB, which is expressed as the BLOBs Anchor boxes and its ordinate mean as

$$x_c = \frac{1}{N}\sum_{i=1}^{N} x_i,\; y_c = \frac{1}{M}\sum_{i=1}^{M} y_i$$

In mathematical terms, the center of N × M BLOB is (x_c, y_c) with x_i and y_i running coordinates.

Object detection consists of indicating where an object is located in the frame. This type of process requires heavy computation effort. Object tracker spots and identifies objects by using their coordinates throughout the frames. Combining both object detection and object tracking is intuitively the route to be considered when seeking to count efficiently and accurately vehicles at a traffic intersection. Somewhat, our tests have shown that combining both object detection and tracking under some specific circumstances does not always produce a satisfactory result, particularly in traffic intersections when object trajectories cross [12]. Therefore, we propose to use the Euclidean distance calculation not only to track objects but also to increase and enhance the precision of assigning the same ID to the same object. Redundancy of tracking methods increases accuracy and precision in counting. Among other interesting tracking methods, object tracking with color matching and QR injection (Quick Response code injection) can considerably improve accuracy in terms of uniquely identify an object. Color matching is an extra dimension added to compare an object color over time and check if it matches. QR injection is a way to embed a code in the binary image to make it unique for blobs to be identified without ambiguity. For the moving direction, we are using a combination of several algorithms to track and count vehicles at traffic intersections.

4.3.1 Centroid Tracking Using Euclidean Distance

In this section, we describe and explain the Euclidean distance algorithm object centroids tracking. Video images are processed frame by frame. Note than a BLOB is a connected pixel delimited by the Anchor box, and the gravity center of the Anchor box is the BLOB/Object Centroid.

Step 1: In the first image, vehicles or objects are detected. Object Centroids and Bounding Boxes are associated with an initial identification number (Figure 4.4). The principle in which vehicles are matched is based on the k-Nearest Neighbors using four steps

where we

1. Calculate *Euclidean Distance* using

$$\text{dist} = \left(\mathrm{N}_{n=1}\left(\mathrm{x}_{1i} - \mathrm{x}_{2i}\right)^2\right)$$

2. Find the Nearest Neighbors
3. Make Predictions
4. Assign ID's

In Figure 4.4, the algorithm has identified two vehicles to track. For each vehicle, the bounding box known as the centroid coordinate is computed and ID#1 and ID#2 assigned respectively to the vehicles. The VW beetle is assigned with ID#1 while the Truck with ID#2.

Step 2: In the next frame, a new object is detected in addition to the two previous vehicles, and the coordinates of its centroids are provided. From the vantage point of the computer, objects detected in a given frame are considered as a set of BLOBs or objects at a given time with no correlation to BLOBs or objects from the previous frame (Figures 4.5 and 4.6). All the effort and the algorithm deployed here is to make the connection between an object in the preview frame with the same object in the current frame.

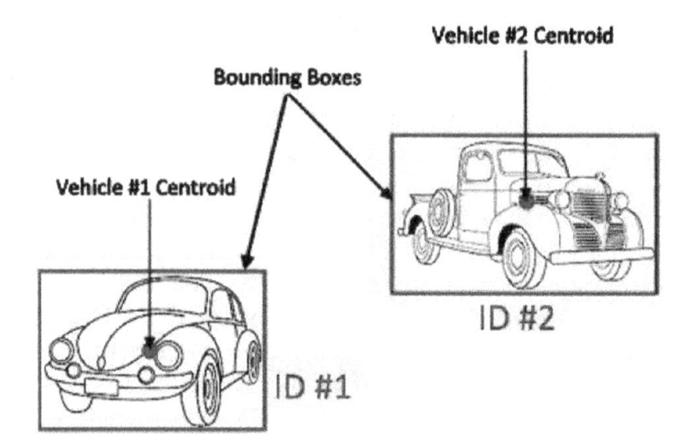

FIGURE 4.4 Initial objects detection with centroids bounding boxes and IDs assignment.

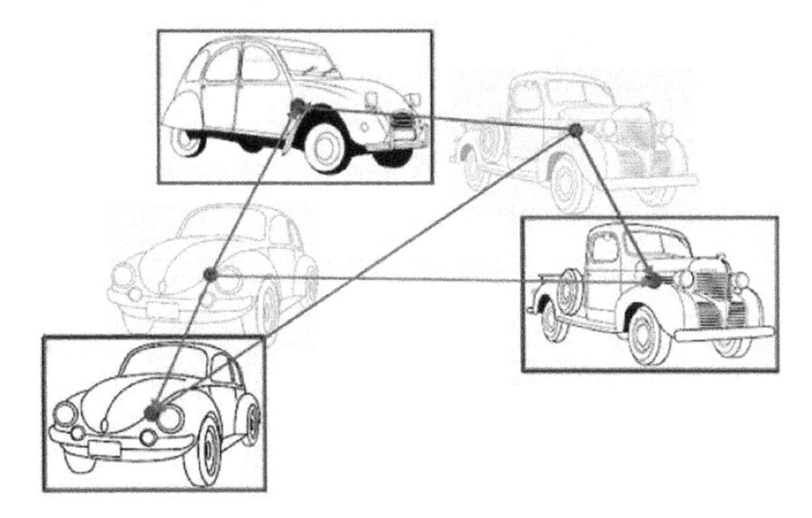

FIGURE 4.5 Euclidean distance calculation between all detected objects and all know centroids from the preview frame.

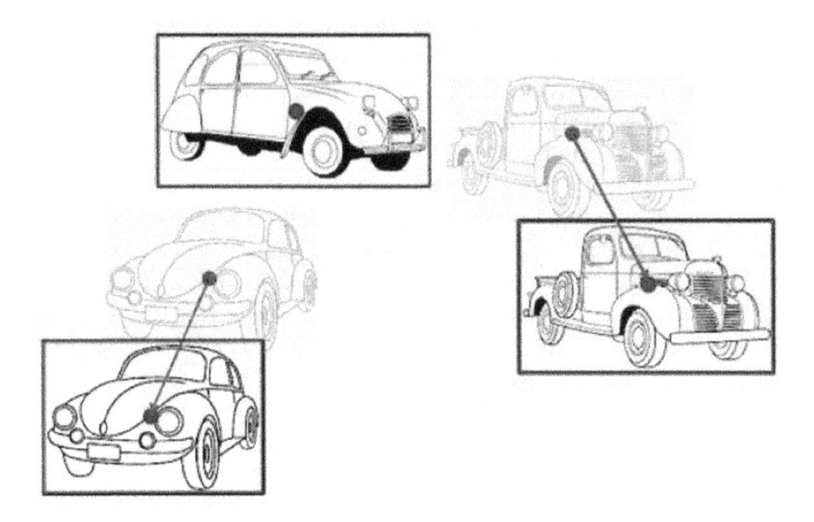

FIGURE 4.6 Matching objects centroids.

From the preview frame, the two initial objects centroid's coordinates are known and their respective IDs have been assigned and stored for future use. The algorithm computes all the Euclidean distances between the known centroids (red dots in Figure 4.5) from the preview frame and all the new centroids (purple dots in Figure 4.5) detected in the current frame.

Step 3: The algorithm compares all the Euclidean distances. The smallest Euclidean distance obtained by computing the distance between a given centroid from the preview frame and the centroid of the current frame is considered to be a match as shown in Figure 4.6. The biggest Euclidean distances are ignored.

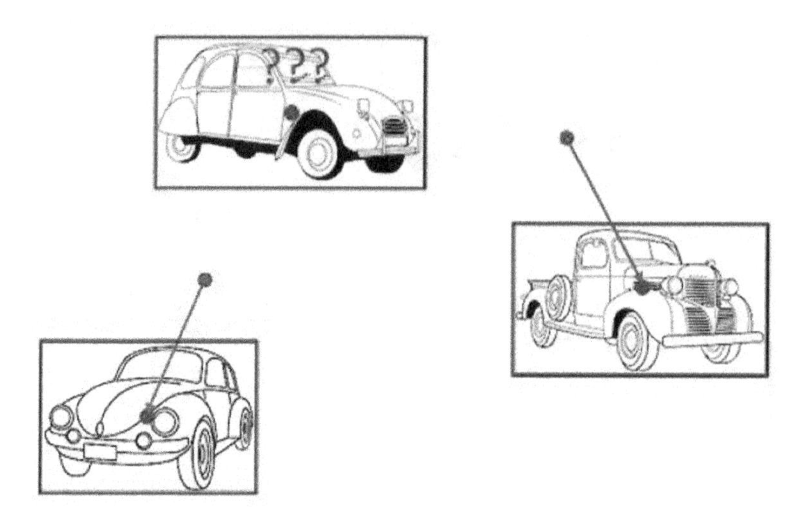

FIGURE 4.7 Objects associated with initial centroid based on smallest Euclidean smallest.

In Figure 4.6, we can clearly see that the smallest Euclidean distance calculated from the pair of centroid representing VW beetle first centroid coordinate to its actual centroid coordinate is the smallest. Similarly, the smallest distance is found to be the distance between the truck initial centroid to its actual centroid. The algorithm makes the inference that the pair of centroids with the smallest distance represents the same object. When an object moves its consecutive, centroid locations are made up of small steps (Figure 4.7).

Step 4: Using the smallest Euclidean distances, all the centroids are associated with a vehicle with their respective IDs. In the process, only new objects will not get matching centroid. In Figure 4.7, we can see that one centroid remains and cannot be paired base on the principle of smallest Euclidean distance. The algorithm assumes that the object must be a new vehicle. Therefore, a new Id is assigned to the unknown new object (Figure 4.8). The algorithm then outputs all the vehicles with their respective.

Centroids of Bounding boxes with their matching IDs number assigned in the preview frame. New objects are assigned with new IDs. A new object is registered by adding its centroid information to the list of tracked objects The implementation involves a process in which a threshold is set to eliminate objects that are missing or not detected after defined consecutive numbers of frames.

4.3.2 The Virtual Line Double Crossing Algorithm (VLDCA)

The postulate of the VLDCA is based on a set of fundamental facts on turning vehicles at a traffic intersection. For instance, an intersection made of two roads intercepting has four arms. In a four arm intersection, a vehicle is considered of making a turn (left turn or right turn) if and only if, at given time t_0, the vehicle is detected on one of the intersection arms; and at the next time t_n, the same vehicle is detected on another intersection arms different from the road arm directly opposed to the arm on which the vehicle was initially detected. A vehicle reentering in an intersection after clearing

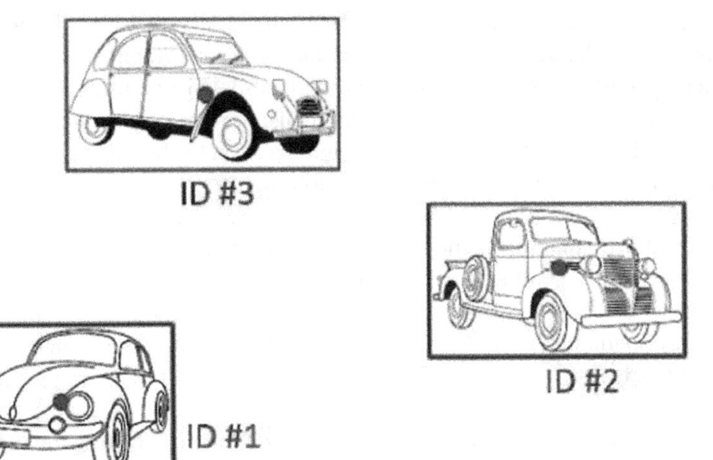

FIGURE 4.8 Object ID assignment.

is considered to be a new vehicle. The VLDCA is based on those postulates with the use of virtual lines to detect what arm of the intersection the vehicle is coming from and where it is heading to. A minimum of two virtual lines should be crossed in order to trigger the process of determining whether a turn has been made or not.

a. Principle:

The coordinate system of a computer screen is a cartesian coordinate system. The origin (0,0) is at the top left of the screen (Figure 4.9).

When the video is loaded, virtual lines are drawn to strikethrough the intersection's different branches (four branches in the case of Figure 4.1). An object centroid is constantly tracked with its coordinates and compared to the virtual lines. A moving object to the left will have decreasing X values and an increasing value when moving to the right. Similarly, when an object centroid Y values decrease, the object is

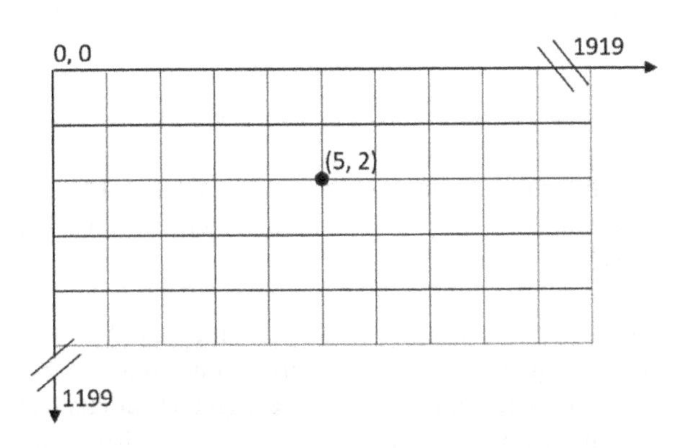

FIGURE 4.9 Coordinate system of a screen (1920 × 1200).

moving up; and inversely when the Y value increases, the object is moving toward the bottom of the screen. A centroid is the center of gravity of a detected object (vehicle, pedestrian, etc.); therefore, the two words will be used interchangeably to describe an object's movement. By continuously processing the centroid coordinate variations, it is possible to determine a moving object's direction. When an object is uniquely identified, turns are detected as follows:

- Eastbound line is the virtual line drawn along the screen rightmost side.
- Westbound line is the virtual line drawn along the screen leftmost side.
- Northbound line is the virtual line drawn along the screen topmost side.
- Southbound line is the virtual line drawn along the screen most bottom side.

A vehicle movement at the intersection is determined by detecting the double virtual line crossing as described below:

1. **Left Turn for a Vehicle Driving Northbound**: The vehicle Centroid comes across South Virtual Line ($Y_s = a_s x + b_s$) first, and then across West virtual line ($Y_w = a_w x + b_w$). At least two lines are crossed here.
2. **Right Turn for a Vehicle Driving Northbound**: The vehicle Centroid comes across South Virtual line ($Y_s = a_s x + b_s$) first, and then across East virtual Line ($Y_e = a_e x + b_e$).
3. **Drive-thru for a Vehicle Driving Northbound**: The vehicle Centroid comes across South Virtual Line ($Y_s = a_s x + b_s$) first, and then across North Virtual Line ($Y_n = a_n x + b_n$).
4. **Left Turn for a Vehicle Driving Southbound**: The vehicle Centroid comes across North Virtual line ($Y_n = a_n x + b_n$) first then across East Virtual line ($Y_e = a_e x + b_e$).
5. **Right Turn for a Vehicle Driving Southbound**: The vehicle Centroid comes across North Virtual line ($Y_n = a_n x + b_n$) first, and then across West virtual line ($Y_w = a_w x + b_w$).
6. **Drive-thru for a Vehicle Driving Southbound**: The vehicle Centroid comes across North Virtual line ($Y_n = a_n x + b_n$) first, and then across South Virtual Line ($Y_s = a_s x + b_s$).

The same principle of double lines crossing detection as shown above goes for vehicles driving Eastbound and Westbound.

Another approach for the virtual line drawing is the use of background subtraction method to determine the real lines of roads and then draw the virtual lines across each branch of the road intersection.

b. **Virtual Lines Initialization**: The virtual line initialization consists of determining the equation lines of the virtual lines drawn across each arm of the intersection. The initialization phase might be automated by using the background subtraction technique to determine the virtual line equations. The coordinates of a pair of points are obtained from each virtual line to be drawn based upon the traffic pattern from the first video frame since the traffic recording is made from a still video camera. A general straight-line equation is: $y = ax + b$ Straight line equation can be computed from the coordinates of two points belonging to that line as follows: Let $P_1(x_1, y_1)$ and $P_2(x_2, y_2)$ points known to be one the virtual line. The line slope (a) is defined as $a = (y_2 - y_1)/(x_2 - x_1)$. Any point belonging to a line satisfies that line equation. So by replacing the slope (a) in the general straight-line equation, we can compute all the unknown values needed to write the line equation.

$$y = ax + b$$

$$a = (y_2 - y_1)/(x_2 - x_1)$$

$$y = [(y_2 y_1)/(x_2 x_1)]x + b$$

Using P_2 to verify the line equation, we can write:

$$y_2 = a(x_2) + by_2 = ax_2 + b$$

$$b = y_2 ax_2$$

By replacing b by its value in the general line equation, we obtain the virtual line equation:

$$y = [(y_2 - y_1)/(x_2 - x_1)]x + [y_2 - ax_2]$$

For verification purposes, let $P_1(2, 5)$ and $P_2(-4, -7)$ be the coordinates of two points obtained by clicking on each side of the segment of road we intend to detect the crossing vehicles. Let us compute the virtual line slope a: $P_1(2, 5)$ and $P_2(-4, -7)$

$$a = (y_2 - y_1)/(x_2 - x_1)$$

$$a = (5 - (-7))/(2 - (-4))a = 12/6$$

Let us use $P_2(2, 5)$ to verify the equation $y = ax + b$ $y_2 = ax_2 + b \Rightarrow 3 = 2\ 2 + b \Rightarrow b = 1$.

Therefore, the virtual line for that segment of the intersection road is $y = 2x + 1$. The slope of the virtual line is $a = 2$. The virtual line intercept is $b = 1$. We can verify the line equation by replacing a and b by their respective value in the virtual line equation $y = 2x + 1$.

Virtual line equations are defined to detect moving object direction when their centroids cross them. To do so, we continuously compute each detected object centroids position relative to the virtual lines. Virtual line equations for a traffic intersection is the set of equations virtually drawn to intercept each branch of road composing the traffic intersection as shown in Figure 4.1.

Virtual line equations intercepting the four-crossing road (road branches) located, respectively, in the East, West, North, and South of the traffic intersection are defined as follows (see Figure 4.1):

$$Y_e = F_e(x) = a_e x + b_e Y_w = F_w(x) a_w x + b_w Y_n = F_n(x) = a_n x + b_n Y_s = F_s(x) a_s x + b_s$$

In case of traffic intersection with more than four crossing roads, additional virtual line equations corresponding respectively to virtual lines north-east, southeast, south-west, and north-west can be added and labeled as follows as linear equations: $Y_{ne} = F_{ne}(x)$, $Y_{se} = F_{se}(x)$, $Y_{sw} = F_{sw}(x)$, and $Y_{nw} = F_{nw}(x)$. While a video is processed by the neural network, objects are detected, classified, and tracked. Relevant outputs below are then collected for computation:

- Object Class (car, truck, pedestrian)
- Object ID (unique identification number)
- Prediction (percentage of certainty that an object belongs to a given class)

We also keep track of object's last three centroid coordinates: (coordinates of the bounding boxes)

1. $o(C_{x0}, C_{y0})$: Centroid coordinate when object is detected the very first time and ID assigned to it by the tracker time t_0.
2. $A(C_{xt}, C_{yt})$: Centroid coordinate a given time t
3. $B(C_{xt} + 1, C_{yt} + 1)$: Centroid coordinate a time t + 1
4. $C(C_{xt} + 2, C_{yt} + 2)$: Centroid coordinate a time t + 2

As the object moves over time, its three last centroid coordinates are updated except its initial centroid (C_{x0}, C_{y0}) when first detected.

Centroid updates algorithm:

$$.\, C_t = (C_{xt}, C_{yt})$$

$$.\, C_{t1} = (C_{xt} + 1, C_{yt} + 1)$$

$$.\, C_{t2} = (C_{xt} + 2, C_{yt} + 2)$$

$$.\, Ct_{new} = \text{Get object current Centroid Coordinates}$$

If Ct_{new} is different of C_{t2}, then

$$\{$$

$$.\ \text{Temp} = C_t$$

$$.\ C_t = C_{t1}$$

$$.\ C_{t1} = C_{t1}$$

$$\}$$

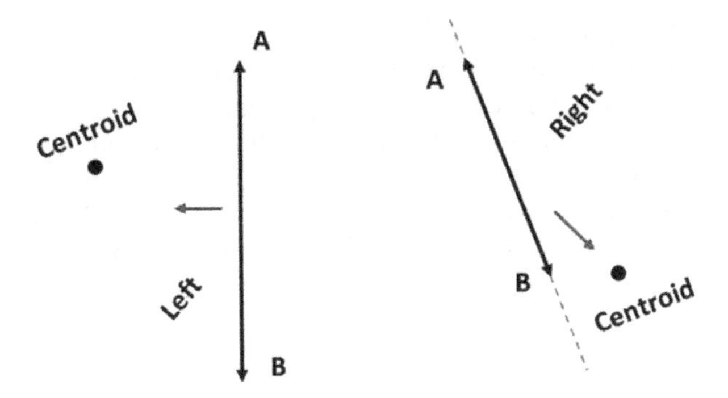

FIGURE 4.10 Centroid direction from a virtual line.

Objects relevant information are kept in a dictionary for future computation. With such a table, we are going to show how our algorithms unfold using mathematical line equations to determine the direction of an object. It consists of using the object centroid coordinates to determine if the object is located on top, bottom left, or right of a virtual line, and the direction of a centroid from a virtual line. The direction of an object centroid C_t from a virtual line consists of computing the direction of the centroid $C_t(x_t, y_t)$ from the virtual line to determine whether the centroid lies to the left of the virtual line or to the right of the virtual line (Figure 4.10).

$$\text{Checking Centroid Coordinate against } Y_e = F_e(x) = a_e x + b_e \text{ and}$$

$$Y_w = F_w(x) a_w x + b_w$$

We apply the cross-product of three points to our object centroids. Let (A_x, A_y), (B_x, B_y), and (C_x, C_y) be the coordinates of three points A, B, and C.

$$\text{diff} = (B_x - A_x)(C_y - A_y) - (B_y - A_y)(C_x - A_x)$$

a. If (diff > 0), then C is located at the left of line b.
b. If diff = 0, then C belongs to the line.
c. If (diff < 0), then C is located at the right side of the line. Similarly, we can apply the cross-product to three consecutive object centroids to determine their position relative to a given virtual line. Let (C_{xt}, C_{yt}), $(C_{xt} + 1, C_{yt} + 1)$, and $(C_{xt} + 2, C_{yt} + 2)$ be three consecutive centroid coordinates of a moving object. Let us compute $\text{diff} = (C_{xt} + 1 - C_{xt})(C_{yt} + 2 - C_{yt}) - (C_{yt} + 1 - C_{yt})(C_{xt} + 2 - C_{xt})$. Similarly: If (diff > 0), then C is located at the left of the virtual line.

If (diff = 0), then C belongs to the virtual line.

If (diff < 0), then C is located at the right side of the virtual line. We also check if a given centroid is above or below a virtual line. The equation of virtual lines are a general equation of a straight line Ax + By + C = 0.

For a straight line $y = f(x) = ax + b$, and a point $P(x_p, y_p)$. If $(y_p - f(x_p) < 0$, then P is under the line represented by $y = f(x) = ax + b$ which means that $[y_p - (ax_p + b)] < 0$ if $(y_p - f(x_p) > 0)$, then P is above the line represented by $y = f(x) = ax + b$ which means that $[y_p - (ax_p + b)] > 0$. When $[y_p - f(x_p)] = 0$, the point $P(x_p, y_p)$ belongs to the line. For a straight line $y = f(x) = ax + b$, and a point $P(x_p, y_p)$ if $(y_p - f(x_p)) > 0$ then P is under the line represented by $y = f(x) = ax + b$ which means that $[y_p - (ax_p + b)] < 0$ if $(y_p - f(x_p)) > 0$, then P is above the line represented by $y = f(x) = ax + b$ which mean that $[y_p - (ax_p + b)] > 0$. When $y_p - f(x_p) = 0$, the point $P(x_p, y_p)$ belong to the line.

4.4 PERFORMANCE EVALUATION

To evaluate the performance, we consider a traffic intersection where a deep learning algorithm continuously detects the vehicles (cars, buses, and motorcycles) and counts using the proposed approach mentioned in Section 4.3.

While evaluating the performance, we used Mean Square Error (RMSE) for predictions. The lower the RMSE the better is the fit. At the same time, particular attention should be made to avoid overfitting which is the consequence of a model that performs well on training data and conversely performs poorly when using evaluation data indicating that the model is memorizing the data and is not able to make prediction on never seen before data. Results are shown in Figures 23–25 for accuracy, recall, and precision.

Different terms used for evaluation are defined as (a) True Positives (TP): The model predicted object as a "Car", but object is known to be a car. (b) True Negatives (TN): The model predicted object as "Not Car", but object is known not to be a car. (c) False Positives (FP): The model predicted object as a "Car", but object is known not to be a car. This is called a "Type I error". (d) False Negatives (FN): The model predicted object as "Not Car", but object is known to be a car. Counting models based on deep learning with high precision might leave out some cars, but it is very unlikely to let objects other than cars through. And a system with high recall might let through some objects not being cars, but it also is very unlikely to miss any object known to be a car (Table 4.1–4.3).

TABLE 4.1
Observed Data – Manual Counting

	Observed Frequencies		Cumulative Rates	
Diagnostic Level	False Positive	True Positive	False Positive	True Positive
1	10	60	0.1887	0.4959
2	0	3	0.1887	0.5207
3	28	3	0.717	0.5455
4	15	55	1	1

TABLE 4.2
Confusion Matrix

Confusion Matrix n = 77		ML System Prediction	
		Predicts: Car/Bus/Bike	**Predicts: Not: Car/Bus/Bike**
Actual truth	Known response: Car/Bus/Bike	True positive:10/3/3 TP	False negative:6/0/0 FN
	Known response: Not: Car /Bus/Bike	False positive:10/28/0 FP 70/31/3	True negative:1/46/74 TN 7/46/74

TABLE 4.3
Accuracy, Recall, and Precision

	Car	Bus	Bike
Accuracy: (TP + TN)/(TP + TN + FP + FN) =	0.792208	0.636364	0.636364
Recall: TP/(TP + FN) =	0.909091	0.909091	0.909091
Precision: TP/(TP + FP) =	0.857143	0.096774	0.096774

4.5 CONCLUSION

In this chapter, we have presented an approach that uses deep learning to effectively detect, track, and accurately count moving vehicles in a traffic intersection. The proposed approach helps to count vehicles (cars, trucks, and motorcycles) in the intersection which is used to make an informed decision about making streets one-way or two-way adaptive traffic light system depending not only on the number of vehicles in the intersection but also based on the time of the day. Deep learning with the use of CNN is the perfect solution for implementing a working monitoring system that effectively detects, tracks, and accurately counts traffic moving objects in an intersection. The profusion of building blocks and its methods and approaches gave the means to effectively tackle the problem of counting turning vehicles in a traffic intersection. As shown in the method, it is clearly possible to track and count turning vehicles in traffic intersection with an accuracy higher than human beings and cost effective. The inability of humans to focus for a long period of time is a serious problem that considerably affects intersection traffic counting. The use of intensive human labor has also tremendous economic implications. The initial implementations using programming languages such as Python, C++, NVIDIA video card with camera, and libraries such as CUDA Toolkit, OpenCV, and Tensor-flow make it possible to achieve the objectives of autonomously counting intersection traffic turning vehicles. Some issues still need to be addressed such as the effect of daylight change, weather, and the threshold distance between cars. Future work should focus on addressing those types of issues.

REFERENCES

1. Danda B. Rawat and Kayhan Zrar Ghafoor. *Smart Cities Cybersecurity and Privacy.* Elsevier, 2018.
2. Stephan Olariu and Michele C. Weigle. *Vehicular Networks: From Theory to Practice.* Chapman and Hall/CRC, 2009.
3. Herbert Hoover. *First National Conference on Street and Highway Safety. Hon. Herbert Hoover, Secretary of Commerce, Chairman. Washington, DC, December 15–16, 1924.* 51 p. Washington, DC: National Capital Press, 1924. 51 p. URL: //catalog.hathitrust.org/Record/102186880.
4. Fei Liu et al. "A video-based real-time adaptive vehicle-counting system for urban roads". *PLoS One* 12.11 (2017), p. e0186098.
5. Zhe Dai et al. "Video-based vehicle counting framework". *IEEE Access* 7 (2019), pp. 64460–64470.
6. Jie Zhou, Dashan Gao, and David Zhang. "Moving vehicle detection for automatic traffic monitoring". *IEEE Transactions on Vehicular Technology* 56.1 (2007), pp. 51–59.
7. Stephen Arhin. *Exploring Strategies to Improve Mobility and Safety on Roadway Segments in Urban Areas*, 2018.
8. Minda Zetlin. *For the Most Productive Workday, Science Says Make Sure to Do This.* Inc.com. Mar. 21, 2019. URL: https://www.inc.com/minda-zetlin/productivity-workday-52-minutes-work-17-minutes-break-travis-bradberry-pomodoro-technique.html (visited on 12/01/2019).
9. Huansheng Song et al. "Vision-based vehicle detection and counting system using deep learning in highway scenes". *European Transport Research Review* 11.1 (2019), p. 51.
10. Joseph Redmon et al. "You only look once: Unified, real-time object detection". en. In: *2016 IEEE Conference on Computer Vision and Pattern Recognition (CVPR).* Las Vegas, NV, USA: IEEE, June 2016, pp. 779–788. ISBN: 978-1-4673–8851-1. DOI:10.1109/CVPR.2016.91. URL: http://ieeexplore.ieee.org/document/7780460/ (visited on 08/16/2020).
11. Qolamreza R. Razlighi and Yaakov Stern. "Blob-like feature extraction and matching for brain MR images". In: *2011 Annual International Conference of the IEEE Engineering in Medicine and Biology Society*, 2011, pp. 7799–7802.
12. Guy M. Lingani, Danda B. Rawat, and Moses Garuba. "Smart traffic management system using deep learning for smart city applications". In: *2019 IEEE 9th Annual Computing and Communication Workshop and Conference (CCWC).* Las Vegas, NV, 2019, pp. 101–106.

5 Toward Rapid Development and Deployment of Machine Learning Pipelines across Cloud-Edge

Anirban Bhattacharjee, Yogesh Barve,
Shweta Khare, Shunxing Bao,
Zhuangwei Kang, Aniruddha Gokhale
Vanderbilt University

Thomas Damiano
Lockheed Martin Advanced Technology Labs

CONTENTS

5.1 INTRODUCTION

5.1.1 Emerging Trends

Internet of Things (IoT), an ecosystem of interconnected devices, such as surveillance cameras, wearable technologies, industrial sensors, smart building appliances, health monitoring systems, and vehicles, generates high volumes of data at high velocity. Predictive analytics applications use these data to derive valuable insights and make informed decisions for a variety of smart application domains. However, building and operationalizing such predictive analytics applications require expertise across machine learning (ML), agile software engineering, and distributed systems.

5.1.2 Challenges and State-of-the-Art Solutions

ML Model Development Challenges: To build robust predictive analytics applications, the ML model developers require expertise on a range of feasible ML algorithms (e.g., linear models, decision trees, or deep neural networks). Moreover, they are also responsible for ensuring high prediction quality of the developed ML models, which require running several experiments to tune the hyperparameters. Training the ML models for big data is time-consuming even with parallelization techniques. With the emergence of ML libraries such as scikit-learn [1] and WEKA [2], the ML model development becomes easier than before. The integration of scalable data frameworks such as Spark [3] and Hadoop [4], or the DL frameworks, such as TensorFlow [5] and MXNet [6], may provide the necessary infrastructure for faster training of the predictive analytics application model; however, it demands expertise in ML, programming languages, and distributed computing.

ML Model Operationalization Challenges: Once the model is trained, the ML model must be integrated in a composable manner with the big data analytics pipeline for serving the prediction requests. This requires integration and interaction with a data ingestion framework, databases, data lakes, and a variety of data visualization services. Inflexibility, poor design quality, and different deployment obstacles impede the operationalization of analytics components for data-driven applications. The DevOps community provides various tools such as Ansible [7] and Terraform [8] to aid the deployment process; however, developing infrastructure-as-code (IAC) requires expertise in the system design and programming concepts.

Resource Management Challenges: Deploying and scaling predictive analytics service components to guarantee its quality of service (QoS) properties is challenging due to the following reasons: (1) compatibility issues of the various components and their data formats; (2) communication overhead of moving data to distant clouds; (3) limited compute capability of the edge devices; and (4) varying number of incoming requests. A requirement-capability analysis must be performed during deployment to find a feasible infrastructure that can deliver the required QoS. Unfortunately, application developers often do not possess the expertise to develop the ML pipeline efficiently and optimized manner and enforce useful application deployment and resource management decisions to maintain QoS. Thus, there is a compelling need for an approach that relieves the predictive analytics application developer from determining the placement of analytics application components, monitoring their resource usage, and controlling different data processing tasks across the cloud-edge devices to enable optimal data control [9–11]. *Serverless computing* shows promise in addressing these issues because it allows application developers to develop the application components without concerns for the intriguing details of the infrastructure. Scrutiny of an IoT data analytics application reveals an application structure that is made up of a collection of loosely-coupled services, such as data ingestion, stream and batch processing, *Machine Learning (ML)-as-a-service*, visualization, and storage [12], where RESTful APIs interconnect individual components. The loosely coupled nature and event-driven nature of these applications make them highly suitable to be hosted using a serverless paradigm. Moreover, the deployment and execution of the trained ML models across the distributed edge-cloud resources using the serverless approach require appropriate and effective auto-scaling mechanisms for individual services to maintain the QoS while optimizing the cost of model execution.

5.1.3 Overview of Technical Contributions

To address the range of the challenges mentioned above, we present a rapid ML pipeline development and deployment framework, called *Stratum* [13].[1] Stratum exploits the benefits of the serverless computing paradigm and relieves the developer of a majority of the deployment and resource management issues, while ensuring their required QoS properties are is met. Specifically, this chapter makes the following contributions:

1. **GUI for Rapid ML Model Prototyping**: Stratum provides an easy-to-use, collaborative, and version-controlled framework for data scientists to quickly build, train, and evaluate the ML model on large datasets. The ML pipeline can be created using intuitive abstractions provided in a graphical environment without writing the code.
2. **Support for Various ML Libraries**: Stratum decouples the data processing and ML algorithms from the specific libraries and frameworks. Once the pipeline is created in the visual modeling environment, the code can be automatically generated in all supported libraries using an underlying domain-specific modeling language (DSML) and its associated generative tools.
3. **Rapid Operationalization of ML Model**: Once the model is trained, the ML pipeline comprises the ML model, and then the data manipulation workflow can be integrated with the data analytics pipeline with minimal user intervention.
4. **Automated Deployment of Application Components in Heterogeneous Environment**: Stratum provides capabilities for automated deployment of application components of the ML pipeline across the cloud-edge spectrum. A new data analytics pipeline can easily be created by binding the desired application components on the various target infrastructure based on user specifications. Thus, Stratum relieves the application developers from the deployment complexities across these heterogeneous environments.
5. **Heterogeneous Resource Monitoring and Application Auto-Scaling**: Stratum provides a resource monitoring and management interface that can be composed with the ML pipeline. Based on the monitored resource performance measurements, the application/service provider can plug their resource management logic to migrate the ML pipeline components across the cloud-edge spectrum.

5.1.4 Organization of the Chapter

The rest of the chapter is organized as follows: Section 5.2 presents a survey of existing solutions in the literature and compares them to Stratum; Section 5.3 uses a

[1] An earlier version of Stratum was presented to showcase the overall capabilities of Stratum. This chapter discusses the design and implementation aspects of the ML pipeline development and deployment in more detail.

motivating case study to elicit the challenges and then describes the problem formulation; Section 5.4 presents the design and implementation of Stratum; Section 5.5 evaluates the usability and efficacy of Stratum with a prototypical case study; and finally, Section 5.6 presents concluding remarks alluding to future directions.

5.2 RELATED WORK

In this work, we compare and contrast Stratum with the existing state-of-the-art solutions for developing, deploying, and managing ML pipeline across cloud and edge computing environments.

Ease.ml [14] is a training platform providing automatic model selection using a declarative programming approach. It relieves users from determining which models to select for a specific task at hand. Ease.ml proposes a resource scheduler, which manages the training job's deployment in a shared cluster. Google Vizier [15] is an ML platform that supports hyper-parameter tuning service. Users within Google can specify the parameter configuration space and the optimization goals. The Vizier service then proceeds with running experimentation trials until it has met the user-specified goals. Similarly, TFX [16] is another ML platform at Google, which provides tools for data preparation and model inference serving. The system is built around the Tensorflow ML framework. Michelangelo [17] is an ML-as-a-service framework deployed at Uber that facilitates building and deploying ML models in the cluster environment. Michelangelo provides an interface to the users to define ML tasks' requirements for the training and inference phases. Alchemist [18] is another internal project at Apple that addresses the training of ML models in a cluster environment. It leverages the Kubernetes container orchestration platform for running the training jobs of the ML models.

To the best of our knowledge, Google Vizier, Michelangelo, and Alchemist are internal tools available for use by the respective organizations and not available in the open-source environment. Concerning commercial offerings, services such as Amazon SageMaker, Microsoft Azure service, and IBM's Watson Studio are also paid services. These end-to-end services running the ML training and deployment pipelines are restricted to their proprietary run-time infrastructures, which can potentially result in vendor lock-in issues for the end-users.

Clipper [19] focuses on the low-latency prediction serving or inference serving aspect of the ML stage. Clipper uses an ensemble of prediction models to select and combine the best models with higher accuracy from different ML frameworks. Unlike Clipper, Rafiki [20] addresses training and inference service deployment in the cloud environment. Rafiki also provides a hyperparameter tuning service for parameter exploration. For inference serving, it gives an ensemble approach to allow for better accuracy of prediction results. MLFlow [21] is an open-source ML platform project that covers end-to-end lifecycle phases of ML development and deployment. It has python-based generic APIs that allow binding to different ML libraries. It also has support for data preparation, training, and deployment of the ML models across heterogeneous service providers such as Microsoft Azure [22], Amazon Sagemaker [23], and Apache Spark [3,24].

InferLine [25] presents prediction pipeline provisioning in a cloud environment. It provides inference serving across a DAG of ML models subject to latency constraints. It provides a hybrid approach for maintaining end-to-end latency constraints by changing model configurations. ML.NET [26] is an open-source ML pipeline framework by Microsoft. ML models can be combined directly into an application codebase. Moreover, predictions can be served by the OS-agnostic platform supported by the .NET Core framework. This feature is useful for prediction serving at the edge computing environments as the application need not communicate with an external service to get prediction results and as the model itself is packaged within the application. DLHub [27] is a platform that supports publishing trained models to model repositories and provides model serving capabilities for ML. DLHub implements an executor model for deploying inference tasks to the serving infrastructures. The currently supported infrastructures in DLHub comprise TensorFlow Serving [16], SageMaker [23], and Parsl-based execution platforms.

Acumos [28] is another open-source effort toward easing packaging, cataloging, and sharing activities of ML models. Acumos provides a marketplace where developers can search and find pretrained models for downloads. It also allows users to publish their custom models to the marketplace for easy sharing. Similarly, ModelDB also provides support for training and managing ML models. It provides a web-based GUI that allows for easy visualization of the ML pipelines. It also supports model versioning and visual exploration of models, and has support for collaborations. Other efforts such as Weka [2], Apache Mahout [29], and Scikit-Learn [1] offer declarative programming means to design and deploy ML pipelines and models. However, these platforms do not provide the means for model versioning and model deployment.

Compared to these existing works, Stratum provides a unified framework that supports design-time tools and deployment tools for model construction and deployment. Also, Stratum handles deployment across a heterogeneous set of platforms spanning from the cloud to edge computing platforms using CloudCAMP [30]. Stratum also provides version support. An easy-to-use visual drag and drop GUI interface is offered by Stratum to design and deploy the ML pipeline. Stratum leverages model-driven engineering (MDE) technologies that facilitate creating custom domain-specific modeling language, automated code generation, and orchestration of models to be deployed on the target infrastructure. Table 5.1 gives an overview of different feature support by Stratum and compares and contrasts with the existing state-of-the-art tool-suites.

In the literature, there exist several model-based approaches to managing end-to-end application performance and resource management autonomically based on the domain-specific modeling language (DSML) [31,32]. However, they do not consider distributing the application across the cloud-edge spectrum like Stratum. Similar to Stratum, which is integrated with our serverless serving platform Barista [33], several platforms leverage virtual machine and container technologies to deploy and scale the application components [34,35]; however, they do not provide end-to-end data analytics deployment and management platform.

TABLE 5.1
Comparing Stratum with Other State-of-the-Art Solutions

	Features									
Frameworks	**ML Model Development**	**Model Training**	**Inference Serving**	**Resource Monitoring**	**Code Generation**	**Resource Management**	**Collaborative Environment**	**Model Versioning**	**DSML**	**Cloud (C)/ Edge (E) Computing**	**Open-Source**
Ease.ML		✓			✓	✓			✓	C	
Vizier		✓							✓	C	
Clipper			✓							C	✓
Michelangelo	✓	✓	✓		✓				✓	C	
Alchemist	✓	✓		✓		✓				C	
Amazon ML Microsoft ML IBM Watson	✓	✓	✓		✓		✓	✓		C/E	
MLFlow	✓	✓	✓							C	✓
InferLine			✓			✓				C	
DLHub			✓					✓		C	
ML.Net	✓									C	✓
Acumos	✓	✓	✓					✓	✓	C	✓
ModelDB		✓						✓		C	✓
Weka Mahout Scikit	✓	✓								C	✓
Rafiki		✓	✓			✓			✓	C	
Stratum	✓	✓	✓	✓	✓	✓	✓	✓	✓	C/E	✓

5.3 PROBLEM FORMULATION

In this section, we use a motivating case study to highlight the key challenges and derive the solution requirements for Stratum.

5.3.1 Motivating Case Study

In this section, we use a motivating case study to highlight the design space. Consider an IoT use case of building a route planning algorithm, where the goal is to predict the optimal route from one point to another in real-time [36]. The traffic cameras are placed all over the city, which can detect the number of cars passing through a particular intersection. Rather than sending all the traffic feeds to the cloud, the traffic cameras equipped with edge devices can detect the vehicles using image recognition service on edge and deliver the result, i.e., the number of cars and car types to the cloud. Thus, edge analytics may reduce communication overhead significantly.

As shown in Figure 5.1, the information from the edge can be sent to the cloud via network protocols such as MQTT or HTTP. The cloud datacenters receive the data using the data ingestion frameworks such as Apache Kafka and then store all the information in a database or data lake. These data can be used to build the route planning model or to update the route planning model based on recent data. Once the route planning algorithm is ready, it can serve the prediction request in near-real-time. The design considerations faced by a developer to build such system architecture are illustrated in Figure 5.1. Manually building such an architecture is time-consuming and error-prone, and requires domain expertise in agile development, system design, and distributed systems. Moreover, building a robust ML model for image recognition and route planning requires ML domain expertise.

5.3.2 ML Model Development

Building predictive analytics requires developing and training ML models–based on historical datasets.

5.3.2.1 Challenges

Next, we highlight the different challenges involved in the design and development of ML models.

1. The ML developer needs to be familiar with a diverse set of ML capabilities, including classification (e.g., logistic regression, naive Bayes), regression, decision trees, random forests, gradient-boosted trees, recommendation (ALS), clustering (K-means, GMMs), DL networks (CNN, LSTM), and many more [37]. Data preparation, feature engineering, ML algorithm selection, and hyperparameter tuning – all of which need to be handled by conducting multiple experiments. A large number of available ML algorithms and their intricacies, along with the parameter tuning mechanisms, require significant ML domain expertise.

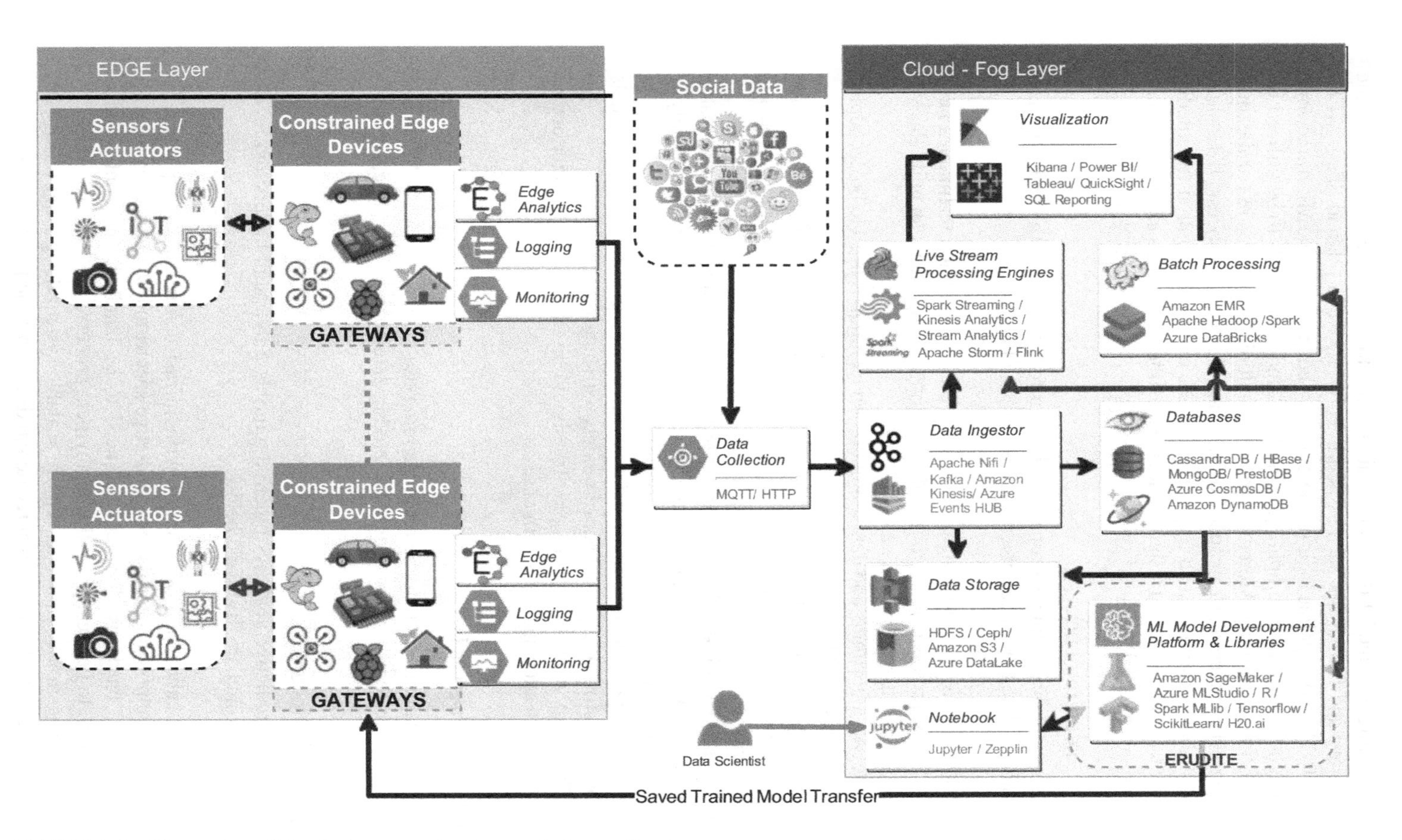

FIGURE 5.1 Illustration of the design space for data analytics application.

2. Building ML model demands coding skills toward a specific ML-specific library and framework such as Scikit-learn [1], Weka [2], and Spark MLlib [24]. The ML pipeline logic becomes tightly-coupled with the ML framework.
3. High-performance computing using GPU and CPU needs to be utilized for ML model training. Moreover, training multiple algorithms in parallel during experimentation can speed up the model selection process. If feasible, distributed training such as distributed DL can be configured to expedite the training process. Designing such experiments requires expertise in data management, distributed system, and software development [5,6,38].

5.3.2.2 Requirements

An ML model development framework is required to address the challenges mentioned above. The core requirements are highlighted below:

1. The diverse set of ML algorithms and data manipulation techniques need to be abstracted away from the ML developer so that they can build the ML model intuitively by using a GUI.
2. The ML algorithms and data manipulation techniques need to be decoupled from existing ML frameworks.
3. A language should be defined to automatically generate the code for the respective framework from the users' high-level ML model and specific choices.
4. The framework should create the experiment environment in an automated manner to speed up the training phase. Requirement-capability analysis needs to be conducted to select optimal infrastructure.
5. The framework should select and save the best model or ensemble of models by evaluating a large number of models with automatic scoring.
6. The components of the ML pipeline should be exposed by an interface so that it can exchange data with ease.

5.3.3 ML Pipeline Deployment

For IoT applications, such as the route planning use case, once the AI/ML model is trained, it needs to be rapidly integrated into big data analytics pipeline, as shown in Figure 5.1.

5.3.3.1 Challenges

Deploying a predictive analytics pipeline requires a thorough understanding of the complexities which are discussed below:

1. The application deployer needs to write infrastructure code to deploy application components across the cloud-edge spectrum [39]. Moreover, all software and hardware dependencies need to be handled, and configuration files need to be created for the deployment process.
2. All application components require scalable system design and infrastructure to handle dynamic resource demands at run-time [40,41].

5.3.3.2 Requirements

The design of the ML pipeline deployment framework necessitates addressing the challenges mentioned above. The core requirements are highlighted below:

1. The intricate infrastructure details and application configuration complexities need to the abstracted away from the deployer so that they can build the big data pipeline using intuitive abstractions.
2. A mechanism to automatically generate the infrastructure code for the respective application components and target infrastructures from the users' high-level system architecture model. The infrastructure code should be validated for correctness during construction.
3. Infrastructure should be designed in such a way so that it can be easily decoupled from one another and can communicate among themselves. The data analytics pipeline needs to provide storage repository APIs, communication services APIs, orchestration APIs, and other platform services APIs to enable the integration of various software components as required by any big data pipeline development.

5.3.4 Infrastructure for Resource Management

To provide low latency for predictive analytics services, sending all requests always to the distant cloud might not be the optimal solution. For example, in the route planning scenario, the vehicle detection model can potentially be placed at the edge unit to minimize the communication overhead incurred between the cloud-edge link.

5.3.4.1 Challenges

Resource management across the cloud-edge spectrum requires addressing the following challenges:

1. The device-to-cloud data round-trip latency is considerable, and hence processing at the resource-constrained edge may be attractive as it reduces latency and makes connected applications more responsive [11,38]. The execution time of the job depends upon the round-trip latency and execution environment. Determining the right hardware and cluster size for executing the job to guarantee the desired QoS is a hard problem.
2. Auto-scaling the application components proactively to handle the dynamic workload is necessary to minimize the operational cost while guaranteeing the QoS.

5.3.4.2 Requirements

The framework requires the following features to address the challenges mentioned above.

1. An integrated but lightweight interface to continuously monitor and log the resource usage of the cluster along multiple dimensions such as GPU, CPU, memory, IO, and network, which is required for proactive and reactive resource management scheduling strategies.

2. The framework should provide an API to interact with the pluggable resource management logic based on user-defined criteria.
3. Based on the decision of the resource manager, the framework should be able to allocate, migrate, and manage the resources across the cloud-edge spectrum seamlessly.

5.4 DESIGN AND IMPLEMENTATION OF STRATUM

This section outlines the design and implementation of Stratum and shows how it satisfies the requirements of Section 5.3. Stratum is built around the core concepts of MDE and generative programming. We developed the interactive GUI for ML model development and deployment, and its generative capabilities using the WebGME modeling environment [42].

5.4.1 Addressing Requirement 1: Rapid AI/ML Model Prototyping Kit

The model-driven ML development framework is expected to benefit both the novice and expert data scientists to prototype their ML models rapidly. The main metamodel, which is the core part of MDE, is shown in Figure 5.2.

5.4.1.1 Overview of the ML Model Development

As shown in Figure 5.3, ML pipelines are composed of data ingestion, data preprocessing, preparation, and transformation strategies, ML algorithms that execute on existing ML frameworks, and evaluation strategies to evaluate the best candidate model. Stratum's development kit provides a standard data exchange format that supports read and write capabilities in different data formats from various data sources such as Amazon S3, Azure DataLake Store, or a scalable datastore (e.g., HDFS-based) via a RESTful API. It also supports querying the data from SQL databases,

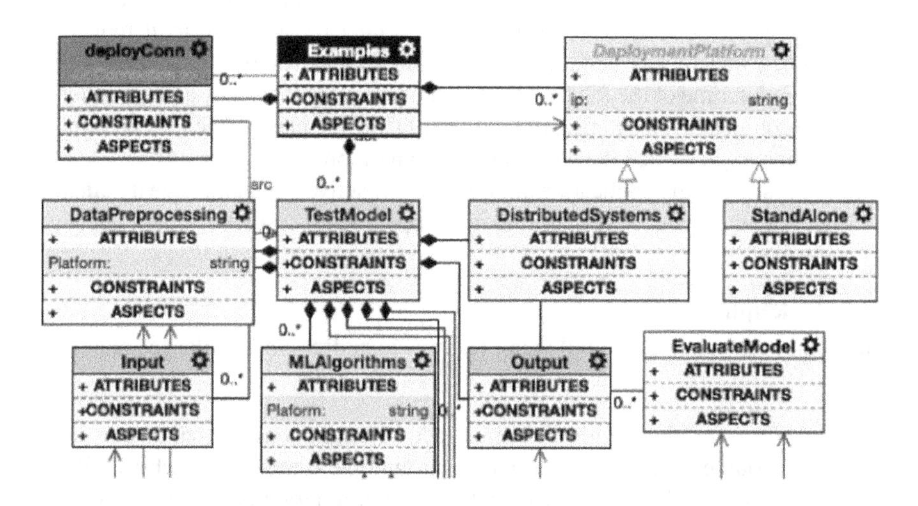

FIGURE 5.2 Main metamodel of ML model development.

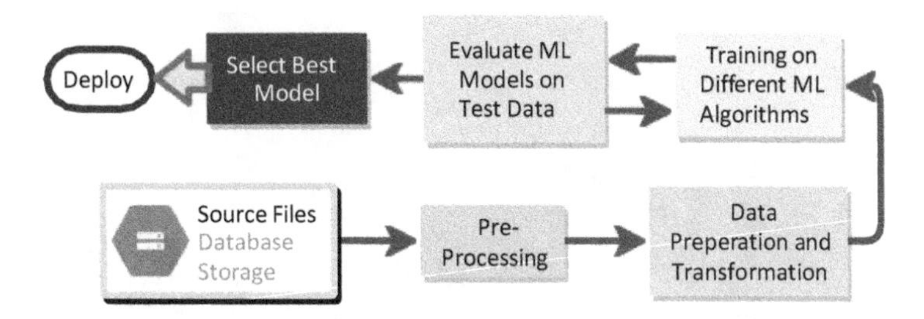

FIGURE 5.3 Sample ML pipeline.

such as MySQL or NoSQL databases such as MongoDB. The data scientist needs to specify the data storage type and location for data ingestion.

As shown in Figure 5.3, data preprocessing and cleansing are also supported in Stratum. The user can specify the strategies to deal with missing values and outliers intuitively in the data preprocessing block. Data preparation and transformation support capabilities such as grouping, scaling, normalization, categorical encoding, and test-train split. Feature extraction and engineering strategies such as PCA and LDA are also supported in this block.

Stratum provides the ML model construction and model evaluation. The user can select the desired ML algorithms from the Stratum GUI to build their ML model. We developed the metamodel for abstracting ML algorithm details and its framework bindings described in Sections 5.4.1.2 and 5.4.1.3. Once multiple ML models are trained, and the best candidate model is selected via evaluation criteria, we store the ML model as a deployable model and integrate it with a Big Data analytics pipeline.

We also integrated Jupyter Notebook and Apache Zeppelin (another notebook-based environment) to provide data scientists the ability to train their models interactively. We fill up the iPython notebook with basic code blocks as per the user-defined model, and if the expert data scientists want to play more with the ML model in the notebook. The data scientist can integrate their modified ML algorithms and data transformation logic in the Jupyter notebook easily rather than developing the ML model from scratch. The data scientist can also directly generate python code if needed.

5.4.1.2 Metamodel for ML Algorithms

We capture various facets of the application and target machine hardware specifications in the metamodel. The metamodel for the supported ML algorithms in Stratum is shown in Figure 5.4. We captured the specifications for a diverse set of ML algorithms, including classification (e.g., logistic regression, naive Bayes), regression, decision trees, random forests, gradient-boosted trees, recommendation (ALS), clustering (K-means, GMMs), and many others.

Using this metamodel, the data scientist can drag relevant ML blocks and define all the parameters such as fit_intercept, normalize, n_jobs for linear regression block or specify the type of layers such as dense, CNN, RNN, their activation functions, and parameter synchronization methods for DL.

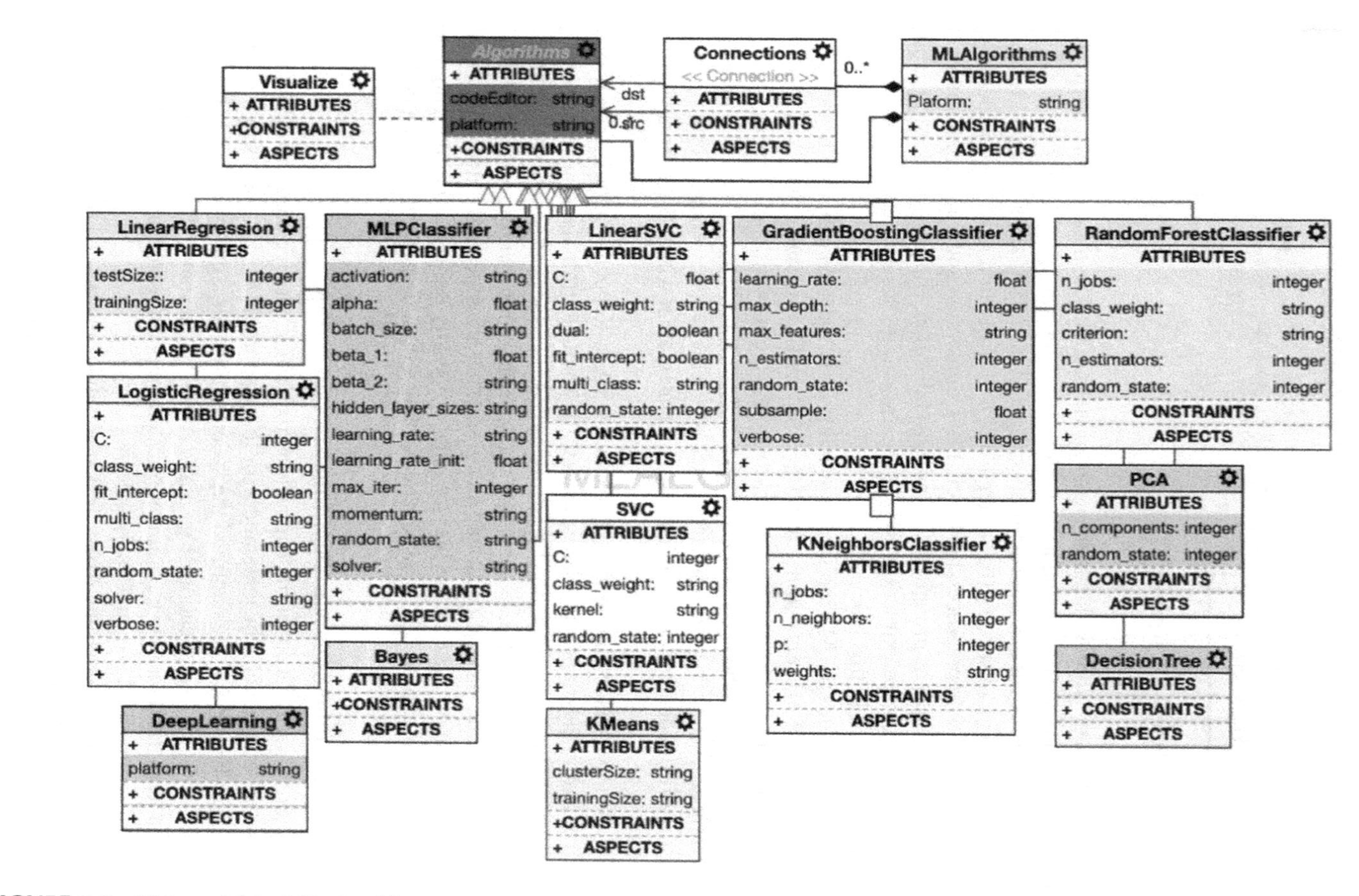

FIGURE 5.4 Metamodel for ML algorithms.

5.4.1.3 Generative Capabilities

The Code Generation Service of Stratum is created by developing plugins for our domain-specific modeling language (DSML). In the WebGME-based GUI, users can express pipelines as a directed acyclic graph (DAG). Each node represents a task such as data preprocessing, training based on different ML techniques, and evaluation of the ML model. The user-defined ML model is built upon Stratum metamodel specifications. The nodeJS- and python-based DSML then realizes the user-defined model and generates the code for specific ML libraries and frameworks such as scikit-learn and Spark MLlib via template-based transformation. The users specify their choice of ML framework. As shown in Figure 5.5, templates for different ML frameworks and their functions are predefined in our DSML. Figure 5.6 shows a sample pipeline prototype for building various regression analysis models for a specific use case using the Stratum DSML. The DAG structure of creating an ML pipeline is flexible, and other ML techniques for the specific dataset can easily be plugged in with the existing pipeline.

The Deployment Platform can be distributed systems or standalone hardware, where the deployment of the ML pipeline on the target machine for training is handled by Stratum ML Pipeline Execution Service, as described in Section 5.4.2.

The ML algorithms are encapsulated in Linux containers and exposed using endpoints. The DSML encapsulates the ML algorithms in Linux containers to support the parallel execution of the pipeline based on the resources' availability. After training the model, we evaluate the model based on different scoring methods such as accuracy, f1 score, precision, r2 score, and mean square error. The ML model developer needs to specify the evaluation method of their choice. Based on the evaluation score, we find the model and save it for prediction jobs. The Stratum framework pushes the saved model into the big data processing workflow.

5.4.2 Addressing Requirement 2: Automated Deployment of Application Components on Heterogeneous Resources

Requirement 2 on automating the deployment on potentially heterogeneous resources is also addressed using Stratum's DSML and its generative capabilities. Figure 5.7 depicts the metamodel for the data analytics and data ingestion aspects of the data analytics pipeline.[2] These metamodels were developed through a combination of (1)

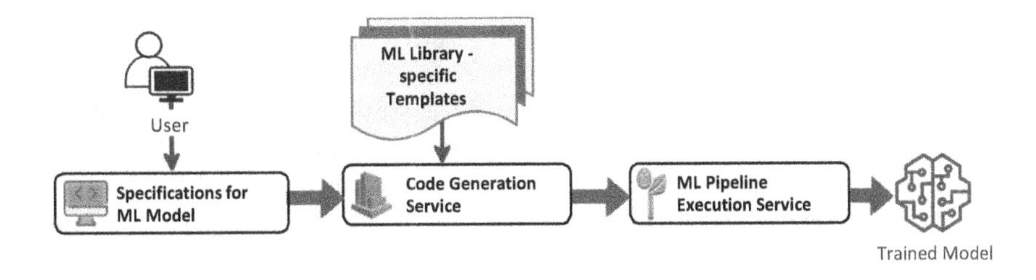

FIGURE 5.5 Stratum workflow for ML model development.

[2] Due to space limitations, we do not show the metamodel for computing resources and storage.

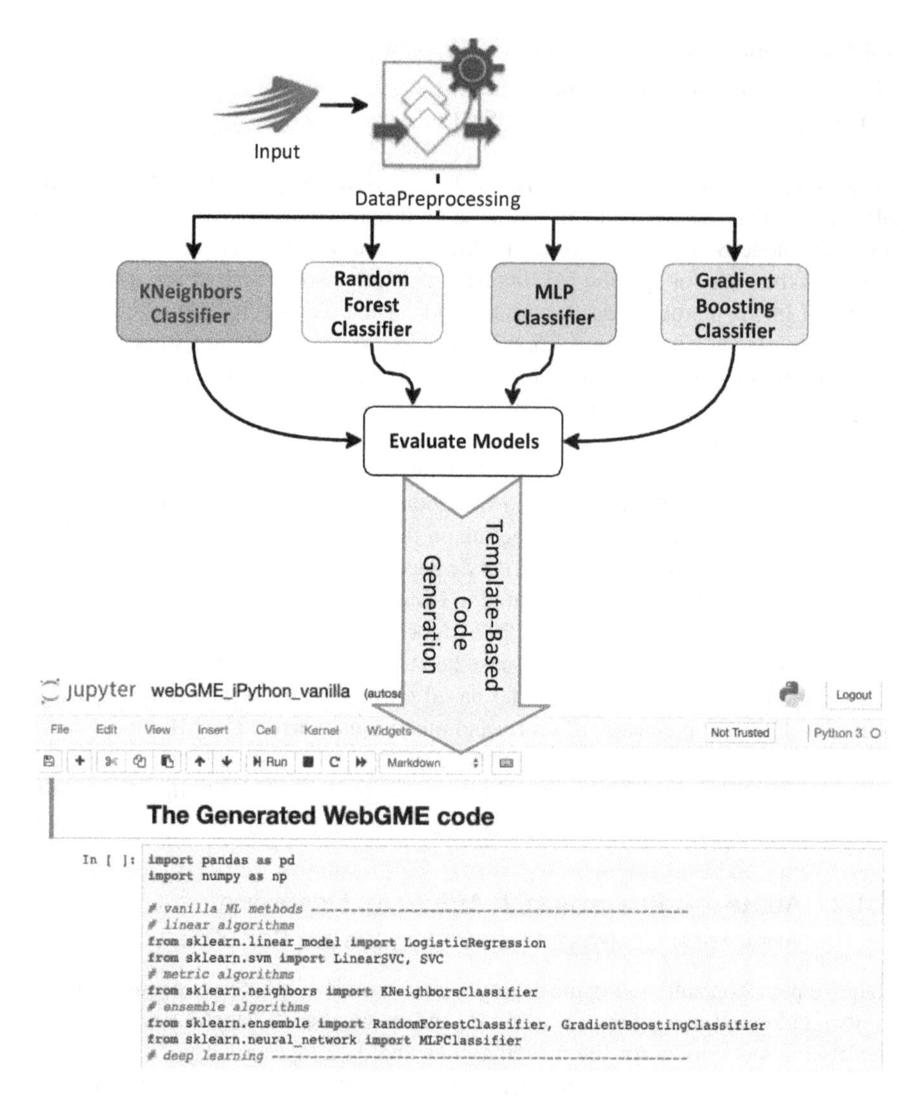

FIGURE 5.6 Sample code generation capabilities of Stratum.

reverse engineering, (2) dependency mapping across heterogeneous hardware, and (3) dependency mapping across different operating systems and their versions. We abstract the specification details from the end-users by identifying the commonalities of the provisioning stacks, which become the high-level reusable building blocks of the deployment and management pipeline that are captured as domain-specific artifacts. The user-defined variability points (e.g., number of machines, predefined business policy) are needed to be specified by the user.

The user models the deployment scenario by using intuitive abstractions. Then, the Stratum *model interpreter* verifies the correctness of the abstract deployment model. A nodeJS-based *code Generator* of Stratum then realizes the user-defined model through one or more interrelated metamodels that capture the syntax and semantics,

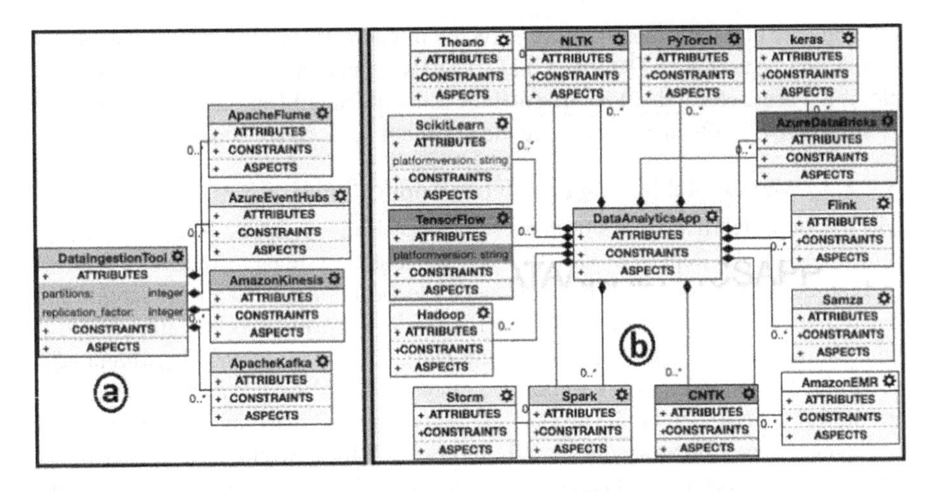

FIGURE 5.7 Metamodel for (a) data ingestion frameworks and (b) data analytics applications.

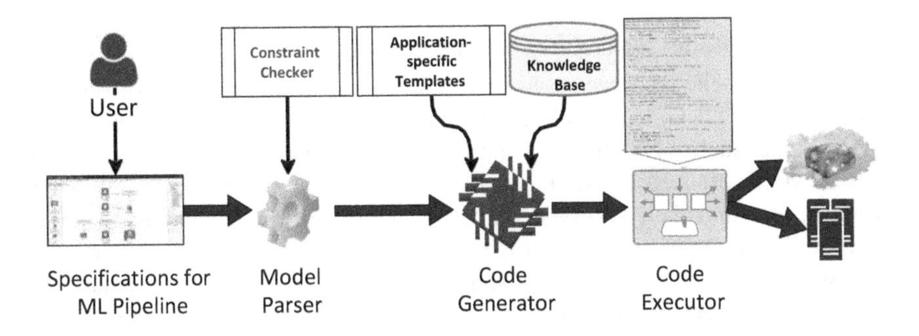

FIGURE 5.8 Stratum workflow for ML model deployment.

and generate the Infrastructure-as-Code (IAC) solution by parsing the user-defined model and deploying it on the target machine using a template-based transformation and knowledge base. Finally, the IAC solution is executed by the *code executor* to deploy the desired data analytics architecture on the target machines across the cloud-edge spectrum, as shown in Figure 5.8. The deployment can be accomplished via the Stratum model-driven approach without writing a single line of code [30]. We capture all these intriguing details in our metamodel using reverse engineering. We enforce all the domain-specific semantics in the visual editor and provide validation of the model during design-time.

5.4.2.1 Metamodel for Data Ingestion Frameworks

Figure 5.7a illustrates the metamodel for data ingestion tools (e.g., RabbitMQ, Kafka), which receive the data from the edge devices, which serve as the publisher. The data is forwarded to the subscribers and can be stored in Data Repositories. The modeling environment includes services for interacting with the Data Repositories and other microservices using RESTful APIs, as shown in Figure 5.1. The user must select the specific data ingestion tool to deploy it on the target platform. We also verify the correctness of the user model based on the semantics defined in the metamodel. For example, if the user selects the AzureEventHubs as their desired Data Ingestion Tool, then Amazon AWS or Open-Stack or bare metal cannot be its target deployment platform.

5.4.2.2 Metamodel for Data Analytics Applications

The live or in-depth analytics frameworks such as Hadoop, Spark, Flink, Samza, and others need to be deployed on the target distributed systems. Moreover, as shown in Figure 5.1, the trained ML model needs to be integrated with the stream and batch analytics frameworks. Henceforth, those frameworks' specifications need to be contained in the Data Analytics parent class. Moreover, as shown in Figure 5.1, the trained ML model needs to be integrated with the stream and batch analytics frameworks. So to deploy the production-ready machine-learning pipeline, we need to capture the specifications for the ML libraries and frameworks such as Tensorflow, sci-kit learn, PyTorch, CNTK, and more, as shown in Figure 5.7b. We also capture the specifications to start the AmazonEMR service on the cluster of Amazon EC2 machines or Azure DataBricks service on the cluster of Azure Virtual Machines.

5.4.2.3 Metamodel for Heterogeneous Resources

As noted before, the resources used can be target hardware such as Raspberry Pis, NVidia Jetson TX1 GPU, any of bare metal CPU machines such as Intel Xeon, or GPU machines such as NVidia Tesla or it could be any cloud platform such as Amazon AWS or Microsoft Azure. The target OS residing on it can be Linux, Windows, Raspbian, etc., and their versions can be different. In the metamodel, we capture hardware details, operating system type, and their version. The package manager of the software packages such as apt and yum depends on the underlying hardware, whereas the software installment process depends on OS type and versions.

5.4.2.4 Metamodel for Data Storage Services

We have captured the storage service specifications in the metamodel also. The subscribers from the data ingestion tools can consume the data in an event-driven way and store it in the storage service. Our storage class contains AmazonS3, HDFS, AzureBlobStorage, and Ceph-based distributed storage services, as shown in Figure 5.9a. The storage service can easily be deployed with user specifications such as folder name and bucket name. The data can also be stored in relational databases such as MySQL or NoSQL databases such as MongoDB and Cassandra. The user has to select the database of their choice, and the database attributes need to be configured by the user by providing database names, user id, password, replication factor, etc. The metamodel for the databases is shown in Figure 5.9b.

5.4.3 Addressing Requirement 3: Framework for Performance Monitoring and Intelligent Resource Management

Once the trained models are deployed and start executing, run-time resource management is required. Stratum supports autoscaling, load-balancing using serverless paradigm, and pinpointing failures.

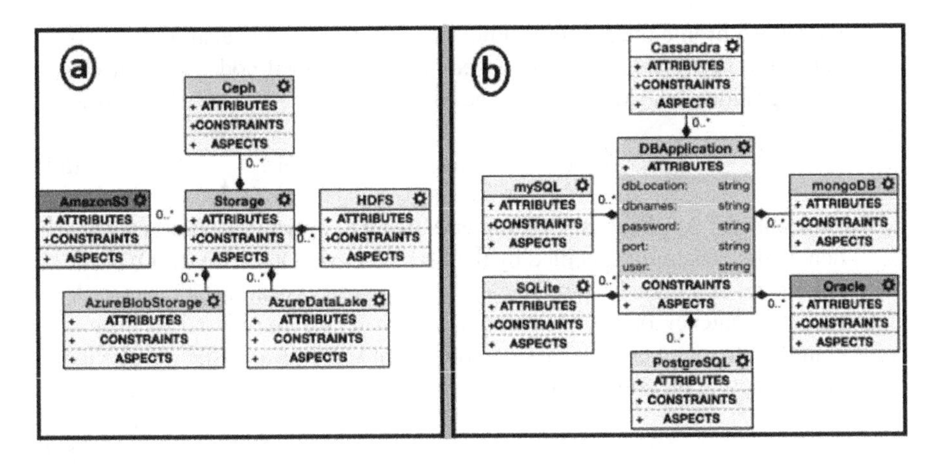

FIGURE 5.9 Metamodel for (a) data storage and (b) databases.

5.4.3.1 Performance Monitoring

It is critical to monitor the system metrics to understand the run-time performance of the infrastructure and the application. Considering the distributed nature of the systems spanning from cloud to edge computing environments, monitoring such systems can become challenging. Also, given the heterogeneity in the computing platforms, there is a need for a monitoring tool that can integrate different architecture-specific monitoring tools in a unified fashion.

To address these challenges, we built our monitoring infrastructure leveraging *CollectD* [43] monitoring daemon and different metric-specific tools such as *Linux Perf* and *nvidia-smi* [44] to monitor various system metrics. We leveraged *rabbitmq* [45] based publish/subscribe system to provide data dissemination in the distributed infrastructure. The collected data is stored in InfluxDB [46], a time-series database for analysis and triggering resource management decisions. The collected system metric data includes GPU-specific metrics such as power consumption, GPU utilization, temperature, and host-specific metrics such as CPU, disk, network, low-level cache utilization, and memory bandwidth utilization.

5.4.3.2 Resource Management

Stratum contains a *Resource Manager* to maintain the QoS of the application components by scaling and migrating the application components. The ML-based data analytics applications' total latency comprises the round trip latency(l_{rt}) of data and the ML model execution time. We profile the ML model execution times on various data and target hardware before actually deploying the model, and consider its 95th percentile execution latency as estimated execution time ($exec_{hw\_id,\ mlmodel}$). The ML prediction model can be deployed in the cloud and edge analytics spectrum for various purposes such as image recognition, speech recognition, and voice assistant.

Migration of ML Prediction Tasks: Before considering edge devices as a potential node for executing predictive analytics, we check if it has sufficient memory to keep the model in memory. If the edge node can host the ML model, we profile the ML model on the edge devices. We also profile 95th percentile network latency(l_{rt}) between edge and cloud node. We consider the migration of the ML model in the edge when the condition below is true.

$$\text{exec}_{\text{edge}}, \text{mlmodel} < \text{exec}_{\text{cloud}}, \text{mlmodel} + l_{rt} \tag{5.1}$$

Auto-scaling of Application Components using Serverless Paradigm: Let x denote the constraint specified by the Service Level Objective (SLO) of the ML model execution latency, and let $\text{exec}_{\text{hw\_id, mlmodel}, p}$ denote 95th percentile execution latency on p CPU cores. For each server configuration, we can compute the number of requests n_req it can serve as a prediction service while meeting the SLOs using p CPU cores.

$$n_{\text{req}} = x/\text{exec}_{\text{hw\_id, mlmodel, p}} \tag{5.2}$$

By monitoring the total number of incoming requests, we can easily calculate the total number of machines (total_incomingrequests/n_{req}) required to handle the workload. Based on the difference between the current state and desired state, we can calculate how many more machines to start. Stratum's *Resource Manager* can deploy the ML model on the required machines and start the prediction service automatically to handle dynamic workloads proactively. Moreover, a batching mechanism is considered to send bulk data at an interval or based on the number of messages, using data ingestion tools like Apache Kafka. Similarly, to handle the incoming message rate, we can autoscale the data ingestion tools also.

5.4.4 Support for Collaboration and Versioning

Stratum core concepts are developed using WebGME, which supports collaboration in a version-controlled environment. We also incorporated the automatic version control in the Stratum framework so that we can recall a specific version of the framework later if required. We save data, code, and attributes of the modeling environment for guaranteeing that every state is reproducible. Because of collaborative editing support, the developers can easily tag branches when they are developing their part of the model, and the branches can be merged easily to integrate the whole model. Figure 5.10 shows an integrated version control for all developers to make any historical state reproducible in the collaborative environment.

5.4.5 Discussion and Current Limitations

Currently, Stratum is capable of generating only Python-based code, and only Scikit-learn and TensorFlow are integrated. However, other languages such as Java and C++ can easily be plugged into Stratum, and other cloud libraries such as Amazon SageMaker and AzureML can be integrated with Stratum very easily. The design of Stratum uses agile methodologies so that it can be extended with ease.

We currently consider that the resources are homogeneous in the cloud, and during scaling, the resources' configuration does not change. The configuration of the

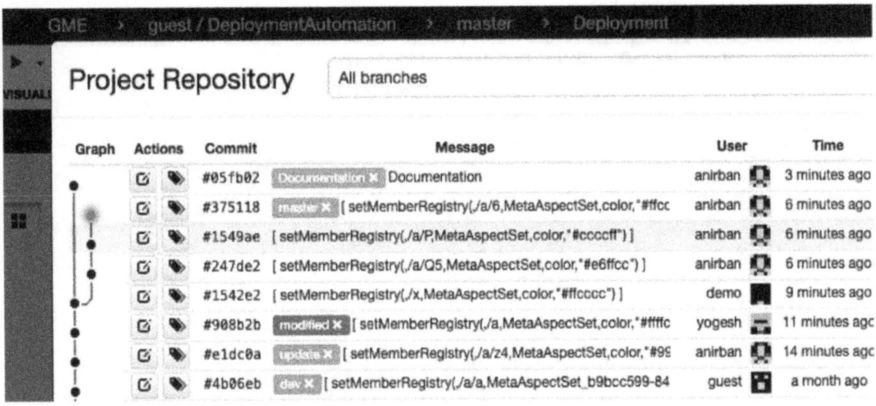

FIGURE 5.10 Integrated version control to reproduce all historical states.

resource is fixed during deployment time based on the user's choice and requirement-capacity analysis. Moreover, the management of resources can be done proactively; however, that problem is out of this chapter's scope.

5.5 EVALUATION OF STRATUM

In this section, we evaluate the simplicity, rapid deployment, and resource management capabilities of Stratum, along with the accessibility, scalability, and efficiency of the ML model development framework of Stratum.

5.5.1 EVALUATING THE RAPID MODEL DEVELOPMENT FRAMEWORK

The Rapid Model Development framework leverages the strength of MDE to provide a visual development environment for machine learning pipelines. It provides a hybrid visual-textual interface for ML model development phases in a version-controlled, collaborative visual environment. The model transformer can distribute different jobs with different ML algorithms over a cluster of connected machines and aids the developer select the best model or ensemble of models based on the user's choice of evaluation methods.

As shown in Figure 5.11, the ML developer can build their machine learning pipeline using the visual interface of Stratum. In the left-hand pane (box 1), all the building blocks are defined using the metamodel. The ML model developer has to drag and drop the required blocks in the design pane (box 2) and must connect the blocks to define the ML pipeline, including preprocessing, ML algorithm selection, hyperparameter tuning, model evaluation, and best model selection criteria. All the selected ML algorithms' attributes, such as max_depth, criterion, and n_estimators, need to be specified by the user (or can take default values) from the right pane (box 3). The name of the attributes are dependent on ML algorithms, and this aspect is captured by reverse engineering. The ML execution framework needs to be mentioned to bind the workflow with a specific library or framework such as Scikit-learn, Spark MLlib, or Tensorflow.

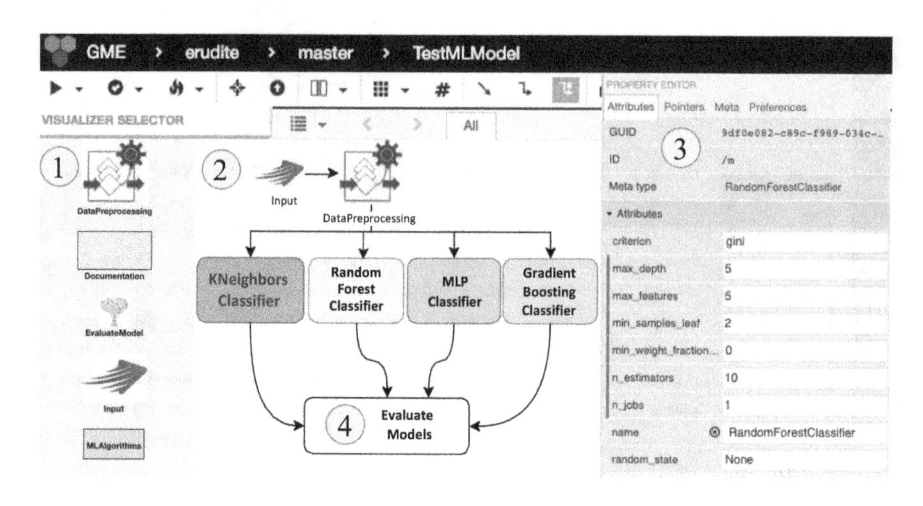

FIGURE 5.11 Usability of the ML model development framework. Box 1 shows the available selection of blocks available to create ML pipeline as shown in Box 2. Attributes can be set using attribute selection panel in Box 3. Box 4 shows model evaluation criteria.

All the ML algorithms are encapsulated in Docker containers, and different algorithms can be executed in parallel to speed up the training and tuning phases. Similarly, in the input building block, the source data type, and path, e.g., database, HDFS needs to be mentioned, and data source type, e.g., csv, Avro, and text, is required. We only support simple data cleaning methods in the data preprocessing block, such as filtering and normalization. In the evaluation building block, the evaluation method needs to be specified, and based on that, Stratum selects the best model.

The model transformer can distribute different jobs with different ML algorithms over a cluster of connected machines and aids the developer in selecting the best model or ensemble of models based on the user's choice of evaluation methods. Box 4 shows the evaluation of ML methods to choose the best model based on training accuracy on a sample dataset. Thus, Stratum helps build the ML model using MDE techniques, and the ML developer does not need to write any code. The framework can also generate the code in the notebook environment as depicted in Figure 5.6 for the expert user, where they can tune the ML model as required. We save the model in the ML framework-specific format and later integrate it with big data processing pipeline.

5.5.2 Evaluation of Rapid Application Prototyping Framework

As depicted in Figure 5.12, using the visual interface of Stratum, the application developer can develop the data analytics application. As described in our previous work [30], the building blocks for application components and infrastructure components need to be dragged from the left panel in the design space. Then all the building blocks need to be connected to build the business application workflow. The 'hostedOn' connection illustrates on which target machine the application

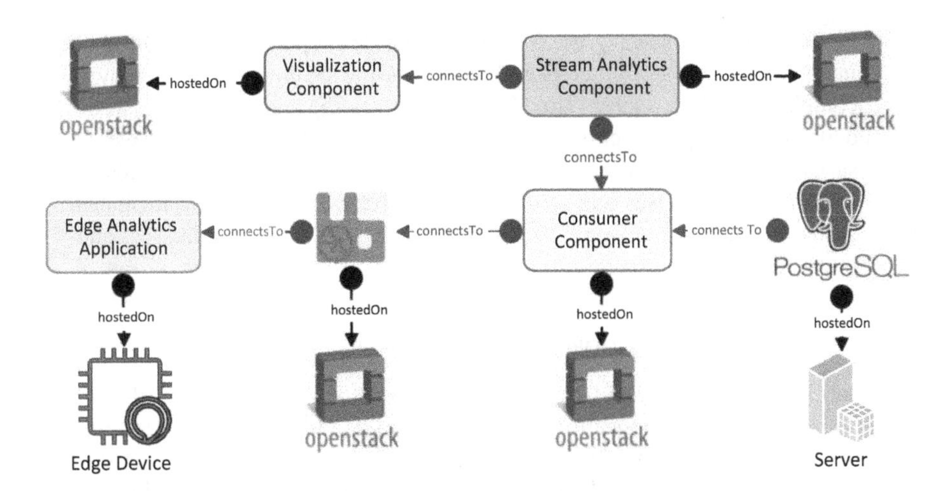

FIGURE 5.12 Example of big data analytics system deployment model across cloud/edge spectrum.

components are deployed, and 'connectsTo' connection represents that the source component needs to be started before destination components because the destination component is dependent on the source component. As shown in Figure 5.8, by parsing the user-defined abstract model tree, the Stratum DSML creates the deployable model (Ansible-specific in our case) and using NodeJS based plugin to execute the deployable model and create the infrastructure of the application as described in Section 5.4.2.

Figure 5.12 illustrates that using the Stratum modeling environment, edge analytics application components can be deployed on Edge devices such as Raspberry Pi machine, and message-broker software, e.g., RabbitMQ (https://www.rabbitmq.com/), can be deployed on cloud VMs, which is maintained by OpenStack. The data consumer application component can be similarly deployed on OpenStack VM, which will consume the data from RabbitMQ in a batch or stream and store it in a MySQL database, which is deployed on top of the bare-metal server. Then, Stream Analytics components, such as Spark Streaming, can be deployed and configured on OpenStack VMs, and a visualization engine like Kibana can be integrated with the workflow. RESTful APIs connect all the application components, so the ML model can easily be pushed into a predictive analytics application during the management phase.

Twitter Sentiment Analysis: We developed a sample big data use case scenario using Stratum for sentiment analysis of the Twitter data stream. Each Raspberry pi is responsible for listening to a particular hashtag and publish the data to the Kafka cluster. The message broker then aggregates the data and publishes the data to be consumed by the Spark engine. We also store the Twitter data in the MySQL database for future batch analytics requirements. The Spark stream processing engine then analyzes the sentiment of each Twitter hashtag and produces the result. The sentiment analysis model is also built using the Stratum model development framework and integrated with the Twitter Big Data processing framework. To deploy a complex workflow such as this, Stratum enables the entire workflow to be deployed on target machines without writing a single line of code, which saves the big data developers hours of manual effort to write hundreds of lines of code. The user needs to specify the infrastructure-specific attributes such as VM type, OS type, network type, access key, IP addresses, and other details.

5.5.3 Performance Monitoring on Heterogeneous Hardware

As described in Section 5.4.3.1, we need to monitor the performance of the infrastructure as well as the application components to take dynamic resource management decisions. To describe the monitoring capabilities of Stratum, we set up the training experiments on NVIDIA GeForce Titan X Pascal GPU machine integrated with Intel(R) Xeon(R) CPU E5-2620 v4. For prediction experiments, we set up a cluster of Intel(R) Xeon(R) CPU E5-2620 v4 machines in the private cloud, Dell OptiPlex 3020 machines in the private cloud, and MinnowBoard with 64-bit Intel Atom devices as edge devices.

DL Model Training for Image Classification: We developed a DL model for image classification using the CIFAR10 dataset. We monitor the accuracy and loss of the custom-developed ML model, as shown in Figure 5.13. The GPU performance metric, such as GPU utilization, GPU memory utilization per core, the power drawn by GPU cores in watts, and the temperature of the GPU machine in Celsius, during the training phase of the ML model can be collected and measured by Stratum framework [13].

Prediction Using Pretrained Image Classification Model: We encapsulate the trained models in Docker containers and build the container for the target hardware. Stratum enables us to monitor the performance of the Docker containers along with host machines on which ML models reside. We collect various metrics of the container from the host, which include execution time, CPU, memory, network, disk utilization along with L2, L3 cache bandwidth, cache miss ratio, and many more.

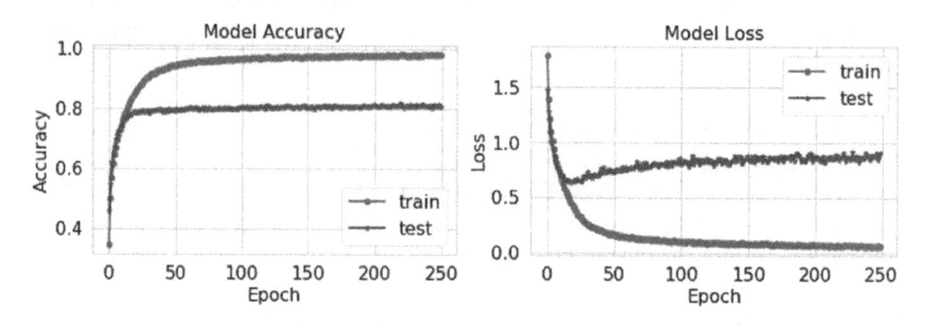

FIGURE 5.13 ML model accuracy and loss trend graph on CIFAR10 dataset (test and train).

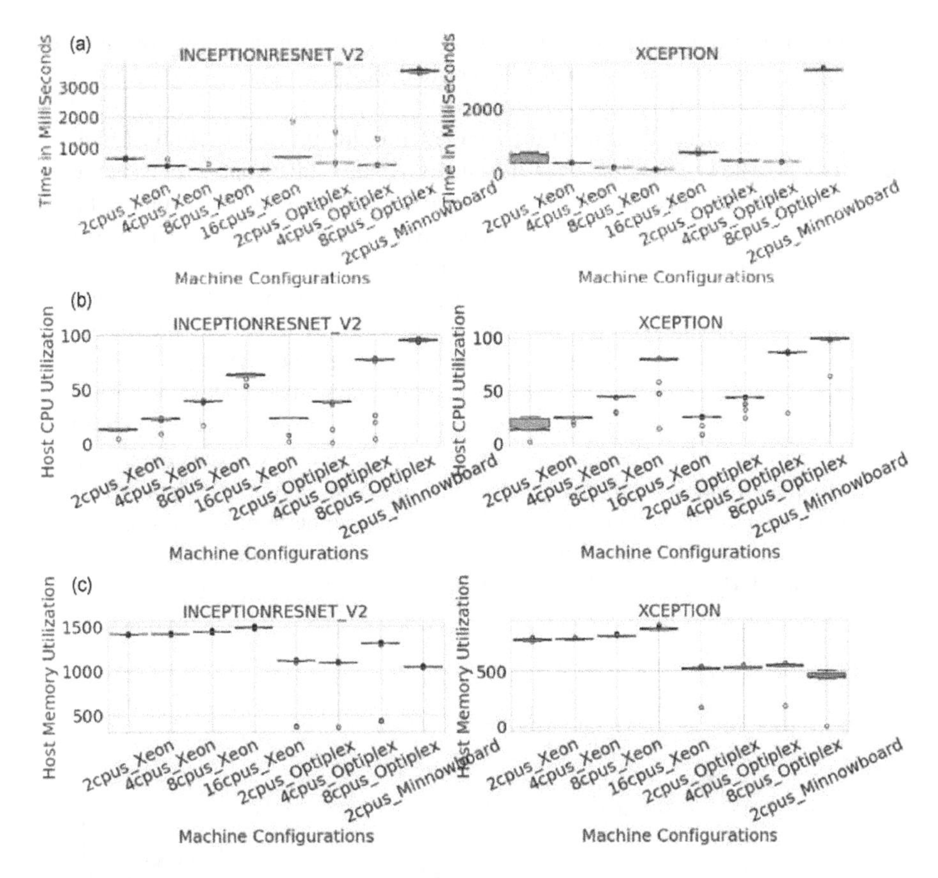

FIGURE 5.14 Performance monitoring of the prediction services (box plot): (a) the execution latency of InceptionResnetV2 and Xception model on different ML containers with variable configurations, (b) host CPU utilization of the ML containers, and (c) host memory utilization of ML containers (in MB).

Figure 5.14 illustrates a glimpse of collected performance metrics during the prediction serving phase using box plot. Figure 5.14a shows the execution latency of Inception- ResnetV2 and Xception models on different ML containers with variable configurations, which are hosted on different bare-metal machines. Figure 5.14b shows CPU utilization of the ML containers from the host machines (e.g., the two cores container only uses around 12% of CPU resources of a 16 cores machine, while the two cores container uses around 100% of CPU resources of two cores machine), and Figure 5.14c shows memory utilization of ML containers (in MB) from the host machines.

5.5.4 Resource Management

As mentioned in Section 5.4.3.2, we profile the prediction service on the specific hardware before deploying it on the cluster of machines. The prediction service is encapsulated in a Docker container, and based on the number of incoming requests (dynamic workload), we scale our system in an event-driven manner. Using the

Docker swarm cluster management tool, we can easily scale up and down the number of ML model containers to guarantee the pre-defined QoS.

Dynamic Resource Allocation for Image Classification Prediction Model: Here, we showed how to manage resources dynamically to optimize resource utilization for incoming request variation using the serverless computing paradigm. The dynamic workload trace we used here is based on NYC thruway toll dataset [47]. This dataset is a motivation to detect car types, license plate detection, and the number of cars passing through the toll booth using an image classification pre-trained model, as described in Section 5.3. As shown in Figures 5.15 and 5.16, we profiled InceptionResnetV2 image classification algorithm on four cores and eight cores virtual machines (VM), which is running on a cluster of Intel(R) Xeon(R) CPU E5-2640 v4 machines. The classification algorithms execute in docker containers, which are pinned on all available cores of the VM. Figure 5.15 depicts that to guarantee 800 minutes execution latency on eight cores machines, how we change the number of machines to handle the dynamic workload [47]; and similarly, Figure 5.16 shows that to guarantee 2 seconds execution

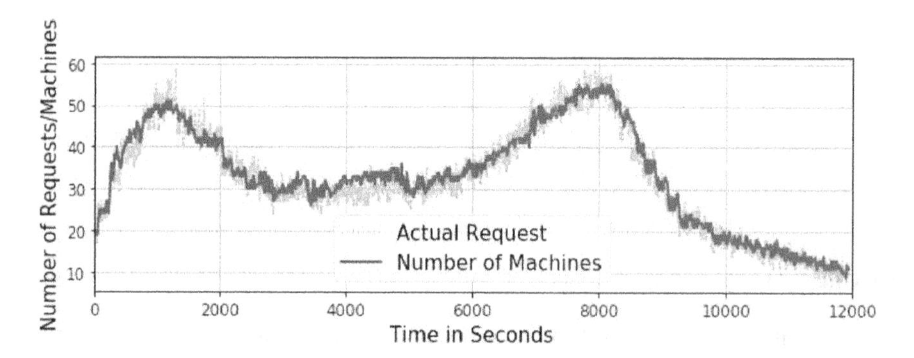

FIGURE 5.15 Number of eight cores machines are required to maintain 800 minutes execution latency of inception resnetv2 model on dynamic workload.

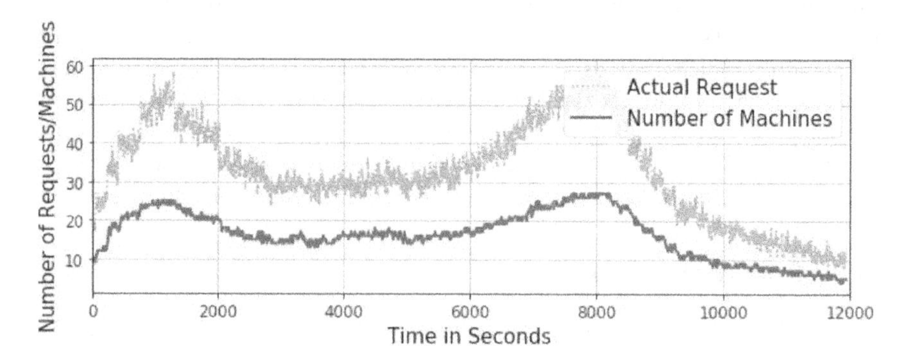

FIGURE 5.16 Number of four cores machines are required to maintain 2 seconds execution latency of inception- resnetv2 model on dynamic workload.

latency on four cores machines, how we change the number of machines. This approach remains the same for any parallelizable prediction services, as shown in our recent work [33]. For the non-parallelizable ML algorithms, we currently run each container on a single core and spawn the required containers every time. In the future, we will look into adaptive batching strategies to handle the problem.

Moreover, to migrate the machine learning application component as described in Section 5.4.3.2, we monitor the round-trip latency to send the data from the edge to cloud, and also the execution latency of the ML model on cloud and edge hardware as shown in Figure 5.14a. If the total execution time (round trip latency + execution latency) is more in the cloud than the edge, we migrate the saved ML model to the edge. We do not provide container migration from cloud to edge because the same Docker container cannot run on edge because of hardware mismatch. We do the requirement-capacity analysis on edge and check the feasibility of deploying the ML model by installing all the dependent software packages on edge to serve the ML model.

5.6 CONCLUSION

As IoT-based analytics becomes increasingly sophisticated, developers are finding themselves lacking the expertise in a wide range of machine learning concepts while also being overwhelmed by the plethora of frameworks, libraries, protocols, programming languages, and hardware available to design and deploy these analytics applications. This chapter presents Stratum, a novel Big Data-as-a-Service for full lifecycle management of IoT analytics applications, to address these highly practical challenges. Stratum systematically integrates several underlying platforms and technologies, hiding most of the details of these underlying artifacts from the users while providing higher-level, intuitive abstractions based on model-driven engineering (MDE), particularly domain-specific modeling and generative programming. The MDE capabilities are offered in our Stratum framework as part of a serverless offering so that the user-supplied specifications can be used to automate the application lifecycle management. Stratum focuses on three key dimensions: automating the deployment of the application, simplifying the machine learning model building process, and ensuring that the models get their desired quality of service properties when the models are used in the inference phase. The Stratum framework is available for download from github.com/doc-vu/Stratum.

ACKNOWLEDGMENTS

This work was supported in part by AFOSR DDDAS FA9550-18-1-0126 and AFRL/Lockheed Martin StreamlinedML program. All authors were affiliated with Vanderbilt University during the time of this work. Any opinions, findings, and conclusions or recommendations expressed in this material are those of the author(s) and do not necessarily reflect the views of AFOSR or AFRL.

REFERENCES

1. F. Pedregosa, G. Varoquaux, A. Gramfort, V. Michel, B. Thirion, O. Grisel, M. Blondel, P. Prettenhofer, R. Weiss, V. Dubourg et al., "Scikit-learn: Machine learning in python," *Journal of Machine Learning Research*, vol. 12, no. Oct, pp. 2825–2830, 2011.
2. G. Holmes, A. Donkin, and I. H. Witten, "Weka: A machine learning workbench," in *Proceedings of ANZIIS'94-Australian New Zealand Intelligent Information Systems Conference*. IEEE, 1994, pp. 357–361.
3. M. Zaharia, R. S. Xin, P. Wendell, T. Das, M. Armbrust, A. Dave, X. Meng, J. Rosen, S. Venkataraman, M. J. Franklin et al., "Apache spark: A unified engine for big data processing," *Communications of the ACM*, vol. 59, no. 11, pp. 56–65, 2016.
4. J. Nandimath, E. Banerjee, A. Patil, P. Kakade, S. Vaidya, and D. Chaturvedi, "Big data analysis using apache hadoop," in *2013 IEEE 14th International Conference on Information Reuse & Integration (IRI)*. IEEE, 2013, pp. 700–703.
5. M. Abadi, A. Agarwal, P. Barham, E. Brevdo, Z. Chen, C. Citro, G. S. Corrado, A. Davis, J. Dean, M. Devin et al., "Tensorflow: Large-scale machine learning on heterogeneous distributed systems," *arXiv preprint arXiv:1603.04467*, 2016.
6. T. Chen, M. Li, Y. Li, M. Lin, N. Wang, M. Wang, T. Xiao, B. Xu, C. Zhang, and Z. Zhang, "Mxnet: A flexible and efficient machine learning library for heterogeneous distributed systems," *arXiv preprint arXiv:1512.01274*, 2015.
7. D. Hall, *Ansible Configuration Management*. Birmingham: Packt Publishing Ltd, 2013.
8. K. Shirinkin, *Getting Started with Terraform*. Birmingham: Packt Publishing Ltd, 2017.
9. M. Satyanarayanan, "The emergence of edge computing," *Computer*, vol. 50, no. 1, pp. 30–39, 2017.
10. P. Ravindra, A. Khochare, S. P. Reddy, S. Sharma, P. Varshney, and Y. Simmhan, "*echo*: An adaptive orchestration platform for hybrid dataflows across cloud and edge," in *International Conference on Service-Oriented Computing*. Springer, 2017, pp. 395–410.
11. S. Shekhar, A. D. Chhokra, A. Bhattacharjee, G. Aupy, and A. Gokhale, "Indices: Exploiting edge resources for performance-aware cloud-hosted services," in *2017 IEEE 1st International Conference on Fog and Edge Computing (ICFEC)*. IEEE, 2017, pp. 75–80.
12. E. Al-Masri, "Enhancing the microservices architecture for the internet of things," in *2018 IEEE International Conference on Big Data (Big Data)*. IEEE, 2018, pp. 5119–5125.
13. A. Bhattacharjee, Y. Barve, S. Khare, S. Bao, Z. Kang, A. Gokhale, and T. Damiano, "Stratum: A bigdata-as-a-service for lifecycle management of IoT analytics applications," in *2019 IEEE International Conference on Big Data (Big Data)*. IEEE, 2019, pp. 1607–1612.
14. T. Li, J. Zhong, J. Liu, W. Wu, and C. Zhang, "Ease. ml: Towards multi-tenant resource sharing for machine learning workloads," *Proceedings of the VLDB Endowment*, vol. 11, no. 5, pp. 607– 620, 2018.
15. D. Golovin, B. Solnik, S. Moitra, G. Kochanski, J. Karro, and D. Sculley, "Google vizier: A service for black-box optimization," in *Proceedings of the 23rd ACM SIGKDD International Conference on Knowledge Discovery and Data Mining*. ACM, 2017, pp. 1487–1495.
16. D. Baylor, E. Breck, H.-T. Cheng, N. Fiedel, C. Y. Foo, Z. Haque, S. Haykal, M. Ispir, V. Jain, L. Koc et al., "Tfx: A tensorflow-based production-scale machine learning platform," in *Proceedings of the 23rd ACM SIGKDD International Conference on Knowledge Discovery and Data Mining*. ACM, 2017, pp. 1387–1395.

17. Uber, Meet Michelangelo: Uber's machine learning platform, 2017. [Online]. Available: https://eng.uber.com/michelangelo/.
18. M. Ma, H. P. Ansari, D. Chao, S. Adya, S. Akle, Y. Qin, D. Gimnicher, and D. Walsh, "Democratizing production-scale distributed deep learning," *arXiv preprint arXiv:1811.00143*, 2018.
19. D. Crankshaw, X. Wang, G. Zhou, M. J. Franklin, J. E. Gonzalez, and I. Stoica, "Clipper: A low-latency online prediction serving system," in *14th {USENIX} Symposium on Networked Systems Design and Implementation ({NSDI} 17)*, 2017, pp. 613–627.
20. W. Wang, J. Gao, M. Zhang, S. Wang, G. Chen, T. K. Ng, B. C. Ooi, J. Shao, and M. Reyad, "Rafiki: Machine learning as an analytics service system," *Proceedings of the VLDB Endowment*, vol. 12, no. 2, pp. 128–140, 2018.
21. M. Zaharia, A. Chen, A. Davidson, A. Ghodsi, S. A. Hong, A. Konwinski, S. Murching, T. Nykodym, P. Ogilvie, M. Parkhe et al., "Accelerating the machine learning lifecycle with MLflow," *Data Engineering*, vol. 41, 2018, p. 39.
22. M. Copeland, J. Soh, A. Puca, M. Manning, and D. Gollob, *Microsoft Azure*, New York, NY: Apress, 2015.
23. K. Venkateswar, "Using amazon sagemaker to operationalize machine learning," 2019.
24. X. Meng, J. Bradley, B. Yavuz, E. Sparks, S. Venkataraman, D. Liu, J. Freeman, D. Tsai, M. Amde, S. Owen et al., "Mllib: Machine learning in apache spark," *The Journal of Machine Learning Research*, vol. 17, no. 1, pp. 1235–1241, 2016.
25. D. Crankshaw, G.-E. Sela, C. Zumar, X. Mo, J. E. Gonzalez, I. Stoica, and A. Tumanov, "Inferline: Ml inference pipeline composition framework," *arXiv preprint arXiv:1812.01776*, 2018.
26. Y. Lee, A. Scolari, B.-G. Chun, M. Weimer, and M. Interlandi, "From the edge to the cloud: Model serving in ml.net," *Data Engineering*, vol. 41, p. 46, 2018.
27. R. Chard, Z. Li, K. Chard, L. Ward, Y. Babuji, A. Woodard, S. Tuecke, B. Blaiszik, M. J. Franklin, and I. Foster, "Dlhub: Model and data serving for science," *arXiv preprint arXiv:1811.11213*, 2018.
28. S. Zhao, M. Talasila, G. Jacobson, C. Borcea, S. A. Aftab, and J. F. Murray, "Packaging and sharing machine learning models via the acumos ai open platform," in *2018 17th IEEE International Conference on Machine Learning and Applications (ICMLA)*. IEEE, 2018, pp. 841–846.
29. V. R. Eluri, M. Ramesh, A. S. M. Al-Jabri, and M. Jane, "A comparative study of various clustering techniques on big data sets using apache mahout," in *2016 3rd MEC International Conference on Big Data and Smart City (ICBDSC)*. IEEE, 2016, pp. 1–4.
30. A. Bhattacharjee, Y. Barve, A. Gokhale, and T. Kuroda, "A model-driven approach to automate the deployment and management of cloud services," in *2018 IEEE/ACM International Conference on Utility and Cloud Computing Companion (UCC Companion)*. IEEE, 2018, pp. 109–114.
31. R. Di Cosmo, A. Eiche, J. Mauro, S. Zacchiroli, G. Zavattaro, and J. Zwolakowski, "Automatic deployment of services in the cloud with aeolus blender," in *Service-Oriented Computing*. Springer, 2015, pp. 397–411.
32. N. Huber, F. Brosig, S. Spinner, S. Kounev, and M. Bahr, "Model-based self-aware performance and resource management using the descartes modeling language," *IEEE Transactions on Software Engineering*, vol. 43, pp. 432–452, 2016.
33. A. Bhattacharjee, A. D. Chhokra, Z. Kang, H. Sun, A. Gokhale, and G. Karsai, "Barista: Efficient and scalable serverless serving system for deep learning prediction services," in *2019 IEEE International Conference on Cloud Engineering (IC2E)*. IEEE, 2019, pp. 23–33.

34. W. Lloyd, S. Ramesh, S. Chinthalapati, L. Ly, and S. Pallickara, "Serverless computing: An investigation of factors influencing microservice performance," in *2018 IEEE International Conference on Cloud Engineering (IC2E), 17-20 April 2018, Orlando, FL, USA*. DOI: 10.1109/IC2E.2018.00039. https://ieeexplore.ieee.org/document/8360324
35. G. Granchelli, M. Cardarelli, P. Di Francesco, I. Malavolta, L. Iovino, and A. Di Salle, "Towards recovering the software architecture of microservice-based systems," in *Software Architecture Workshops (ICSAW), 2017 IEEE International Conference on*. IEEE, 2017, pp. 46–53.
36. L. Figueiredo, I. Jesus, J. T. Machado, J. R. Ferreira, and J. M. De Carvalho, "Towards the development of intelligent transportation systems," in *ITSC 2001. 2001 IEEE Intelligent Transportation Systems. Proceedings (Cat. No. 01TH8585)*. IEEE, 2001, pp. 1206–1211.
37. A. L'heureux, K. Grolinger, H. F. Elyamany, and M. A. Capretz, "Machine learning with big data: Challenges and approaches," *IEEE Access*, vol. 5, pp. 7776–7797, 2017.
38. A. Bhattacharjee, A. D. Chhokra, H. Sun, S. Shekhar, A. Gokhale, G. Karsai, and A. Dubey, "Deep-edge: An efficient framework for deep learning model update on heterogeneous edge," in *2020 IEEE 4th International Conference on Fog and Edge Computing (ICFEC)*, 2020, pp. 75–84.
39. M. Zimmermann, U. Breitenbücher, and F. Leymann, "A tosca-based programming model for interacting components of automatically deployed cloud and IoT applications." in *ICEIS (2)*, 2017, pp. 121–131.
40. H. Gupta, A. Vahid Dastjerdi, S. K. Ghosh, and R. Buyya, "ifogsim: A toolkit for modeling and simulation of resource management techniques in the internet of things, edge and fog computing environments," *Software: Practice and Experience*, vol. 47, no. 9, pp. 1275–1296, 2017.
41. S. Yi, C. Li, and Q. Li, "A survey of fog computing: Concepts, applications and issues," in *Proceedings of the 2015 workshop on mobile big data*. ACM, 2015, pp. 37–42.
42. M. Maróti, R. Kereskényi, T. Kecskés, P. Völgyesi, and A. Lédeczi, "Online collaborative environment for designing complex computational systems," *Procedia Computer Science*, vol. 29, pp. 2432–2441, 2014.
43. Collectd, Collectd - the system statistics collection daemon, 2018. [Online]. Available: https://collectd.org/.
44. Nvidia, Nvidia system management interface, 2018. [Online]. Available: https://developer.nvidia.com/nvidia-system-management-interface/.
45. Rabbitmq, Messaging that just works — rabbitmq, 2018. [Online]. Available: https://www.rabbitmq.com/.
46. Influxdb, Influxdb – time series database, 2018. [Online]. Available: https://www.influxdata.com/time-series-platform/influxdb/.
47. DATA.GOV, Data catalog. [Online]. Available: https://catalog.data.gov/dataset/nys-thruway-origin-and-destination-points-for-all-vehicles-15-minute-\intervals-2017-q4-46887.

6 Category Identification Technique by a Semantic Feature Generation Algorithm

Somenath Chakraborty
The University of Southern Mississippi

CONTENTS

6.1 INTRODUCTION

Image processing is the most fundamental area of research where new innovative technologies coming day by day and hence automatic, more generalized, robust, and efficient methods are always in demand in this area. In modern society, everything is digitally connected. Due to the recent advancement in the Internet of Things (IoT) where everything around us needs to be integrated within a digital control system. Consequently, different kinds of systems need to be correlated first, and then the objective is to connect every device into a common system that can easily monitor and control both locally and remotely. As a result, it is an integration of multiple technologies, ingenious systems, embedded systems, wireless sensor networks, and all possible kinds of automation like home automation or city automation.

With this immense potential of this field, it is obvious we need to have all sorts of advanced methodologies in every field. This chapter describes the semantic feature generation in the image analysis domain with the application description in the image categorization field.

Research of image analysis which deals with the extraction of meaningful information for object category identification is increasing day by day. The most notable challenges vary a lot. As there are many alternative conceptions of color representation in a digital image, for instance, HSL (hue, saturation, lightness), HSB (hue, saturation, brightness), and HSV (hue, saturation, value) mainly come into a similar category of color description where hue and saturation amounts of different kinds of web unstructured images varied a lot with the same category of items, for instance, you have an image of an animal but with varying degree of hue and saturation. So, if the model is based on color criteria, then it is ready hard to put them into the same cluster. This ever-increasing growth of available content makes it more difficult for visual understanding problems of visual information-related applications, which in turn established the urgent need for tools to intelligently analyze, index, and manage the available visual content in a manner capable of satisfying end user's real information needs effectively. It finds many practical applications in many diversified areas ranging from pedestrian detection, object category classification and recognition, face detection and recognition, scene text extraction, vulnerable object identification, industrial verification, advanced robot guidance software for robotic vision equipment, driver assistance, autonomous vehicles, and many more.

There are different kinds of image feature generation algorithm exist in the domain of image processing, but mostly they are not semantic. The feature generation algorithm mostly relies on different image characteristics like color, content properties, and structure properties. However, this work is mainly based on semantic feature generation; hence, it is a more robust feature generation process. To differentiate the category of the content of an image is a crucial task and a challenging research topic for so many reasons. This chapter proposes a semantic approach that is robust enough to categorize different object classes by use of feature extraction algorithm coupled with deformable part-based model and support vector machine. Semantic analysis is a special kind of task that needs to be done in such a way that declarations and statements are semantically correct which ensures that their meaning is coherent with how to control structure and data types are supposed to be used.

6.2 LITERATURE REVIEW

In this section, we have a description of different kinds of well-known researchers in this field and compare the advantages and disadvantages of their methodologies with the application. To start with, Dasiopoulou et al. [1] present a detailed understanding of the state-of-the-art technology of semantic processing of color images. They describe semantic processing as the bridge between low-level visual features that can be automatically extracted from the visual content and the high-level concepts capturing the conveyed meaning.

Semantic approaches are of two types. These are as follows:

- Data governs approach.
- Knowledge governs approach.

The data governs approach is a set of models where some global objective function leads to the computation of values directly from the data. No hierarchy of meaningful intermediate derivations not needed. Consequently, the internal structure becomes abstract to the user's point of view.

On the other hand, the knowledge governs approach is based on signal to symbol paradigm where there is a descriptor for an intermediate level. Indeed, computational vision–related analysis cannot proceed in one single step from signal domain analysis to spatial and semantic analysis. In another paper, Hanif et al. describes [2] a system for text detection and localization. They take text segment images of 32 × 16 or its multiples. They process three different kinds of property of the sample images like Mean Difference Feature (MDF), Standard Deviation (SD), and Histogram of Oriented Gradient (HOG) with a feature set of 39 features (7 MDF, 16 SD, and 16 HOG). The dimension of this feature vector is 151. They proposed a modified AdaBoost algorithm (CAdaBoost) to train the classifier. Felzenszwalb et al. [3] proposed a deformable part-based model, which is very effective for object detection. The improved part-based model presented in the papers [4,5] provides enhancements in their approach. That paper introduces new ideas with Dalal, and Triggs's Histogram of Oriented Gradient (HOG) [6] feature descriptor with support vector machine (SVM) classifier but they build up a deformable which signifies that the object can be part wise divided and detected so that the part can be change positions which match the practical scenario of different objects and later the parts are integrated to detect the object as a whole. It is the most fundamental improvement which can be applied with a variety of object classes. However, the basic concept comes from Fischler and Elschlager's approach [7], which is 44 years old paper describing the fundamental concept of representation and matching of pictorial structures. There is also a pathbreaking concept comes from Lowe's [8] Scale Invariant Feature Transform (SIFT), which outlines a very successful feature generation algorithm that is a very strong affine invariant in nature. The basic technique behind SIFT is scale space. There are other algorithms like, a combination of Principal Component Analysis and Scale Invarient Feature Transform (PCA-SIFT) [9], A SIFT Descriptor with Global Context (GSIFT) [10], A SIFT descriptor with color invariant characteristics(CSIFT) [11], Speeded up robust features (SURF) [12], and Affline-Scale Invarient Feature Transform (ASIFT) [13], which are the different variants and improvements of SIFT. Moreover, a very good scene text detection technique (MHOG) is described in the paper [14]. These techniques are used for different kinds of image analysis applications, such as pedestrian detection, face detection and recognition, human detection, category detection, scene text detection and recognition, and many other applications.

A recent advancement of deep neural network or deep learning algorithms has a huge impact, as well as providing good results make this research field more interesting. With the help of deep learning, researchers can cope up with new challenges that come up with huge datasets or big data analytics or IoT challenges. Deep learning is used in many areas as well as in image classification and recognition domain. Litjens et al. [15] present different deep learning methodologies that are used in medical image analysis and help to develop medical prognosis system which is capable of automatic detection and prediction of different diseases based on image analysis. Various Deep Learning algorithms like ResNet [16], ImageNet [17,18], Mask R-CNN [19], TensorMask [20] that uses Mask RCNN, and Faster R-CNN [21] that uses real-time detection by a modified faster deep neural network architecture named as R-CNN [22–24] are coming up with the use of convolution neural network, recurrent neural network, and a mixture of different model architecture. All these deep neural network architecture use a lot of dense hidden layers which can process complex features within an image or video frame and able to detect and recognize accordingly.

6.3 PROPOSED APPROACH

The proposed approach presented here is twofold. One using semantic feature with SVM and another using a multilayer perceptron.

The general overview of the proposed approach is pictorially represented by the following block diagram (Figure 6.1).

Hereafter taking the images we first go for gamma correction and color normalization. Then comes the feature generation part. In the feature generation, the proposed approach uses Semantic Feature Generation Algorithm (SFGA), which is described in later sections. Then those semantic features are fed to a multi-class SVM to generate the statistical multi-class model upon training, cross validation, and testing. The dataset used for this approach changes from 70% to 85% for training and the rest 30%–15% for testing. Then the SVM outputs the class labels where confusion metrics is created and precision and recall are calculated.

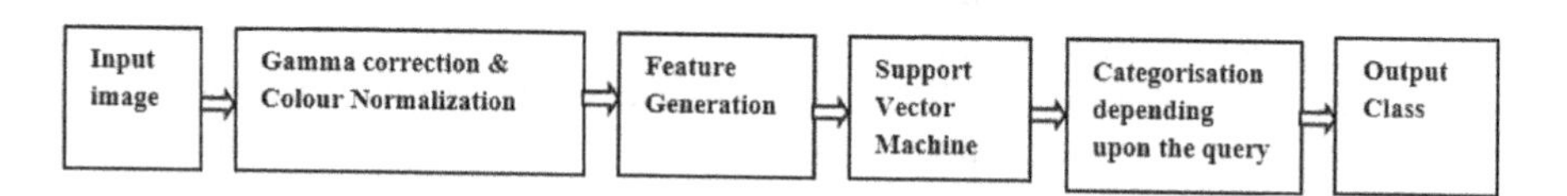

FIGURE 6.1 Block diagram representation of proposed approach.

6.3.1 Image Feature Generation

Here the image feature generation process uses the approach in a semantic way to bridge the gap between high-level image descriptor annotations to knowledge-driven system. First, we need to study the spatial relationship for image understanding.

Semantic data processing is the process of extracting global scene semantic and next to the recognition of the perceptual entities depicted. The feature extraction model combines with a knowledge-driven systems like neural networks, expert systems, fuzzy logic, ontology, decision trees, static and dynamic Bayesian networks, factor graphs, and Markov random fields. These are frequently used mechanisms in recent times for storing and enforcing high-level information.

In our approach, we build up a semantic image extraction mechanism coupled with SVM. The following diagram details the system working representation of the model we proposed in this paper (Figure 6.2).

As represented in the structure, different kinds of ontology have been used to build the knowledge base, and by using semantic rule, we build up the knowledge architecture. Incorporating prior knowledge is a prerequisite for the retrieval to allow understanding of the semantics of user search, and subsequently, aligning and matching with the content annotation semantics. Proprietary knowledge representation solutions, although effective within the predefined usage context, impose serious limitations considering reusability and interoperability. Mainly, the repositories of the semantic gap between the high-level conceptual user queries and the low-level annotations automatically extracted from visual content, improve the process of analysis which aim at this time at automatic extraction of high-level semantic-based annotations. Ontology-based descriptors have been developed that provide support for expressing rich semantics, while at the same time ensuring the formal definition framework is required for making these semantics explicit. Also, ontology alignment, merging, and modularization are receiving intense research interest leading to methodologies that further establish and justify the use of ontology as a knowledge

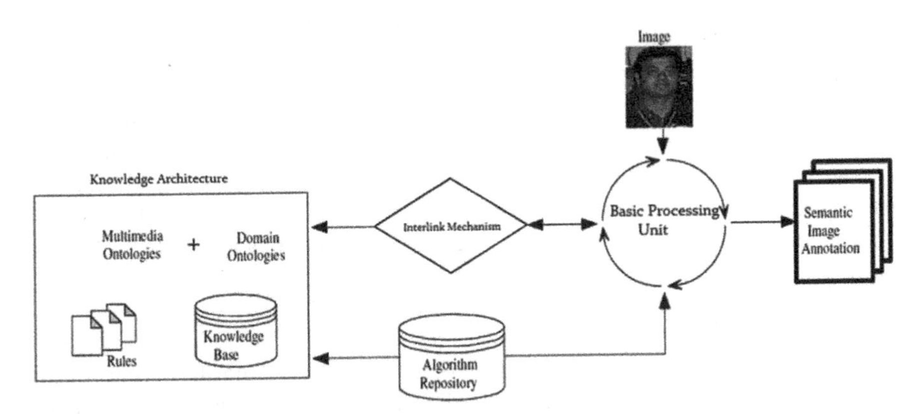

FIGURE 6.2 System architectural flow of semantic analysis.

representation system. There are two layers of semantic used in this system. The first layer semantic is the layer semantics of visual content, i.e., the semantics of the actual conveyed meaning, and the next layer of semantic is the semantics referring to the media itself, within that different kind of ontology is involved in the analysis process. More specifically, domain ontologies are used to model the conveyed content semantics concerning specific real-world domains. Here, we present an ontology-based framework for knowledge-assisted domain-specific semantic image analysis. The assisted knowledge involves qualitative object attributes and quantitative low-level features generated by training as well as low-level processing methods. Rules are used to describe how systems for image analysis should be used, depending on object attributes and low-level features, for the detection of objects corresponding to the semantic concepts defined in the ontology. The added value comes from the coherent architecture achieved by using ontology to describe both the analysis process and the domain of the examined visual content.

6.3.2 The MPEG-7 Visual Descriptors

The ISO/IEC MPEG-7 standard [25], which is a multimedia content analysis and description standard, allows interoperable searching, indexing, filtering data, and browsing of audiovisual (AV) content and focuses on the non-textual description of multimedia content aiming to provide interoperability among applications that use audio-visual content descriptions. It includes color descriptors, shape descriptors, texture descriptors, and motion standardized descriptors. Each of the descriptors has special descriptor schemes (DS) and descriptors (D) attached which are controlled by the data description language (DDL).

6.3.3 Color Descriptors

Color content of an image is a very crucial feature as it is robust to viewing angle, translation, and rotation of the regions of interest. These kinds of descriptors are also very helpful for image retrieval analysis and search processing. The MPEG-7 color descriptors [26] consist of histogram related descriptors, a dominant color descriptor, and a color layout descriptor (CLD). The presentation of the color descriptors begins with a description of the color spaces used in MPEG-7. The color spaces supported are the monochrome, RGB, HSV, YCbCr, and the new HMMD [26]. The color space descriptors are of many kinds depending upon the information it uses for the analysis process. Examples are Scalable Color Descriptor (SCD), Color Layout Descriptor (CLD), and Color Structure Descriptor (CSD).

6.3.3.1 Scalable Color Descriptor (SCD)

The fundamental idea behind the Scalable Color Descriptor (SCD) is to derive descriptor which is indistinguishable to color histogram generation. The difference between them is SCD that uses Haar filter on the histogram and expresses it in the frequency domain. Thus, this approach provides a scalable description because the length of the description can be varied according to the precision as required by the system. SCD descriptors sent data from low band to high band. This approach is quite indistinguishable to quantization in the frequency domain of image or video codec, SCD reserves more bits for "low frequency" bins, and vice versa.

6.3.3.2 Color Layout Descriptor (CLD)

The principal idea behind Color Layout Description (CSD) is that this descriptor procedure represents an image by color accumulation and the local spatial distribution of colors. The procedure of CSD histogram uses a moving structuring window (SW), which shifts one row or one column at a time, to analyze which colors are present in it, and then updates those color bins by only adding one. No matter how many equal value color pixels exist or not.

6.3.3.3 Color Structure Descriptor (CSD)

In Color Structure Descriptors (CSD), the nonlinear quantization is performed by the binary comparison method and folding skills. Then histogram accumulation happens using a structuring window approach over the image. It consists of three parallel local histograms observing (LHO) blocks, which are used to indicate which colors are present in each structuring window. After completion of the color indicating three structure window (SW), their results are summed and sent to CSD histogram accumulation. After histogram accumulation is finished, the final operation is the nonlinear histogram quantization.

6.3.4 Texture Descriptors

The uses of texture descriptors come when the same color visual information varies with a different texture. Consequently, the analysis of texture descriptors plays a vital role in deriving valuable texture features. Texture descriptors are capable of recognizing repeated patterns in an image and analyzing the energy distribution in the frequency domain where frequency domain analysis is needed. The texture descriptors also vary depending upon the information they use for analysis like Edge Histograms Descriptor (EHD), Homogeneous Texture Descriptor (HTD), and Local Edge Histogram (LEH).

6.3.5 Shape Descriptors

Distinguishing categories depending upon the shape is an important aspect of visual information understanding. The shape of an object may be a very expressive feature when used for similarity search and retrieval. MPEG-7 proposes three shape descriptors [27]: Region-Based Shape Descriptor, Contour-Based Shape Descriptor, and 2-D/3-D Shape Descriptor.

The matching with different descriptors can be performed in the following way. Let D_1 and D_2 be two dominant color descriptors.

$$D_1 = \left\{\left\{ c_{1i}, p_{1i}, v_{1i} \right\}, s_1 \right\}i\}, \text{ Where } i = 1, 2, 3, \ldots, N_1$$

$$D_2 = \left\{\left\{ c_{2i}, p_{2i}, v_{2i} \right\}, s_2 \right\}i\}, \text{ Where } i = 1, 2, 3, \ldots, N_2$$

Ignoring variances and spatial coherencies (which are optional), the dissimilarity between them may be defined as follows:

$$R^2(D_1, D_2) = \sum_{i=1}^{N_2} P_1 i^2 + \sum_{j=1}^{N_2} P_2 i^2 - \sum_{i=1}^{N_1}\sum_{j=1}^{N_2} 2al_i, 2jplip2j$$

In this equation, a_{ij} is the similarity coefficient between two colors c_k and c_l, defined by:

$$a_{k,l} = \begin{cases} 1 - d_{k,l}/d_{max} & \text{if } d_{k,l} \leq T_d \\ 0 & \text{if } d_{k,l} > T_d \end{cases}$$

and $d_{k,l}$ is the Euclidean distance between the two colors c_k and c_l, T_d is the maximum distance for two colors and $d_{max} = aT_d$. More details of the modification parameters and others can be found at [28].

6.4 UNDERSTANDING MACHINE LEARNING

Machine learning is the process of finding different kinds of patterns in data using different shorts of statistical and mathematical tools. After finding the patterns, the knowledge can be further used to identify or recognize the similarity with unknown test data and sometimes used to classify and predict different kinds of real-life problems. So, it is a process of building a model that can automatically perform different kinds of identity, recognize, analyze, predict, decide, etc. which the knowledge learned by the system or machine. The new innovation of machine learning, which are gaining popularity are in computer video gaming, simulation modeling, and augmented reality is known as Reinforcement Learning(RL) where agents learn through performing different kinds of experiments and learn over time without having any knowledge of the system.

Basically, there are four types of machine learning models used in machine learning domain.

- Supervised learning model
- Unsupervised learning model
- Semi-supervised learning model
- Reinforcement learning model

6.4.1 Supervised Learning Model

In this form of machine learning, computer is fed with known databases known as the training phase wherein this phase using different kinds of statistical and mathematical models, system understands the desired patterns which need to classify and perform operations on test data.

Mathematically, the process can be described as a technique in which you have input variables (x) and an output variable (Y) and you use a model to learn the mapping function from the input to the output.

$$Y = f(x)$$

The objective is to approximate the mapping function so that when you have new test input data (x) that you can predict the output variables (Y) for that data.

Though it looks very simple with this simplest form of the mathematical equation, essentially different kinds of feature generation statistical approaches are used to exact patterns from a given data. Also, the size of the training dataset should be standardized and large enough so that the model can learn effectively.

Like in a class room environment where the teacher's objective is to teach the information to the students then students also analyze that information to gain knowledge and then after acquiring the knowledge, the students can solve similar kinds of problems related to that information.

Supervised learning can be further classified into two categories:

- Classification
- Regression

6.4.1.1 Classification

Classification is a supervised learning technique where the machine classifies the output variable as a typed category. It could be binary classification like "Red" or "Green" or multivariable classification where more than two categories can be identified.

The application of classification could be like we can build a classification model to categorize bank loan applications as either safe or risky. The system analyzes a huge set of applications dependent on different parameters and label 1 for safe and 0 for risky. Now, whenever a new application comes, it checks all those parameters, automatically classifies the label, and categorizes the application accordingly.

There are diffident kinds of classification algorithms that are very popular in the machine learning domain.

They are as follows:

- Linear classifiers
- Support vector machines
- Decision trees
- K-nearest neighbor
- Random forest
- Bayesian belief networks

6.4.1.2 Regression

Regression is a predictive statistical process where the model builds in such a way that it is capable to find the patterns between dependent and independent variables. Most of the variables are independent variables, and there should be one dependent variable which we need to predict using the recreation model. Examples such as to predict a continuous number like sales, income, and test scores. The equation for basic linear regression can be written as

$$\hat{y} = w[0] \times x[0] + w[1] \times x[1] + \cdots + w[i] \times x[i] + b$$

where x[i] is the feature(s) for the data, and w[i] and b are parameters that are developed during training. For simple linear regression models with only one feature in the data, the formula looks like this:

$$\hat{y} = wx + b$$

where w is the slope, x is the single feature, and b is the y-intercept. The regression model predictions are represented by the line of best fit. For models using two features, the plane will be used and for a model using more than two features, a hyperplane will be used.

There are different kinds of regression model used in machine learning domains are

- Linear regression
- Logistic regression
- Polynomial regression
- Stepwise regression
- Ridge regression
- Lasso regression
- ElasticNet regression

6.4.2 Unsupervised Learning Model

Unsupervised learning model is the type of machine learning model where pre-existent label training data are not needed and machines are expected to look for patterns depending upon the similarity present in the dataset. So, in this case, the objective is to look through the data and find all kinds of possible pattern exist in the dataset.

It is a kind of learning analogous to "Learning without a teacher".

Mathematically, it a technique where a set of N observations (x_1, x_2, …, x_N) of a random p-vector X having joint density Pr(X). The goal is to directly infer the properties of this probability density without the help of a supervisor or previously trained model providing correct answers or degree-of-error for each observation.

The dimension of X is sometimes much higher than in supervised learning, as here no training model is used which is previously reduced in dimension and the properties of interest are often more complicated than simple location estimates. These factors are somewhat mitigated by the fact that X represents all the variables under consideration, one is not required to infer how the properties of Pr(X) change, conditioned on the changing values of another set of variables.

Some applications of unsupervised machine learning techniques include:

a. **Clustering**:

Clustering allows you to automatically split the dataset into groups according to similarity. Often, however, cluster analysis overestimates the similarity between groups and doesn't treat data points as individuals. For this reason, cluster analysis is a poor choice for applications like customer segmentation and targeting.

b. **Anomaly Detection**:

Anomaly detection can automatically discover unusual data points in your dataset. This is useful in pinpointing fraudulent transactions, discovering faulty pieces of hardware, or identifying an outlier caused by a human error during data entry.

c. **Association Mining**:

Association mining identifies sets of items that frequently occur together in your dataset. Retailers often use it for basket analysis, because it allows analysts to discover goods often purchased at the same time and develop more effective marketing and merchandising strategies.

d. **Latent Variable Models**:

Latent variable models are commonly used for data pre-processing, such as reducing the number of features in a dataset (dimensionality reduction) or decomposing the dataset into multiple components.

6.4.3 Semi-Supervised Learning Model

Semi-supervised learning model is a machine learning model where we mix supervised learning techniques with unsupervised learning techniques.

In this kind of learning model generation, we have a large amount of input data (X) and only some of the data is labeled (Y).

In most cases of application, the bigger chunk of the dataset is unlabelled. It helps to reduce complexity and in certain cases where dimension reduction is very crucial for fast execution of the model, we use this kind of technique.

6.4.4 Reinforcement Learning Model

The reinforcement learning model is a kind of machine learning model where we do not need any kind of previous knowledge to meet our objectives. Here, software agents which are basically different kinds of actions or set of programs help to understand and gain knowledge in the environment. A set of learning rules is defined which in most cases a combination of exploration and exploitation by which machine gains the required knowledge to perform its objective task. Here exploration is different kinds of random event generation by which the software agent interacts with the environment and learns different kinds of knowledge. Exploitation signifies that whatever knowledge the software agent learns over a period, it can be used to identify unknown events.

This kind of learning model is very useful in computer video game design environment and different kinds of software simulation generation processes.

6.5 SUPPORT VECTOR MACHINE (SVM)

SVM is a supervised learning technique that is very efficient and fast and can perform well without demanding high resources.

SVM is a supervised discriminative classifier formally defined by a separating hyperplane. In other words, given labeled training data, the algorithm outputs an optimal hyperplane that categorizes new examples. In two-dimensional space,

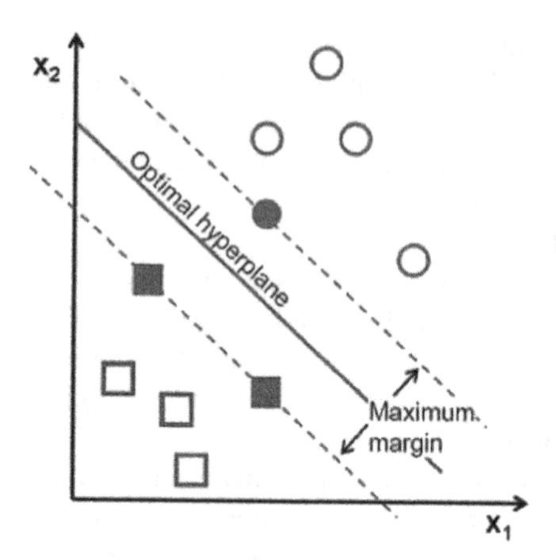

FIGURE 6.3 Support vector machine.

this hyperplane is a line dividing a plane into two parts wherein each class lay on either side. SVM is a practical implementation of a hypothetical classifier named as maximal-marginal classifier.

The objective of the SVM algorithm is to find a hyperplane in an N-dimensional space, where N is the number of features that distinctly classify the data points (Figure 6.3).

In this project for image identification technique, after generation of image feature values for each class of the category, we need to feed those values into SVM [29], which is basically a multi-class classifier capable of identifying the different category of objects depending upon the query given or feature value fed to the system. Each class corresponds to a particular category. For each class of the category, the SVM needs to be trained with the same size as the image feature class. In our case, it is 40 × 50 classes of vectors. Depending upon the values, we prepare the confusion matrix by which precision and recall are calculated. We obtained the two measures precision (p) and recall (r) of our extraction results. Expressions used for computation of p and r values are as follows:

$$p = \frac{P \cap T}{P} \text{ and } r = \frac{P \cap T}{T}$$

where P is the area of the minimum enclosing rectangles of all connected components extracted by the proposed strategy and T is the area of all the rectangles provided as the ground truth together with the sample images.

6.5.1 Tuning Parameters: Kernel, Regularization, Gamma, and Margin

Kernel: The learning of the hyperplane in linear SVM is done by transforming the problem using some linear algebra. This is where the kernel plays a role. For linear kernel, the equation for prediction for a new input using the dot product between the input (x) and each support vector (x_i) is calculated as follows:

$$f(x) = B(0) + \text{sum}\left(a_i \times \left(x, x_i\right)\right)$$

This is an equation that involves calculating the inner products of a new input vector (x) with all support vectors in training data. The coefficients B_0 and a_i (for each input) must be estimated from the training data by the learning algorithm.

The polynomial kernel can be written as $K(x, x_i) = 1 + sum(x \times x_i)^d$ and exponential as $K(x, x_i) = exp(-gamma \times sum(x - xi^2))$.

Regularization Parameter: The regularization parameter describes the SVM optimization how much we want to avoid misclassifying each training example.

For large values of C, the optimization will choose a smaller-margin hyperplane if that hyperplane does a better job of getting all the training points classified correctly. Conversely, a very small value of C will cause the optimizer to look for a larger margin separating hyperplane, even if that hyperplane misclassifies more points.

Gamma: The gamma parameter defines how far the influence of a single training example reaches, with low values meaning "far" and high values meaning "close". In other words, with low gamma, points far away from plausible separation lines are considered in the calculation for the separation line. By contrast, high gamma means the points close to a plausible line are considered in the calculation.

6.6 EXPERIMENTAL RESULTS

After extracting the semantic features using different kinds of semantic descriptor as described above, the features are then fed to multi-valued linear SVM which is actually based on maximal-marginal classifier, labeled the multi-class level according to the class category level present in the training dataset.

The proposed approach was implemented on windows platform, using OpenCV and MATLAB 7.6.0(R2008a).

The experimental results were obtained using the PASCAL VOC [30] best practices dataset specifically captured separately for Training and Testing. For the "Train file" here, we have taken 1740 positive images, i.e., image samples that have desired class of objects in it, and 1910 negative images, i.e., image samples that do not have the desired class of objects in it. As a result, the total image sample size for the train category is 3650. For the "Test file", we take 1240 positive sample images and 2180 negative sample images. As a result, the total image sample size in this category is 3420. We need to do the same process for each class of the desired image object.

The comparative results are presented in Table 6.1 along with object detection results on a few images (Figures 6.4 and 6.5).

TABLE 6.1
Comparative Performance Analysis of Text Detection Approaches

Algorithms	Precision (p)	Recall (r)
Felzenszwalb et al. [3]	0.69	0.61
Arbeláez et al. [31]	0.63	0.59
Carreira et al. [32]	0.73	0.60
Dalal et al. [6]	0.74	0.60
Winder et al. [33]	0.51	0.67
Proposed method	0.78	0.62

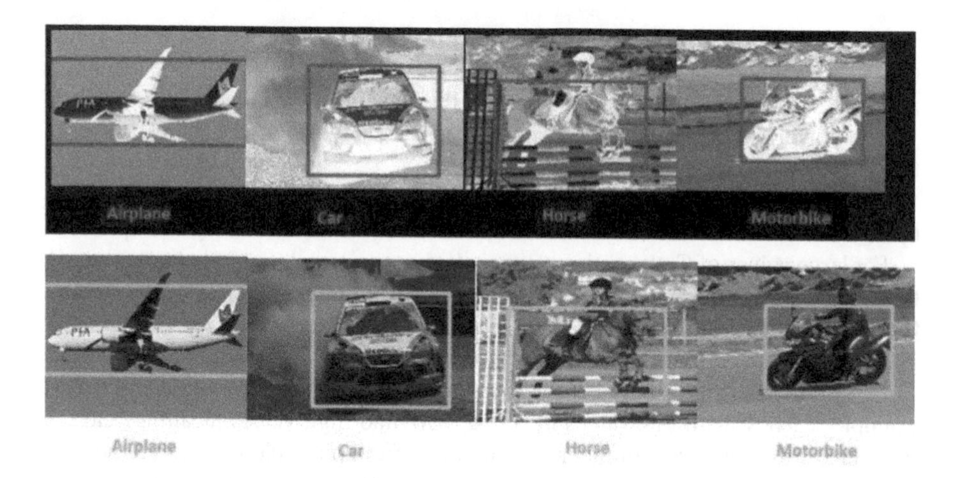

FIGURE 6.4 Visualization for different categories of images.

FIGURE 6.5 ImageNet [17,18] database example pictures.

6.7 CONCLUSION AND FUTURE WORK

This project work proposed a semantic feature generation approach for the detection of different kinds of objects with distinguishable characteristics. This helps immensely in dimension reduction and complexity that makes the model fast and efficient. Then these semantic features are labeled by the help of multi-class SVM. This classification model can be applied in many fields like license plate detection of car, camera capture Image Scene Text Detection, and web-generated images. We have a plan to use this in all possible areas of application.

REFERENCES

1. S. Dasiopoulou, E. Spyrou, Y. Avrithis, Y. Kompatsiaris, and M.G. Strintzis, "Semantic processing of color images," in *Color Image Processing: Methods and Applications*, R. Lukac and K. N. Plataniotis, ISBN 9780849397745 Published October 20, 2006 by CRC Press
2. S. M. Hanif and L. Prevost, "Text detection and localization in complex scene images using constrained AdaBoost algorithm," *10th International Conference on Document Analysis and Recognition*, 2009. DOI:10.1109/ICDAR.2009.172.
3. P. Felzenszwalb, R. Girshick, D. McAllester, and D. Ramanan, "Object detection with discriminatively trained part-based models," IEEE Transactions on Pattern Analysis and Machine Intelligence, 32(9): 1627–1645, Sept. 2010, DOI:10.1109/TPAMI.2009.167.
4. P. Felzenszwalb, R. Girshick, and D. McAllester, "Cascade object detection with deformable part models," in *CVPR*. IEEE, 2010.
5. P. F. Felzenszwalb, R. B. Girshick, and D. McAllester, "Discriminatively trained deformable part models, release 4." http://people.cs.uchicago.edu/pff/latent-release4/.
6. N. Dalal and B. Triggs, "Histograms of oriented gradients for human detection," in *CVPR*, 2005.
7. M. A. Fischler and R. A. Elschlager, "The representation and matching of pictorial structures," *IEEE Transactions on Computer*, 22(1):67–92, January 1973.
8. D. Lowe, "Distinctive Image Features from Scale-Invariant Keypoints," *IJCV*, 60(2): 91–110, 2004.
9. Y. Ke and R. Sukthankar, PCA-SIFT: "A more distinctive representation for local image descriptors," *Proc. Conf. Computer Vision and Pattern Recognition*, pp. 511–517, 2004.
10. E.N. Mortensen, H. Deng, and L. Shapiro, "A SIFT descriptor with global context," in *Computer Vision and Pattern Recognition (CVPR 2005)*, 20–25 June 2005. IEEE, Vol. 1, 184–190, 2005.
11. A.E. Abdel-Hakim and A.A. Farag, "CSIFT: A SIFT descriptor with color invariant characteristics," in *Computer Vision and Pattern Recognition (CVPR 2006)*, 17–22 June 2006. IEEE, Vol. 2, 1978–1983, 2006.
12. H. Bay, T. Tuytelaars, and L.V. Gool, "SURF: Speeded up robust features," in *Computer Vision – ECCV 2006 9th European Conference on Computer Vision*, 7–13 May 2006. Springer, Part II, 404–417, 2006.
13. J.M. Morel and G. Yu, "ASIFT: A new framework for fully affine invariant image comparison," *SIAM Journal on Imaging Sciences*, 2(2):438–469, 2009.
14. S. Chakraborty and S. K. Bandyopadhyay. "Scene text detection using modified histogram of oriented gradient approach," *International Journal of Applied Research*, 2(7):795–798, 2016.
15. G. Litjens, T. Kooi, B. E. Bejnordi, A. A. A. Setio, F. Ciompi, M. Ghafoorian, J. A.W.M. van der Laak, B. van Ginneken, and C. I. Sánchez, "A survey on deep learning in medical image analysis," *Medical Image Analysis*, 42:60–88, 2017, ISSN 1361-8415, DOI:10.1016/j.media.2017.07.005.
16. K. He, X. Zhang, S. Ren, and J. Sun, "Deep residual learning for image recognition," in *2016 IEEE Conference on Computer Vision and Pattern Recognition (CVPR)*, Las Vegas, NV, 2016, pp. 770–778. DOI:10.1109/CVPR.2016.90.

17. J. Deng, W. Dong, and R. Socher, "Imagenet: A large-scale hierarchical image database," in *Proceedings of the IEEE Conference on Computer Vision and Pattern Recognition*, pp. 248–255, Miami, FL, USA, June 2009.
18. http://www.image-net.org/.
19. K. He, G. Gkioxari, P. Dollár, and R. Girshick, "Mask R-CNN," in *2017 IEEE International Conference on Computer Vision (ICCV)*, Venice, 2017, pp. 2980–2988. DOI:10.1109/ICCV.2017.322.
20. X. Chen, R. Girshick, K. He, and P. Dollar, "TensorMask: A foundation for dense object segmentation," in 2019 IEEE/CVF International Conference on Computer Vision (ICCV), Seoul, Korea (South), 2019, pp. 2061–2069. DOI:10.1109/ICCV.2019.00215.
21. S. Ren, K. He, R. Girshick, and J. Sun, "Faster R-CNN: Towards real-time object detection with region proposal networks," *IEEE Transactions on Pattern Analysis and Machine Intelligence*, 39(6):1137–1149, 1 June 2017. DOI:10.1109/TPAMI.2016.2577031.
22. Y. Wei, W. Xia, M. Lin, J. Huang, B. Bi, J. Dong, Y. Zhao, and S. Yan, "HCP: A flexible CNN framework for multi-label image classification," *IEEE Transactions on Pattern Analysis and Machine Intelligence*, 38(9):1901–1907, 2016.
23. F. Schroff, D. Kalenichenko, and J. Philbin, "Facenet: A unified embedding for face recognition and clustering," in *Proceedings of the IEEE Conference on Computer Vision and Pattern Recognition*, pp. 815–823, Boston, MA, USA, June 2015.
24. A. Esteva, B. Kuprel, R. A. Novoa, J. Ko, S. M. Swetter, H. M. Blau, and S. Thrun, "Dermatologist-level classification of skin cancer with deep neural networks," Nature, 542(7639):115–118, 2017.
25. S. F. Chang, T. Sikora, and A. Puri, "Overview of the mpeg-7 standard," *IEEE Transactions on Circuits and Systems for Video Technology*, 11(6):688–695, 2001.
26. B. Manjunath, J. Ohm, V. Vasudevan, and A. Yamada, "Color and texture descriptors," *IEEE Transactions on Circuits and Systems for Video Technology*, 11(6):703–715, 2001.
27. M. Bober, "Mpeg-7 visual shape descriptors," *IEEE Transactions on Circuits and Systems for Video Technology*, 11(6):716–719, 2001.
28. MPEG-7, "Visual experimentation model (xm) version 10.0." ISO/IEC/ JTC1/SC29/ WG11, Doc. N4062, 2001.
29. K. Crammer and Y. Singer. "On the algorithmic implementation of multi-class SVMs," *JMLR*, 2001.
30. M. Everingham, L. Van Gool, C. K. I. Williams, J. Winn, and A. Zisserman. "The PASCAL visual object classes (VOC) challenge," *IJCV*, 88:303–338, 2010.
31. P. Arbel´aez, B. Hariharan, C. Gu, S. Gupta, L. Bourdev, and J. Malik. "Semantic segmentation using regions and parts," in *CVPR*, 2012.
32. J. Carreira and C. Sminchisescu, "CPMC: Automatic object segmentation using constrained parametric mincuts," *TPAMI*, 34:1312–1328, 2012.
33. S. A. Winder and M. Brown, "Learning local image descriptors," in *Computer Vision and Pattern Recognition, 2007. CVPR'07.* IEEE Conference on, 2007, pp. 1–8.

7 Role of Deep Learning Algorithms in Securing Internet of Things Applications

Rajakumar Arul
Amrita School of Engineering

Shakila Basheer
Princess Nourah bint Abdulrahman University

Asad Abbas
University of Central Punjab

Ali Kashif Bashir
Manchester Metropolitan University

CONTENTS

7.1 INTRODUCTION

The Internet of Things (IoT) is changing our thoughts over sensing and sharing, and it has already made a huge impact in our day-to-day life from agriculture to wearable gadgets. In addition to the inestimable values being utilized, IoT leaves us with a hotbed of security threats. The last few decades of conventional internet computing threats were associated with information leaks and service loss. However, threats to IoT adds a physical security risk along with our old bucket of threats. IoT works on various platforms and portfolios, where each needs to be addressed with our dedicated security solutions. Regarding new security solutions, all the threats of the former computing era should be revisited.

Ensuring the security features like confidentiality, integrity, and reliability while exchanging the data between the things on the internet are the responsibilities of strong Crypto mechanisms. In a normal IoT scenario, data is exchanged between the entities which can be input for other entities on the internet or it can be a report for the end-user to verify. This shared data is highly sensitive or an individual's private information. Thus addressing these privacy issues is a critical issue. Crypto mechanism protects this sensitive information when it is transmitted, when it is stored, and when it is processed. IoT's special unique characteristic is its heterogeneity and battery power. Hence all the available authentication solutions won't address the threats which are imposed from high power systems.

Communication in IoT can be classified into two categories namely domestic and foreign [1]. In domestic communication, access to the public network is necessarily not needed. Domestic communication authentication mechanisms like Bluetooth or ZigBee needs a password for pairing before transfer. Application models for domestic communication are wearable devices, smart furniture, and smart home appliances. Here communication happens within a closed network of home or organization. On the contrary, in a foreign communication scenario, a wearable device or a sensor on a vehicle or in remote places that is in mobility needs access to the public network to transfer data to distant objects. This range of communication is more prone to attacks.

In IoT ecosystem, even improperly configured devices can also cause a threat to the whole framework, which can be a compromise of a device, infra, and communication network. Based on a study, less than 10% of IoT carries nonpersonal information, remaining carries personal information. Only 30% of the IoT devices encrypt the data to the network. Only four out of ten devices that provide user interfaces are safe from attacks like XSS-Persistent cross-site scripting. Since these IoT devices are heavily dependent on wireless and cellular networks, regular threats like Eavesdropping, Denial of service, authentication bypass, and role bypass are very common in IoT. Hence, in this chapter, we propose a design by involving DL algorithms for common threat mitigation in IoT. Our design takes the advantage of DL algorithms that are deployed in the cloud environment to process the network logs. When the logs are mined, there is a high chance of finding the corelation in the network behavior, and based on that, we can segregate the vulnerable traffic. This can help us in early detection of the attacks, and also the same can be used to mitigate the attacks.

The remainder of this chapter is organized as follows. Section 7.2 presents the literature survey of security threats in IoT and the recent mitigation strategies for the same. Section 7.3 provides the machine learning (ML) algorithms that are popular and can be used with IoT classification. Section 7.4 discusses the DL algorithms that can be modeled for attack detection. Section 7.5 describes the proposed ideology for threats and vulnerabilities mitigation in IoT. Finally, Section 7.6 concludes the study.

7.2 LITERATURE SURVEY OF SECURITY THREATS IN IoT

In this section, a detailed literature survey has been done in the various aspects of IoT terminals with respect to the security concerns and the recent research solutions for the same. As the number of IoT terminals is increasing with the revolution of technology, security risks that incur are also high [2,3]. Some of the predominant attacks [4] are shown in Table 7.1.

Jamming is a process followed by intruders in the networks to cause degradation or denial in service by decreasing the signal-to-noise ratio at receiver sides through the transmission of interfering wireless signals [5,22]. There are two types of disruption that occur in a wireless network: one is intended (jamming), and other is unintended (interference). The unintended interference may be caused due to the overlap in the frequency either by the wireless node itself or through the neighboring device. On the other hand, intentional interference is usually conducted by an attacker who intends to interrupt or prevent communications in networks. Jamming is possible in all levels of networks [22]. Jamming is common in IoT terminals, and there are many recent research solutions proposed to overcome these. Namvar et al. proposed a novel anti-jamming strategy for OFDM-based IoT systems to protect IoT devices against malicious radio jammer [5]. Tang et al. proposed a jamming mitigation strategy based on the Hierarchical Security Game by involving reactive jamming techniques [2].

TABLE 7.1
Predominant Attacks in IoT

Attacks	Characteristics
Jamming [5]	Makes the devices nearby unstable leading to denial of service
Spoofing [6,7]	Hiding the original identity to avail unauthorized service/access
Replay attack [8]	Using the old message sequence to steal sensitive information
Buffer reservation attack [4]	Using the temporary storage buffer records to regenerate the message sequence
Attacks based on authentication [9–20]	Gaining access to unauthorized services with fake credentials/certificates
Insecure interface [4]	Established connection between the terminal is open to attackers
Insecure firmware [8]	Easily accessible device drivers and firmwares lead to faulty information exchange
Privacy threats [21]	Information leaks that lead to privacy concerns

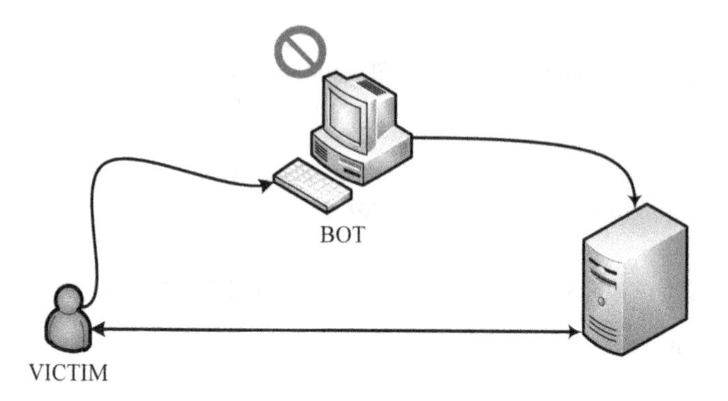

FIGURE 7.1 Process of spoofing.

Spoofing is one of the predominant attacks that active attackers follow [6]. Spoofing is common in all types of networks and even in hardware-based scenarios as shown in Figure 7.1. In this process, the attacker hides the original identity and acts as a known identity in a network. Spoofing is mainly done to acquire sensitive information and important credentials, and to gain control over a specific target [7]. Also, attackers use spoofing techniques to overhear sensitive conversations and to gain unauthorized access to confidential data. Spoofing in IoT terminals is highly critical as it may give access to the entire network. Arafin et al. proposed a novel spoofing prevention scheme using the hardware oscillators by using the drifts in the frequencies. Koh et al. have developed a unique location spoofing detection algorithm for geospatial tagging and location-based services in the IoT [1].

A replay attack is a form of passive network attack in which a valid data is maliciously repeated or delayed. This is done by intercepting the data that is transferred in the past and retransmitting it after a stipulated time with an intention to acquire information as shown in Figure 7.2. Jan et al. explain replay attacks with respect to IoT terminals as one of those attacks which can be done with the intention to fool the receiver by retransmitting the data sent to the same terminal with a different timestamp. This not only reveals the confidential information, but also provides a clue for the keys that will be generated in the future for various purposes. A few threats even challenge the privacy of the network. For instance, let's consider an IoT-based light bulb, where the attacker can easily operate the bulb without the user's permission by enforcing a replay attack. Rughoobur et al. have proposed a replay attack mitigation

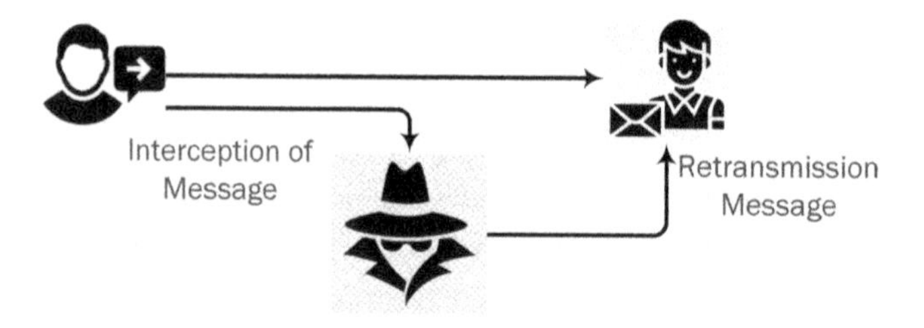

FIGURE 7.2 Replay attack.

strategy by using strong digital signatures with timestamps, and some of the other known techniques include generating an OTP during each transmission and sequencing of messages and nonacceptance of duplicated messages [8].

A receiving node requires to reserve buffer space for reassembly of incoming packets, and an attacker can easily exploit it by sending incoming packets. This attack results in denial-of-service in an IoT device and an inevitable attack strategy. Arış et al. propose an investigation of how buffer reservation attacks are affecting IoT devices and have suggested many possibilities to overcome such attacks [23,24].

The devices that are connected in the IoT and the users in IoT need to be authenticated through specific key management systems. Any loophole in security at the network layer may put the entire network at risk, and the vulnerable communication may expose the network to a large number of vulnerabilities. These vulnerabilities can arise due to constrained resources and overhead of DTLS (Datagram Transport Level Security) requirements that need to be minimized. The authentication may be mutual authentication which involves both the sender and receiver have to authorize each other. This will purposefully avoid rouge attacks. Venkatesan et al. propose a method of securing IoT endpoints from this attack by introducing Secure Volts which is a multikey-/multi-password-based mutual authentication mechanism [3,16–20,25–36].

The IoT endpoints are hardware devices that are fabricated for specific purposes. As such, the devices cannot interact with other devices in the network, and hence it requires an interface to establish a connection between other terminals. The interface not only provides connectivity, but also connects the supplementary devices to the internet. Since wireless interface may be over the air, there is a high risk for passive attacks in this interface. Similarly, if it's a wired one, physical protection of the same will require additional overheads. An insecure interface may invite all privacy-related threats as well as active threats that may be vulnerable to the entire IoT ecosystem [22].

Several attacks can be triggered by accessing the data stored on the endpoint which in turn can be used to violate identity and location-based privacy concerns. Since most of the data from the IoT terminals are stored in the cloud, the devices that establish the connection should be verified and authenticated at as many places as possible. Because on the varied spectrum of device integration in IoT, there is a need for a certificated special firmware that can be used to verify these devices. The Application Programming Interface (API) that is developed with languages like JSON, XML, SQL, and XSS is prone to attacks and needs to be tested carefully. Similarly, the software/firmware updates need to be carried out in a secure manner. Salah et al. proposes a Blockchain smart contract that has the ability to provide decentralized authentication rules and logic to be able to provide single and multiparty authentication to IoT Devices [5,10].

Thus, when million devices are connected to an IoT platform, there should be some technique or strategy to prevent attacks on the connected devices. Through our literature, we have inferred that the number of attacks on the IoT terminals is growing in an exponential way which in turn invites many research solutions to mitigate the common attacks. Most of the recent solutions rely on the attack type and attacking strategies for attack detection. By considering all the drawbacks of the existing solutions, we propose an ideology that takes the advantage of the DL frameworks for attack detection in this chapter.

7.3 ML ALGORITHMS FOR ATTACK DETECTION AND MITIGATION

In this section, various ML algorithms and DL frameworks are discussed with the deployment scenario to attack detection and mitigation. As there are many learning algorithms in practice, a broader spectrum of algorithms is listed here.

7.3.1 Linear Regression

Linear regression is a type of supervised machine learning algorithm that maps an input to an output based on input-output pairs. When dealing with a single input, it is called a single linear regression, while two or more inputs lead to multiple linear regression as shown in Figure 7.3. Different techniques can be used to prepare or train the linear regression equation from data, the most common of which is called Ordinary Least Squares. Linear regression is widely used in price predictions, stock trading, etc. Ieno et al. address the disadvantages of this model in that it assumes the existence of a linear relation between dependent and independent variables, which is not always true.

Distributed Denial of Service (DDoS) is a common attack of the IoT terminal where the legitimate users are denied services (network resources, information systems, etc.) by the actions of malicious actors in the network. Linear regression can be used to deploy a mechanism to detect the DDoS attack, by establishing a linear relationship between the dependent and the independent variable as discussed by Chen [37].

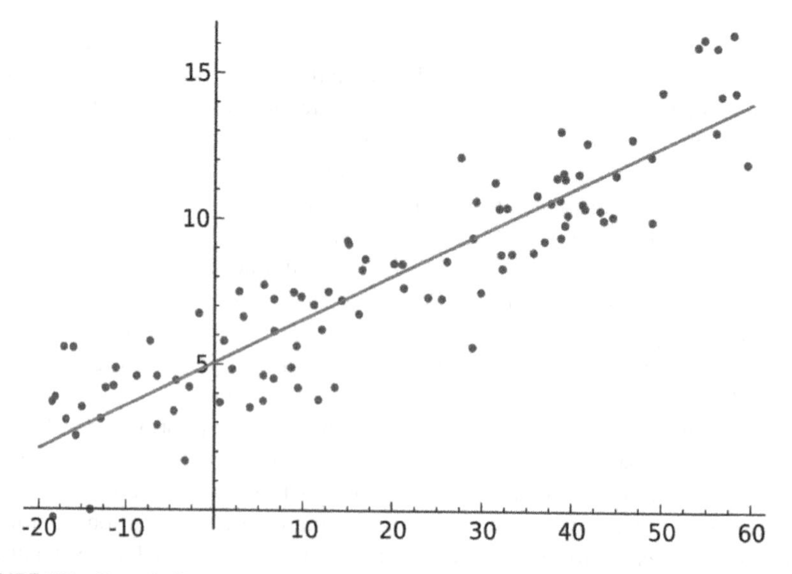

FIGURE 7.3 Sample linear regression.

7.3.2 Principal Component Analysis

The Principal Component Analysis (PCA) is a technique to bring out strong patterns in a dataset by suppressing variations. The algorithm is used to clean data structures to explore and efficiently analyze it. This technique is used in domains like facial recognition, computer vision, and image compression. The major disadvantage of PCA is that one must standardize their data before implementing it; otherwise, PCA will not be able to find the optimal principal components.

PCA can be viewed as an attractive solution for anomaly detection in IoT networks with its reduced complexity. Anomalies may include cardiac episodes, mechanical failures, hacker attacks, or fraudulent transactions but the suppression process helps in yielding favorable results. Dang Hai Hoang and Ha Duong Nguyen have discussed the various existing PCA models for anomaly detection in the IoT network and also have depicted further developments from their side [23].

7.3.3 Q-Learning

It is an off-policy reinforcement learning algorithm that seeks to find the best action to take given the current state as depicted in Figure 7.4. "Off policy" refers to the algorithm's attempt to learn from actions outside its current policy. It is usually used for higher-dimensional problems such as industrial applications, but it comes with rather a large time complexity. Its most popular use was in AlphaGo and Google's DeepMind.

Q-learning-based authentication in IoT devices can be used to detect DoS attacks, jamming, spoofing, and malware attacks.

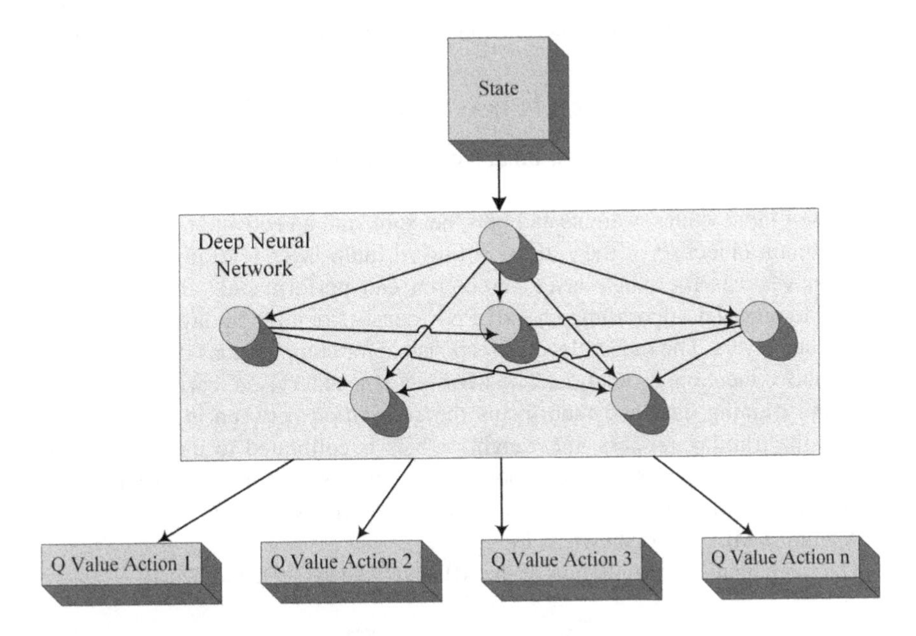

FIGURE 7.4 Q-learning procedure.

7.3.4 K-Means Clustering

K-means clustering is one of the most popular unsupervised ML algorithms. In K-means, learning of data starts from the randomly chosen centroids which in turn act as the starting point of each cluster. The same process is iterated until we obtain an optimized centroid. K-means is popularly used in identifying crime localities and analysis of Fantasy League Statistics.

The disadvantages include the formation of different clusters when using different initial partitions and the inefficiency in the global cluster, but the most important one being the difficulty in predicting the K-value. An implementation of K-means in python is as follows:

```
from sklearn.cluster import KMeans
Kmean = KMeans(n_clusters=2)
Kmean.fit(X)
```

A neural network with K-means clustering algorithm can easily apply techniques for malware detection and access control (security technique to regulate the access to the computing environment).

Table 7.2 describes the security level of various ML algorithms and their adaptability toward the specific attacks. Further reading on how ML can be adopted for threat detection can be found at [9,38].

7.4 DL ALGORITHMS AND IoT DEVICES

In this section, we will briefly discuss some of the important DL algorithms and their applications [39] toward the detection and mitigation of attacks in IoT terminals.

7.4.1 Multilayer Perceptron Neural Network (MLPNN)

Perceptron: The perceptron is a binary classification algorithm that classifies the input into two segments. As such perceptron is a single-layer neural network that consists of input values, weights and bias, net sum, and an activation function [40].

The main objective of the transition toward multi-layer is to involve nonlinear function whereas the single-layer perceptron can perform only linear functions. Similar to other DL algorithms, the MLPNN consists of an input layer, output layer, and hidden layers. The hidden layer is responsible for computations that are involved in the above-mentioned function. The MLPNN is a supervised learning model that takes the training data and then forms the correlation between input and output. During the training process, the weight or bias is calibrated to make the desired output which in turn refines the decision-making process. The backpropagation technique is used for the refinement of decisions in the MLPNN.

A Sample MLPNN is shown as Figure 7.5.

The mathematical formulation of the MLPNN can be seen as Eq. 7.1:

$$C_K = f\left(\sum_y A_{yz} P_y\right) P_y = f\left(\sum_x A_{xy} P_x\right) \tag{7.1}$$

TABLE 7.2
Security Level of ML Algorithms with Attacks

Attacks	Linear Regression	PCA	Q-Learning	K-Means Clustering
Information leakage	○	○	△	△
Spoofing	△	○	△	△
Replay attack	△	◇	△	○
Buffer reservation	△	◇	△	◇
Insecure interface	◇	◇	○	△
Insecure firmware and privacy	◇	○	○	△

△ - *Fully secure*

○ - *Partially secure*

◇ - *Not secure.*

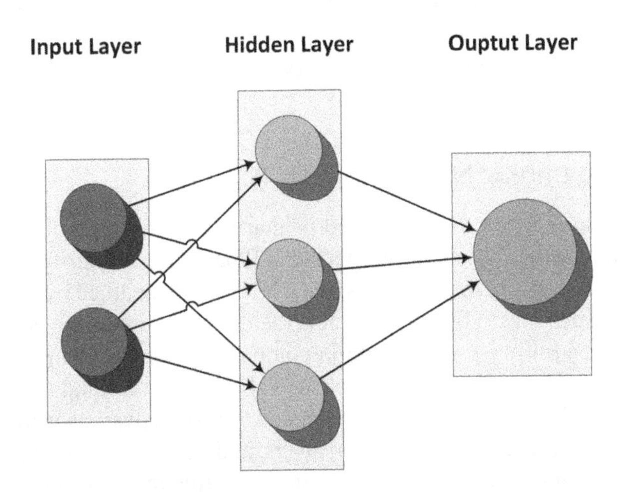

FIGURE 7.5 Sample MLPNN.

where A denotes weight, Z is the output layer, y is the hidden layer, and x is the input layer.

For the purpose of refinements, the weight can be returned or adjusted using Eq. 7.2:

$$A = A_{old} + \eta\delta_b \text{ where } \delta = C_{target} - C \quad (7.2)$$

Similarly, sigmoid (S-shaped) function properties can be modified by choosing an appropriate function. The most common choice for it is Eq. 7.3:

$$C = f(m) = \frac{1}{1+e^{-m}} \quad (7.3)$$

and the derivative of sigmoidal function is:

$$f(m) = f(m)(1-f(m)) \quad (7.4)$$

The weights that are initialized can be modified by following certain rules as per our criteria.

Generally, weight change from any unit y to unit z by gradient descent that is called as Generalized Delta Rule (GDR, or Backpropagation)

$$\Delta_a = a - a_{old} = -\eta\frac{\partial E}{\partial a} = \eta\delta_b \quad (7.5)$$

So, the weight change from the input layer unit x to hidden layer unit y is:

$$\Delta A_{xy} = \eta\delta_y B_x \text{ where } \delta_y = P(1-P_y)\sum_z A_{yz}\delta_z \quad (7.6)$$

The weight change from the hidden layer unit y to the output layer unit z is:

$$\Delta A_{yz} = \eta\delta_z P_y \text{ where } \delta_z = (C_{target,z} - C_z)C_z(1-C_z) \quad (7.7)$$

7.4.2 Convolutional Neural Network (CNN)

CNN is a neural network that is formed primarily for image processing [41]. CNN can be used for classification and clustering [42] based on similarity in object recognition process. In most of the scenarios, the CNN is accompanied by a visual element that includes image and video analysis. It is the main reason for the popularity of the DL paradigm, and it plays a vital role in computer vision problems which involves self-drivings cars, autonomous robots, etc. CNN [43] is not limited to image processing problems, and they can be further extended to text analysis as well.

CNNs consist of several convolutional layers, and each layer is equipped with filters to perform convolutional operations. The input for this model is a two-dimensional matrix, i.e., an image transformed to a 2d matrix, that can be used to identify and

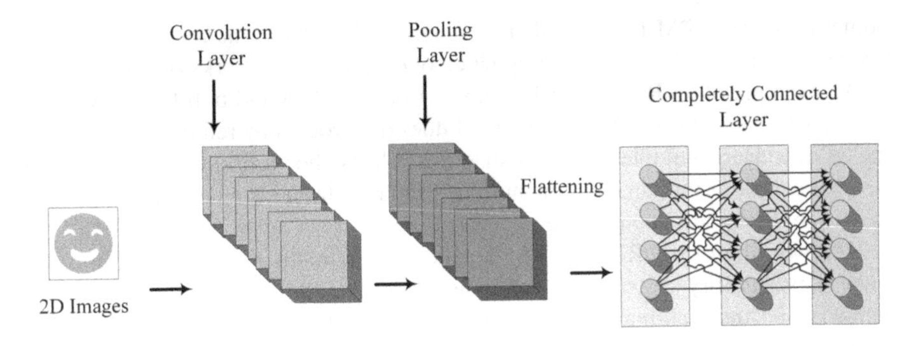

FIGURE 7.6 Convolution neural network.

classify the object. The rectified linear unit layer is responsible for the element-wise operations that in turn produce rectified feature maps. The next layer is the pooling layer that takes the map as input to reduce the dimensionality of the feature map. Through the flattening process, the resultant 2d matrix will be converted into a single linear vector. Then comes the fully connected layer which takes the linear vector as input for classification and identification.

The sample CNN is depicted in Figure 7.6.

One aspect to be noted in the convolution layer is the role of ReLU's. These units do not require input normalization as it may lead to saturation.

Denoting $a^i_{x,y}$ the activity of a neuron computed by applying kernel i at position (x, y) and then applying the ReLU nonlinearity, the response-normalized activity $b^i_{x,y}$ is given by the expression

$$b^i_{x,y} = a^i_{x,y} \Big/ \left(k + \alpha \sum_{j=\max(0,i-n/2)}^{\min(N-1,i+n/2)} \left(a^j_{x,y}\right)^2 \right)^{\beta} \tag{7.8}$$

where the sum runs over n "adjacent" kernel maps at the same spatial position, and N is the total number of kernels in the layer. The constants k, n, α, and β are hyperparameters whose values are determined using a validation set.

7.4.3 Restricted Boltzmann Machine (RBM)

Geoffrey Hinton introduced the RBM [44] for various purposes. RBM is a stochastic neural network that can learn from probability distribution over a set of inputs. RBM is capable of performing classification, collaborative filtering, dimensionality reduction, feature learning, regression, and topic modeling. In general, the RBM is made of two layers: first the visible input layer and the second hidden layer. Each component in the network is called a node, and all the computations happen here. All the nodes are connected between the layers but the nodes on the same layer are independent of one another. The name restricted BM comes as it restricts the interlayer communication, i.e., no node in the same layer can be connected for any communication.

Sometimes, this RBM is also called *symmetrical bipartite graph* as there are only two layers. This RBM is the building block of deep belief networks (DBN).

An example of an RBM model is shown in Figure 7.7a and b. It basically works as an unsupervised learning model, and it does the process by refining the values in the hidden layer. Multiple layers are also possible as shown in the figure where the restriction resides the same but the number of hidden layers can be increased from 1 to n as per the requirement of the system.

The working of the RBM purely depends on the mathematical operations that involve probability distribution functions. Initially, the input layer receives the input x, and the weightage (w) is multiplied with the input parameter (x) while entering the hidden layer. Inside the hidden layer, there is a parameter that is added to the computed value, i.e., bias(b) and then, the same is fed into an activation function f(x).

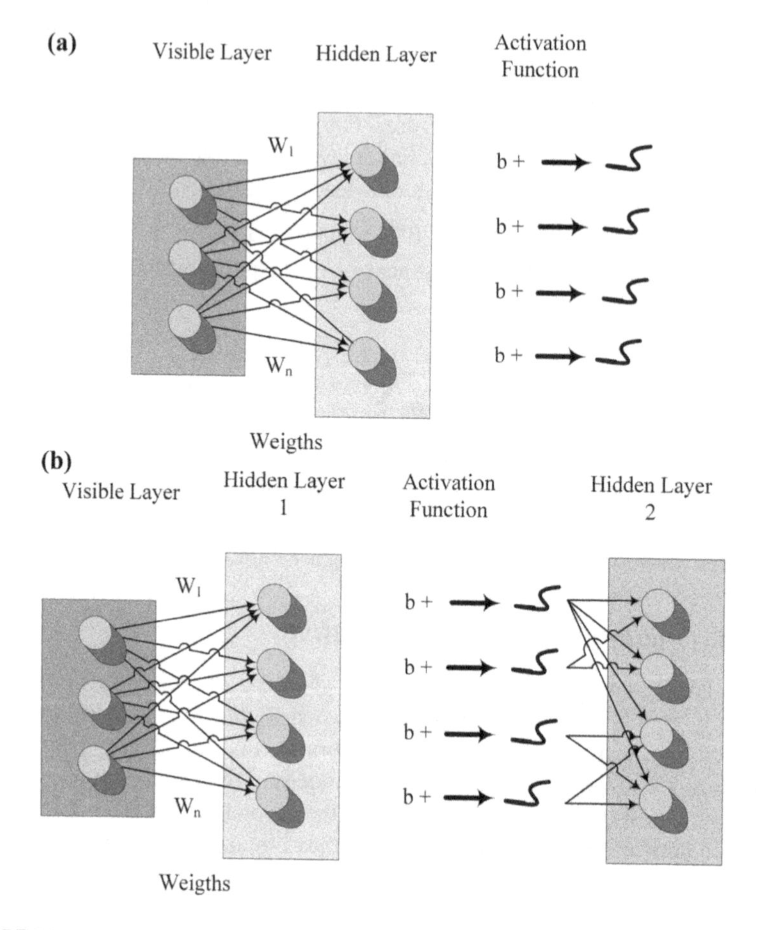

FIGURE 7.7 (a) Single hidden layer RBM, (b) Multiple hidden layer RBM.

The key mathematical refinements of the RBM can be listed as

$$E(v,h) = -\sum_{i,j} V_i h_j w_{ij} \tag{7.9}$$

where E (v, h) is the energy with configuration v on the visible units and h on the hidden units

V_i is binary state of visible unit i
h_j is binary state of hidden unit j
w_{ij} is weight between units i and j

$$-\frac{\partial E(v,h)}{\partial w_{ij}} = V_i h_j \tag{7.10}$$

Overall process of the RBM can be inferred from Eq. 7.11, where we will start with a training vector on the visible units and then alternate between updating all the hidden units in parallel and updating all the visible units in parallel.

$$\frac{\partial \log p(v)}{\partial w_{ij}} = \langle v_i h_j \rangle^0 - \langle v_i h_j \rangle^\infty \tag{7.11}$$

The general form of an RBM is

$$\Delta w_{ij} = \varepsilon\left(\langle v_i h_j \rangle^0 - \langle v_i h_j \rangle^1\right) \tag{7.12}$$

As this RBM takes a training dataset that is unsupervised, this model can be used to train uncertain network traffic that is generated by IoT terminals. The traffic that is fed to the input layer must have all the data segments that are inclusive of genuine and attack requests. The number of hidden layers in the proposed model should be based on a scenario, and the input layer remains 1 in all the cases. The weights (w) multiplied with the input are randomly initialized, so there may be a deviation in the reconstruction of the original data during the training phase and that can be solved through exhausting training with similar datasets. If one such model is designed, it can be used to avoid attacks such as replay, rouge, and denial of service.

7.4.4 Deep Belief Network (DBN)

A DBN [45] is a generative graphical neural network that is constructed basically using RBM's to solve the vanishing gradient problem and as an alternative to back-propagation. It can be viewed as a stack of RBM's that are linked one after the other with the capacity of exchanging information.

In the DBN, the setup consists of one input layer and many hidden layers. Unlike other models, in DBN, each hidden layer acts as a visible layer to the upcoming hidden layers. In other words, each layer will be an RBM layer as specified in Figure 7.8. The key aspect of the DBN is that each layer owns the entire input set, and hence the refinement level of the training will be high and efficient when compared to

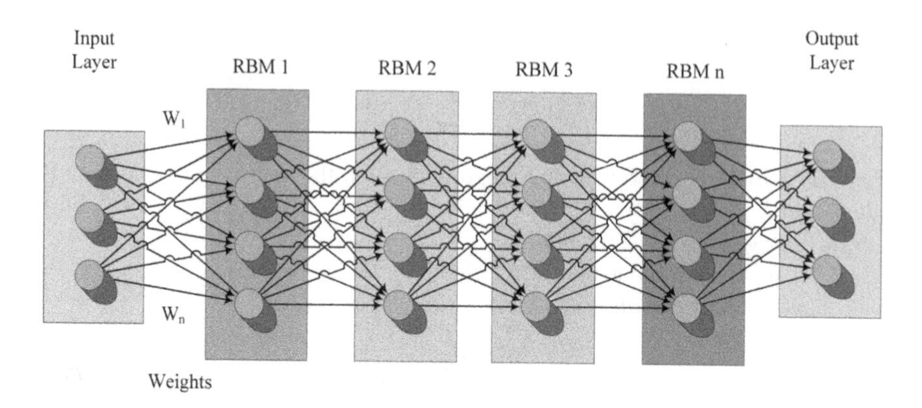

FIGURE 7.8 Deep belief networks.

other models like CNN [46]. The DBN works globally with the entire input provided, unlike other models which perform only with refined values that are passed between the layers. One of the main advantages is that it can produce all the possible values of the specified scenario. This is a greedy layer-wise unsupervised training process where we will try to make refinements in the whole process to achieve a supervised learning criteria.

Special aspect of the DBN is that, except the first and last layer, all the layers will have a dual role.

These deep belief networks will be highly helpful in increasing the accuracy of the attack mitigation strategies. As we know, the accuracy of attack detection is a crucial parameter for active attacks which can evade the whole network. By using a model like DBN, it is possible to detect the attacker even before the attack sequence starts.

Table 7.3 shows the list of DL algorithms and the recommendation for those algorithms to be used against the type of attacks. Through our research, we could notice that GAN outperforms in most of the cases thereby making an appropriate algorithm for attack mitigation. However, for the issues that are related to privacy concerns which involve images and location tracking, RBM and DBN are the best recommended solutions. Similarly, for the physical layer–based attacks, CNN became the recommended algorithm. Even though we can make a difference among the available algorithms, the usage of each algorithm is purely based on the scenario and the available infrastructure and architecture.

7.5 REQUIREMENTS OF SECURED IoT SYSTEM

In this section, we propose the following requirements to achieve an efficient secure IoT device:

- Device and data security – The data that is transmitted to and from IoT devices should be encapsulated with a decent encryption algorithm.
- Rigid authentication mechanisms should be followed for devices to maintain confidentiality and integrity of data.

TABLE 7.3
Attack Detection in IoT and the Recommended DL Algorithms

Algorithms \ Attack	Jamming	Spoofing	Replay Attack	Buffer Reservation Attack	Attacks Based on Authentication	Insecure Interfaces	Insecure Firmware	Privacy Threats
CNN	✓	×	✓	×	✓	×	✓	✓
MLPNN	✓	✓	×	✓	✓	✓	×	×
RBM	×	✓	✓	×	✓	×	✓	✓
DBN	×	✓	✓	×	✓	×	✓	✓
GAN	✓	✓	✓	×	✓	✓	✓	×

- Security is crucial in all fields, and when it comes to the internet, it's obvious that security enhancements play a vital role in the design of architecture. Hence, implementing and running security operations at an IoT scale should be possible.
- The devices that are manufactured/fabricated should meet the minimum requirements of the standards that are standardized by reputed consortiums.
- Additional security features can reduce the performance of the end devices, and in certain scenarios, they can drain even the battery. In such scenarios, there should be sufficient security measures to avoid attacks by taking advantage of the property.

Ideal IoT security solutions need to implement the following restrictions:

- **Device Trust**: Establishing and managing device identity and device integrity
- **Data Trust**: Data that is sent/received from the devices need to be protected for privacy and end-end data security
- **Operationalizing the Trust**: Automating and interfacing to the standards-based, proven technologies/products
- An ideal IoT system would be the one that withholds all the above criteria

7.6 IDEOLOGY

Through the detailed literature, it's obvious that the number of security threats is growing with the advancement of technology. In connection with that, we have come up with a multitier architecture that can work interoperably with most of the IoT endpoints. The proposed architecture consists of hybrid architecture that takes the advantages of the DL frame to detect and mitigate the most common attacks in the IoT domain by integrating the benefits of cloud computing.

As shown in Figure 7.9, we divide the entire IoT infrastructure into three segments. The first segment corresponds to IoT endpoints, while the second segment provides the network connectivity to IoT devices. The third tier is the most important tier where all the logs of the entire process are stored. We propose a hybrid architecture involving the cloud computing paradigm where all the logs are used as training data to make the inference of attacks and the normal traffic. In our proposed design, the neural network/DL framework is deployed in the cloud environment which has the potential to track the ongoing traffic and can recommend the precautionary measures that have to be taken to avoid future attacks. The algorithm to be used for specific attacks can be selected based on Table 7.1, and it depends on the scenario and the type of attack.

In Figure 7.10, the message sequences that are generated from the IoT terminal are directly received by the ISP host and the service will be provided to the client only after a series of checks made at the upper tier. In this way, the security of the system becomes rigid at the same time, and it increases the overhead time which can be ignored as the level of security increases drastically. The cloud tier makes the high-level computation that is required for the DL framework, possible through the compute engines.

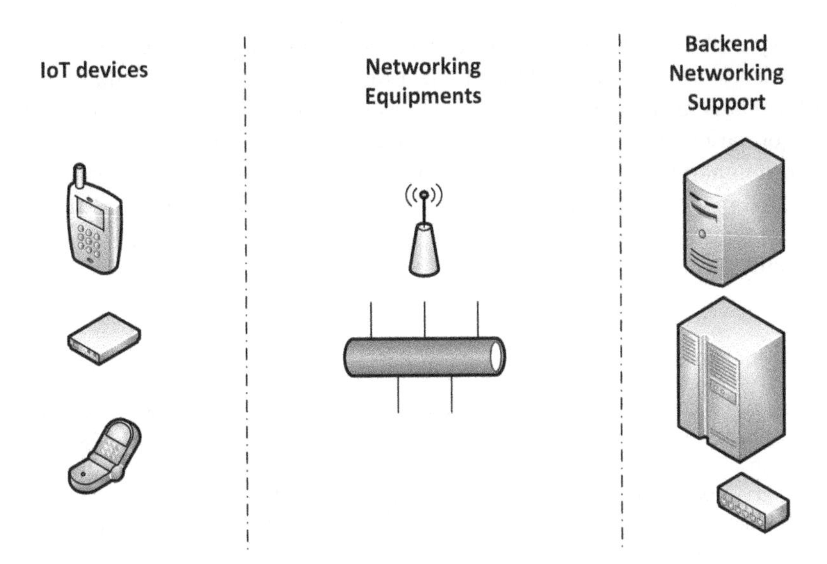

FIGURE 7.9 IoT system connection.

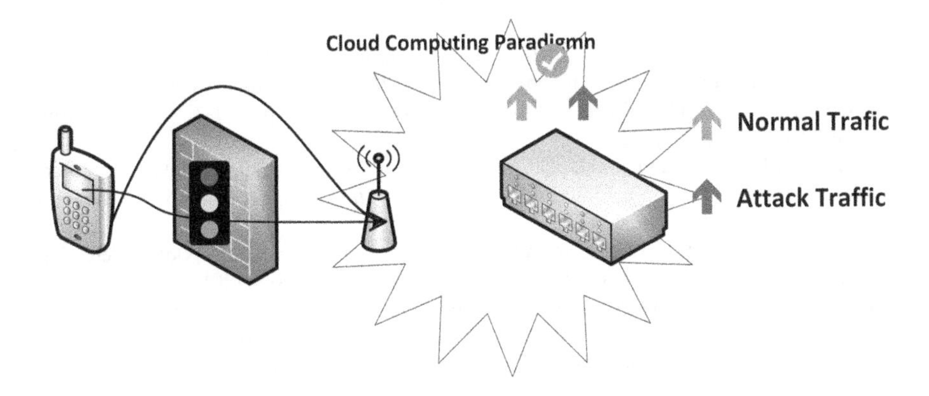

FIGURE 7.10 Cloud-integrated paradigm for attack detection.

Thus, when DL algorithms are deployed on the network logs, it provides the flexibility to learn the network behavior as well as the type of attack traffic. Once the behavior is learned by the algorithm in the cloud, we can easily segregate the attack requests and the accuracy of the attack detection will obviously improve.

7.7 SUMMARY

In the last few years, the emerging domain for the IoT has been attracting significant interest and will continue for the years to come. In spite of rapid evolution, still there are several difficulties and challenges which required our attention. In this literature, we concisely reviewed security loopholes in the IoT endpoint by defining few attacks that an IoT device might face, and then we analyzed security characteristics by deeply exploring the weakness of IoT devices and stating why these attacks take place. Also, we have compared the efficient ML and DL algorithms suitable to prevent these attacks in the future. Through our research, we have estimated the adoptability of various DL algorithms and frameworks for specific attack scenarios. At last we summarized several security requirements that are needed for an ideal IoT device. All in all as the development of the IoT booms it will bring in more serious security problems, which are always the focus and the primary task of the research community.

REFERENCES

1. Alaba, F.A., M. Othman, I.A.T. Hashem, and F. Alotaibi. "Internet of Things security: A survey." *Journal of Network and Computer Applications* 88 (2017): 10–28.
2. Khan, M.A., and K. Salah. "IoT security: Review, blockchain solutions, and open challenges." *Future Generation Computer Systems* 82 (2018): 395–411.
3. Rahman, A.F.A., M. Daud, and M.Z. Mohamad. "Securing sensor to cloud ecosystem using internet of things (IoT) security framework." In *Proceedings of the International Conference on Internet of things and Cloud Computing*, pp. 1–5. 2016.
4. Zhang, Z.-K., M.C.Y. Cho, and S. Shieh. "Emerging security threats and countermeasures in IoT." In *Proceedings of the 10th ACM Symposium on Information, Computer and Communications Security*, pp. 1–6. 2015.
5. N. Namvar, W. Saad, N. Bahadori and B. Kelley. "Jamming in the Internet of Things: A game-theoretic perspective." In *2016 IEEE Global Communications Conference (GLOBECOM)*, Washington, DC, 2016, pp. 1–6. DOI:10.1109/GLOCOM.2016.7841922.
6. Arafin, M.T., D. Anand, and G. Qu. "A low-cost GPS spoofing detector design for internet of things (IoT) applications." In *Proceedings of the on Great Lakes Symposium on VLSI*, 2017, pp. 161–166.
7. Koh, J.Y., I. Nevat, D. Leong, and W.C. Wong. "Geo-spatial location spoofing detection for Internet of Things." *IEEE Internet of Things Journal* 3, no. 6 (2016): 971–978.
8. Rughoobur, P., and L. Nagowah. "A lightweight replay attack detection framework for battery depended IoT devices designed for healthcare." In *2017 International Conference on Infocom Technologies and Unmanned Systems (Trends and Future Directions) (ICTUS)*, pp. 811–817. IEEE, 2017.
9. Bashir, A.K., R. Arul, S. Basheer, G. Raja, R. Jayaraman, and N.M.F. Qureshi. "An optimal multitier resource allocation of cloud RAN in 5G using machine learning." *Transactions on Emerging Telecommunications Technologies* 30, no. 8 (2019): e3627.
10. Shafiq, M., Z. Tian, A.K. Bashir, X. Du, and M. Guizani. "CorrAUC: A malicious Bot-IoT traffic detection method in IoT network using machine learning techniques." *IEEE Internet of Things Journal* (2020).
11. Basu, D., A. Jain, R. Datta, and U. Ghosh, "Optimized Controller Placement for Soft Handover in Virtualized 5G Network." In *2020 IEEE Wireless Communications and Networking Conference Workshops (WCNCW)*, Seoul, Korea (South), 2020, pp. 1–8. DOI:10.1109/WCNCW48565.2020.9124902.
12. Basu, D., R. Datta, U. Ghosh, and A.S. Rao. "Load and latency aware cost optimal controller placement in 5G network using sNFV." In *Proceedings of the 21st International Workshop on Mobile Computing Systems and Applications (HotMobile '20)*. Association for Computing Machinery, New York, NY, USA, 106, 2020.
13. Basu, D., U. Ghosh, and R. Datta. "Adaptive Control Plane Load Balancing in vSDN Enabled 5G Network." *arXiv* (2020).

14. El-Latif, A.A., B.A. El-Atta, S.E.V. Andraca, H. Elwahsh, M.J. Piran, A.K. Bashir, O.Y. Song, and W. Mazurczyk. "Providing end-to-end security using quantum walks in IoT networks." *IEEE Access* 8 (2020): 92687–92696.
15. Shafiq, M., Z. Tian, A.K. Bashir, X. Du, and M. Guizani. "IoT malicious traffic identification using wrapper-based feature selection mechanisms." *Computers & Security* 94 (2020): 101863.
16. Shafiq, M., Z. Tian, A.K. Bashir, A. Jolfaei, and X. Yu. "Data mining and machine learning methods for anomaly and intrusion traffic classification: A survey." *Sustainable Cities and Society* 60 (2020): 102177.
17. Zheng, Z., T. Wang, J. Weng, S. Mumtaz, A.K. Bashir, and C.S. Hussain. "Differentially private high-dimensional data publication in internet of things." *IEEE Internet of Things Journal* 7 (2020): 2640–2650.
18. Arul, R.K., G. Raja, A.O. Almagrabi, M.S. Alkatheiri, C.S. Hussain, and A.K. Bashir. "A quantum safe key hierarchy and dynamic security association for LTE/SAE in 5G scenario." *IEEE Transactions on Industrial Informatics* 16 (2019): 681–690.
19. Zhang, D., Y. Liu, L. Dai, A.K. Bashir, A. Nallanathan, and B. Shim. "Performance analysis of FD-NOMA based decentralized V2X systems." *IEEE Transactions on Communications* 67 (2019): 5024–5036.
20. Alaba, F.A., M. Othman, I.A.T. Hashem, and F. Alotaibi. "Internet of Things security: A survey." *Journal of Network and Computer Applications* 88 (2017): 10–28.
21. Hwang, Y.H. "IoT security & privacy: Threats and challenges." In *Proceedings of the 1st ACM Workshop on IoT Privacy, Trust, and Security*, pp. 1–1. 2015.
22. Tang, X., P. Ren, and Z. Han. "Jamming mitigation via hierarchical security game for IoT communications." *IEEE Access* 6(2018): 5766–5779.
23. Hoang, D.H., and H.D. Nguyen. "A PCA-based method for IoT network traffic anomaly detection." In *2018 20th International Conference on Advanced Communication Technology*, 2018.
24. Arış, A., S.F. Oktuğ, and S.B.Ö. Yalçın. "Internet-of-things security: Denial of service attacks." In *2015 23nd Signal Processing and Communications Applications Conference (SIU)*, pp. 903–906. IEEE, 2015.
25. Alladi, T., V. Chamola, B. Sikdar, and K.-K. Raymond Choo. "Consumer IoT: Security vulnerability case studies and solutions." *IEEE Consumer Electronics Magazine* 9, no. 2 (2020): 17–25.
26. Frustaci, M., P. Pace, G. Aloi, and G. Fortino. "Evaluating critical security issues of the IoT world: Present and future challenges." *IEEE Internet of things journal* 5, no. 4 (2017): 2483–2495.
27. Hosen, S.M.S., S. Singh, P.K. Sharma, U. Ghosh, J. Wang, I.-H. Ra, and G.H. Cho. "Blockchain-based transaction validation protocol for a secure distributed IoT network." *IEEE Access*, 8 (2020): 117266–117277.
28. Malik, A., D.K. Tosh, and U. Ghosh. "Non-intrusive deployment of blockchain in establishing cyber-infrastructure for smart city." In *2019 16th Annual IEEE International Conference on Sensing, Communication, and Networking (SECON)*, Boston, MA, USA, 2019.
29. Singh, A.P., N.R. Pradhan, S. Agnihotri, N. Jhanjhi, S. Verma, U. Ghosh, and D. Roy. "A novel patient-centric architectural framework for blockchain-enabled healthcare applications." *IEEE Transactions on Industrial Informatics* (2020).
30. Román-Castro, R., J. López, and S. Gritzalis. "Evolution and trends in IoT security." *Computer* 51, no. 7 (2018): 16–25.
31. Oracevic, A., S. Dilek, and S. Ozdemir. "Security in internet of things: A survey." In *2017 International Symposium on Networks, Computers and Communications (ISNCC)*, pp. 1–6. IEEE, 2017.

32. Joshitta, R.S.M., and L. Arockiam. "Security in IoT environment: A survey." *International Journal of Information Technology and Mechanical Engineering* 2, no. 7 (2016): 1–8.
33. Farris, I., T. Taleb, Y. Khettab, and J. Song. "A survey on emerging SDN and NFV security mechanisms for IoT systems." *IEEE Communications Surveys & Tutorials* 21, no. 1 (2018): 812–837.
34. Alkurdi, F., I. Elgendi, K.S. Munasinghe, D. Sharma, and A. Jamalipour. "Blockchain in IoT security: A survey." In *2018 28th International Telecommunication Networks and Applications Conference (ITNAC)*, pp. 1–4. IEEE, 2018.
35. Guo, Z., Y. Shen, A.K. Bashir, M. Imran, N. Kumar, D. Zhang, and K. Yu. "Robust spammer detection using collaborative neural network in internet of thing applications." *IEEE Internet of Things Journal* (2020).
36. Qiao, F., J. Wu, J. Li, A.K. Bashir, S. Mumtaz, and U. Tariq. "Trustworthy edge storage orchestration in intelligent transportation systems using reinforcement learning." *IEEE Transactions on Intelligent Transportation Systems* (2020).
37. Chen, E.Y. "Detecting TCP-based DDoS attacks by linear regression analysis." In *Proceedings of the Fifth IEEE International Symposium on Signal Processing and Information Technology*, 2005.
38. Arul, R., R.S. Moorthy, and A.K. Bashir. "Ensemble learning mechanisms for threat detection: A survey." In *Machine Learning and Cognitive Science Applications in Cyber Security*, pp. 240–281. IGI Global, 2019.
39. Iwendi, C., P.K.R. Maddikunta, T.R. Gadekallu, K. Lakshmanna, A.K. Bashir, and M.J. Piran. "A metaheuristic optimization approach for energy efficiency in the IoT networks." *Software: Practice and Experience* (2020).
40. Gardner, M.W. and S.R. Dorling. "Artificial neural networks (the multilayer perceptron)—A review of applications in the atmospheric sciences." *Atmospheric Environment* 32, no. 14–15 (1998): 2627–2636.
41. Lawrence, S., C.L. Giles, A.C. Tsoi, and A.D. Back. "Face recognition: A convolutional neural-network approach." *IEEE Transactions on Neural Networks* 8, no. 1 (1997): 98–113.
42. Shakil, M., A. Fuad Yousif Mohammed, R. Arul, A.K. Bashir, and J.K. Choi. "A novel dynamic framework to detect DDoS in SDN using metaheuristic clustering." *Transactions on Emerging Telecommunications Technologies* (2019): e3622.
43. Kalchbrenner, N., E. Grefenstette, and P. Blunsom. "A convolutional neural network for modelling sentences." *arXiv preprint arXiv:1404.2188* (2014).
44. Sutskever, I., G.E. Hinton, and G.W. Taylor. "The recurrent temporal restricted boltzmann machine." In *Advances in Neural Information Processing Systems*, pp. 1601–1608. 2009.
45. Mohamed, A.-R., G.E. Dahl, and G. Hinton. "Acoustic modeling using deep belief networks." *IEEE Transactions on Audio, Speech, and Language Processing* 20, no. 1 (2011): 14–22.
46. Lee, H., R. Grosse, R. Ranganath, and A.Y. Ng. "Convolutional deep belief networks for scalable unsupervised learning of hierarchical representations." In *Proceedings of the 26th Annual International Conference on Machine Learning*, pp. 609–616. 2009.

8 Deep Learning and IoT in Ophthalmology

Md Enamul Haque and Suzann Pershing
Stanford University

CONTENTS

8.1 INTRODUCTION

Internet of Things (IoT) has been playing an important role in modern life. It has taken us closer to the things that we wanted to see in the future. It has made us more connected than ever with respect to using not only mobile devices but also all types of digital devices around us. This technology has introduced connected home [1], smart cars [2], precision agriculture [3], smart healthcare [4], and smart retail solutions [5].

On the other hand, deep learning (DL) 6 technology is a subfield of machine learning (ML) which introduced numerous modern-day applications used within IoT-enabled devices. IoT and DL together can achieve high-performance healthcare for patients by identifying early signs of diseases, providing preventive healthcare suggestions, and informing precision healthcare. In this chapter, we provide an intuitive architecture for ophthalmology built on top of IoT and DL, which can devolve primary eye care to remote patients as an end-to-end solution.

8.1.1 Chapter Roadmap

The rest of the chapter is organized as follows. We present introductory details of IoT and DL and their applications in Sections 8.2 and 8.3, respectively. We demonstrate our proposed universal DL and IoT-based ocular services in Section 8.4. Section 8.5 presents the privacy and security of the proposed architecture. Section 8.6 presents a brief discussion on the research challenges related to DL and IoT-based ocular services. Finally, Section 8.7 concludes the chapter with future directions of DL and IoT in ophthalmology.

8.2 INTERNET OF THINGS

The significant improvements in scope and quality of digital communications have accelerated our lives by providing the ability to sense and monitor our surrounding environment including human and objects of various kinds. IoT is one of the outcomes from such digital communications technology, which consists of devices with both homogeneous and heterogeneous configurations. Devices with homogeneous configurations refer to various types of computational items such as smartphone and washing machine. These devices run on separate operating systems because of their heterogeneous hardware profiles. By contrast, homogeneous hardware profile-based devices such as phones of different brands run on similar operating systems. The device configurations play an important part in communicating among multiple machines, which was difficult prior to the availability of the IoT infrastructure.

The major components of IoT infrastructure include Internet protocols, sensor network integration, cloud computing, and other applications. Mainly, IoT devices can communicate with each other by transferring required data along with proper communication protocols [7]. This feature has enabled us to employ the power of IoT protocols on devices and to enrich our daily lives with smart healthcare, smart home, and smart transportation. Although IoT infrastructure has been operating for quite some times, the ubiquity is not yet prominent among the users of different ages, cultures, and geographic locations. Most of the users who have familiarity with IoT and its applications have either background in technology or work in the industries that are actively adopting the technology. This is an important caveat relevant to ophthalmology, since the ophthalmic patient population disproportionately consists of older adults, who are often less technologically facile.

Recent studies suggest that the number of smart devices will increase to 50 billion by the end of 2020 [8,9], which will have significant implications for the way we communicate using technology. We also notice communications across different settings between humans and devices such as human-to-human, human-to-things, and things- to-things [10]. Traditional human-to-human communication refers to natural processes such as speech, facial expression, and touch. However, human-to-human communication in the IoT infrastructure also encompasses collaborative data sharing through sensors attached to or implanted in the human body. Things-to-things are the most common types of connectivity among devices such as smart phone and the thermostat used in the home. Human-to-things communication is also becoming more widespread, mainly connecting data received from humans to another device. A simple example of such a case would be data extracted from smart "wearables" (smart watch, implanted chip, etc.) and remote smartphones of physicians. There are numerous examples of human-to-things communications in agriculture, smart home, transportation, and healthcare which improve the existing quality of services.

8.2.1 Applications in Healthcare

IoT infrastructure contributes to a significant number of applications in the healthcare domain, including but not limited to patient monitoring, drug development and delivery, assistive technology, and telesurgery. IoT is also referred to as Internet of Medical Things (IoMT) in several studies because of the wide range of specific usage in the healthcare domain [11].

Patient monitoring is one of the most ubiquitous and easily accessible services among all the healthcare applications supported by the IoT infrastructures. Usually, wearable devices such as smart watches and attached/implanted sensors are tagged with users' smartphones or other remote devices which are controlled by authorized persons [12]. The wearable IoT devices collect different health-related data such as blood glucose levels, oxygen level, cardiogram, and daily activity data. Cloud storage is used to reserve the data to apply analytics before sending them to the physicians, health insurance providers, and other authorized parties. The remote monitoring system may also be used to track elderly patients who need extensive care and vigilant monitoring due to frailty or limited mobility. The wearable devices on these patients can send data securely on cloud storage, where pattern analysis-based algorithms can identify whether the patient has a fall, for example, and ambulance services may be dispatched without involving multiple levels of communications among human agents. As a result, the IoT device– based patient monitoring systems allow several nonessential services to be performed remotely to extend parallel healthcare to more patients. Consequently, IoT may reduce overall costs and expand access to care by minimizing unnecessary patient visits to healthcare systems and utilizing available resources efficiently.

Further, IoT technology helps improving drug developments for pharmaceutical companies and delivery to the end users. According to Pharmaceutical Research and Manufacturers of America (PhRMA),[1] pharmaceutical companies spend an average of 10 years and approximately $2.6 billion to develop new medications. Such massive expense includes the cost of failed experiments and clinical trials. IoT can improve overall drug development time and cost by reducing research efforts, optimizing the production process, and administering patient protocols during clinical trials. A clinical trial for a new medicine usually takes nearly 70% of the total drug development time before it receives approval from the Food and Drug Administration (FDA) for mass manufacturing. The FDA approves manufacturers to develop medicines based on data generated during clinical trials and other preclinical studies. Therefore, it is very important to optimize clinical trial design, patient recruitment, and monitoring,

---

[1] https://www.phrma.org/

to develop the most successful and cost-effective new medications and technologies. IoT can provide accelerated support to achieve this goal. For example, recent research on replicating human physiology on a chip has enabled drug developers to observe the effect of new drugs on these artificial organs without waiting for longer periods [13].

Furthermore, medication production and delivery can be monitored by IoT sensors and system parameters uploaded to the cloud for advanced analytics using artificial intelligence (AI) programs. AI performs in-depth analysis on uploaded data and provides insights on individual components of the system. For instance, AI can predict system failure in advance and provide warnings or suggest temperature adjustment within the machinery. AI may also reduce clinical trial time by allowing easy administration and monitoring of trial participation, which is possible due to the wearable devices and cloud technologies integrated to IoT. For instance, ingestible pills that are equipped with IoT sensors may be used to monitor the dose of a medicine taken by patients and trial subjects and also to verify medication adherence. In addition, drug refills (including eye drops) could be automated from medication bottles which are IoT-sensor enabled and warn patients when supply is low.

Assistive systems are another avenue where IoT plays a significant role. Older adults aged 65 years and above are at higher risk of falls due to comorbid conditions such as dementia, frailty, and low vision from eye disease. IoT devices can be used on these patients to collect tracking and health-related data to better support the healthcare system as well.

Remote surgery (also known as telesurgery) is another application of IoT-enabled healthcare services [14,15]. This is an extension of robotic surgery in which patients are operated on by physicians from remote locations with IoT-enabled surgical robots. One major requirement for telesurgery is uninterrupted Internet and secured connectivity to ensure successful surgeries [16,17].

8.2.2 Role of Cloud Computing

Apart from the physical layer connectivity of different digital devices, cloud data storage and processing capability are important components of all healthcare applications. As most applications require the tracked and monitored data to be analyzed using AI programs, the limited hardware profiles of the sensors are not yet capable of processing all the information locally. Therefore, a secondary storage media is required where complex computational models can be executed and the outcome is sent to appropriate parties. Hence, the role of cloud computing and its additional components (such as edge and fog computing) are termed as the enabler of IoT infrastructure to function properly [18,19].

Recent studies also suggest to incorporate different combinations of cloud, fog, and edge computing for healthcare applications. Tyagi et al. proposed a conceptual framework built on cloud computing and IoT infrastructure for better healthcare delivery [20]. Fernandez et al. discuss the challenges of healthcare delivery architecture built on IoT and cloud computing [21]. Gia et al. present a case study on ECG (Electrocardiogram) feature extraction using IoT and fog computing [22]. Klonoff presents the impact of edge and fog technologies over cloud computing architectures applied on data collected from diabetes-tracking devices [23]. Cloud, fog, and edge computing-based IoT infrastructure will be crucial elements for ophthalmology research and clinical care, where high-quality images of interior and exterior eyes are processed for advanced diagnostics.

8.3 DEEP LEARNING

DL is a subfield of ML and AI methods [24]. This learning method is mostly popular for processing images to extract patterns and tag relevant labels to each image. It is also useful in identifying patterns from text data by extracting significant features and using them for classification and clustering at the later stages of the pipeline. As improvements in DL research continue, we notice significant progress in computer vision-based applications such as smart cars, face recognition, biometric identification, and pedestrian detection. DL models consist of a special type of neural network named as convolutional neural network (CNN) which are specialized in extracting useful patterns from input data.

A typical DL model consists of several convolution layers in the feature generation and selection steps. At each convolution layer, a defined filter or kernel is applied to the pixels of input images based on the sliding window approach. In the classification layer, the output of the final convolution layer is first encoded and pooled before passing through the softmax classifier. The softmax classifier for a vector v, containing image pixels, is defined as follows:

$$\hat{y} = \arg\max \frac{\exp(v_i)}{\sum_{i=1}^{K} \exp(v_i)} \tag{8.1}$$

where $\hat{y}$ denotes the predicted class label of the image where $\hat{y} \in [1 \ldots K]$. K refers to the number of diagnosis classes present in the training data.

8.3.1 Applications in Healthcare

DL solutions have a broad impact on healthcare applications. DL is especially relevant for image processing tasks required in the healthcare domain. Ophthalmologists can potentially use DL-enabled digital systems as tools to identify patterns, diagnose different eye conditions, predict outcomes, and support decision-making. For example, Figure 8.1 presents a sample image processing task where a fundus photo (demonstrating the retina of an eye) is classified using several CNN, encoding, pooling, and softmax layers. Initially, the image is passed through several convolutional layers to extract useful features of the image. The features are then encoded and passed through the pooling layer that select values based on a defined filter. Finally, the softmax layer performs the classification task by providing probability scores for all classes present in the training image.

Similar to image classification to identify eye diseases, there are numerous other applications of DL such as modeling disease progression [25], precision health [26], and developing predictive models for optimal treatment and disease course early in

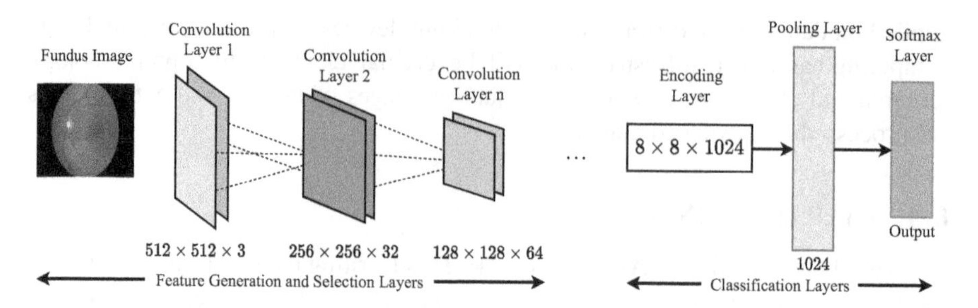

FIGURE 8.1 DL framework for ocular image classification models. The dimensions of intermediate layers are set as an example to demonstrate the transformation of image pixels.

diagnosis [27]. Next, we provide a detailed discussion on the usage of DL in the context of designing an end-to-end solution with potential for the provision of universal primary eye care with the aid of IoT infrastructure.

8.4 DL- AND IoT-BASED INFRASTRUCTURE

In the previous sections, we presented a brief overview of DL and IoT infrastructures in healthcare applications. In this section, we describe DL and IoT-based infrastructure as applicable to primary eye care.

Both DL- and IoT-based infrastructures can be used to aid physicians and ophthalmologists in diagnosing and treating different types of eye diseases. DL-based methods usually process retinal fundus images to automate the identification of eye diseases including glaucoma, age-related macular degeneration (AMD), and diabetic retinopathy [28–30]. Several studies use DL to diagnose specific diagnostic subsets, such as preperimetric glaucoma, where glaucomatous damage is not yet traditionally detectable from visual field testing [31,32]. Other eye diseases and disease progression can be detected using CNN, a variant of DL [33]. Retinal disorders, such as diabetic retinopathy, may be detected using models developed on 1.2 million eye images as training data and neural networks for classification of test images [34,35]. In summary, DL methods are widely used for eye image processing, classification, and clustering because of capability to extract relevant features from the images, which is not otherwise possible using typical feature engineering techniques.

IoT is an enabler of DL technologies on different platforms with the help of cloud, fog, and edge computing. As most wearable devices do not have complex computation capability in terms of power and processing hardware, the data is first offloaded to the cloud for extensive processing and then the outcome is forwarded to designated parties.

Although there had been an increasing trend of incorporating technology to aid physicians in eye disease diagnosis and management, the integrated platform is yet to be realized so that the patients from remote and underprivileged areas are able to benefit from access to and improvement in eye care. In this book chapter, we present an integrated and universal eye-care solution with potential to improve ocular diagnosis, early disease detection, and access to care in underserved communities.

This process will also introduce the capability of eye-care service continuation during pandemics such as COVID-19 where people are required to maintain social distancing and sheltering in place within their home [36].

8.4.1 Proposed Architecture

In this section, we present an architecture for ophthalmic diagnostics involving DL and IoT technologies. We have already seen that the applications of DL and IoT are prevalent in healthcare and ophthalmic research. However, we need an integrated solution where patients can receive eye care and consultation based on their specific conditions and needs. We present our proposed architecture in Figure 8.2 using four components and subcomponents. We include existing technologies and propose to integrate them within different components in this architecture. This architectural template can be extended to specific eye treatment, etc. to meet requirements from individual vendors and service providers. An architecture such as this cannot replace all elements of eye care, including complex medical decision-making and surgical procedures such as incisional glaucoma surgery. However, by increasing access to and accelerating initial screening and consultation, it can lead to an overall improvement in eye care. In addition, over time we envision that complex eye treatments may also be performed remotely using robot-assisted and telesurgery technologies.

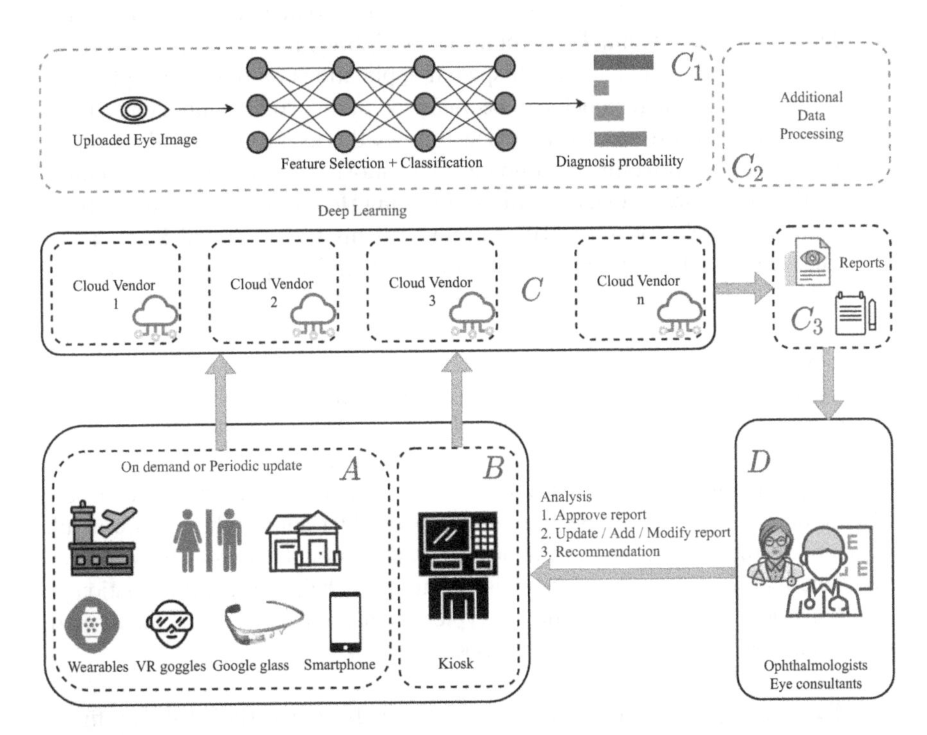

FIGURE 8.2 A framework involving DL and IoT technology to provide primary ophthalmic services to remote users.

In addition to surgical simulation and preparatory training, VR goggles and wearables such as Google Glass could also be used to assist in patient encounters and remote surgery supervision.

8.4.2 Components of the Architecture

Our proposed architecture consists of the following four main components. In this part, we describe detailed functions and business models that can be used within each component mentioned in Figure 8.2.

A. **Active User End**: The first component is composed of different user interfaces which are connected to the cloud vendors or service providers. We denote the component as A where users' main requirement includes voluntary clinical data sharing and on-demand service. Users will be able to share their health data including eye images through specialized hardware and software applications outside the traditional clinical patient care setting. Specialized hardware such as smartwatches, VR goggles [37,38], and Google Glass are easily accessible to users in moderate to higher income groups. These wearables are suitable for collecting health data including blood pressure, glucose level, eye movements, and heart rate. Smartphones and VR goggles could be equipped with the capability to capture visual acuity, and other devices such as contact lenses, intraocular lenses, or other technology could enable intraocular pressure measurement as well. Users can install special applications on their smartphones to upload eye image which can be analyzed in the cloud components. The availability of different IoT devices allows the users to share their data while they are traveling or at home. The data-sharing scheme can be controlled from users' devices to allow enforcing Health Insurance Portability and Accountability Act (HIPAA) requirements in the United States and General Data Protection Regulation (GDPR) in European Union and the European Economic Area.

All the devices in component A need to be connected to cloud vendors using Internet connectivity to ensure that the processed data securely reach component D, where expert opinion is added with the automated analysis. The data uploading strategy to the cloud is either on-demand or periodic. On-demand data uploading allows users to avail specific and one-time consultation from component D. On the other hand, periodic data uploading may be more suitable for older adults who require regular evaluation of their health and eye issues. Periodic data uploading to the cloud will enable the specialists in group D to advise the users of their eye care information and plan. A service fee for uploading data is typically charged to customers based on the type of data and upload strategy they choose.

B. **Passive User End**: The second component, B, is a variation of the first component which is an alternative for on-demand data uploading to the cloud. This passive user end is usually composed of kiosks similar to ATM machines. Kiosks can be equipped with special eyeglasses, cameras, or other imaging devices, as well as data receiving ports. Upon paying a fee at the kiosk, patients could choose from different services provided by the service provider. The list of services might include various predictive modeling of eye diseases such as the presence of cataract, glaucoma, age-related macular degeneration, and diabetic retinopathy, available treatment, and the likelihood of benefit from treatment. The kiosk option would allow users of all income groups to take the opportunity of preventive eye care and consultation from remote locations. Component B is introduced to provide services to a broader group of people who cannot afford the expensive wearable devices. However, cost relative to human exam, limitations in automated imaging technology, and sensitivity of imaging equipment may limit feasibility. Components A and B encompass data generator type and uses of the IoT device technology.

C. **Cloud Vendors**: Component C uses DL and additional data processing to generate reports from user-uploaded images and monitoring data. It is the main component where different service providers can operate. We show that there are n cloud providers each having three common subcomponents. We primarily include three major subcomponents C_1, C_2, and C_3, denoting DL, data processing, and report generator modules, respectively. The cloud providers can include additional modules within their cloud storage to offer advanced ocular diagnostics to the users.

The majority of the ocular data received from user components may include digital eye images, OCT (Optical Coherence Tomography) [39], and visual fields [40]. The DL component will use the images as input and generate reports at the report generation subcomponent, C_3. Figure 8.2 shows that uploaded eye images are passed through a deep neural network for selecting features and classifying images. Initially, the learning modules are provided with a large training database consisting of ocular images having different tagged diagnostics. This process is referred to as a supervised image classification method where training data is provided to the learning algorithm to obtain an optimal hypothesis to be applied on test data [41,42]. The diagnosis labels can be both binary and multi-class. In the figure, we demonstrate that the DL model performs a four-class diagnostic on the uploaded image. The final outcome from the DL model would provide different diagnosis probability.

An additional data processing module, C_2, is responsible for including other types of tracking and uploaded data such as blood pressure, eye movement using a time-series format, and glucose level. The module incorporates DL model output and additional data to create a summary report using different statistical models including but not limited to Bayesian, causal, and probabilistic graphical models [43–47].

The report generation module prepares the outcome of DL and additional data processing module in a human readable format and forwards to the consultant component D. In addition, users might request this report directly sent to them when they first upload the data and request for services. Overall, the cloud vendor part is an important component of our proposed architecture where actual data processing is performed. This can be shared by multiple service providers as well, provided that they follow the appropriate data privacy and sharing protocols. For instance, Amazon web services (AWS)[2] or Google Cloud Platform might provide the cloud computing, DL, and additional data processing services to different service providers who build their custom solutions. However, users would have the option to choose from multiple service providers for their ophthalmic reports and consultation, either at kiosks or from their own IoT-enabled devices.

D. **Consultation**: Component D consists of different entities including physicians, ophthalmologists, and ancillary providers such as optometrists. The report generator module from the cloud vendor sends the detailed report to consultant(s) when users submit requests for consultation. The users can either select from a list of consultants from different specialist areas or let the DL module decide after analyzing the uploaded data.

Consultants either approve the automated report so that it can be forwarded to the user or make additional recommendations such as the need for emergency department or clinic visit, or urgent surgery. The recommendation is then fed to the DL module in parallel to the user so that the DL model can learn additional knowledge from the consultants. This process helps the system to continuously update its hypotheses based on new available beliefs. As the service providers primarily collect digital images from a user, other clinical history data will not be available in their databases. However, users can be offered an option to provide their historical medical information, similar to how traditional face-to-face clinics ask patients to fill out forms indicating their medical history and consent to requests for prior medical records. Users' historical data will allow the consultants to provide improved and detailed reports back to both the system and the user. Finally, consultants would have the flexibility to define their consultation hours in the system, allowing them to potentially work with multiple service providers.

8.4.3 Types of Ocular Disease Diagnoses

The advanced medical image capturing technology and DL-based image processing with the help of Graphical Processing Units (GPU) has introduced a significant improvement in diagnosing various diseases [48–52]. Most importantly, the advent of AI and DL-based image processing has made routine clinical practice easier and more affordable [53]. We observe that the benefits of AI-based healthcare have primarily been assumed by developed countries. However, as recent progress of AI has made the technology more affordable, going forward it may be used in both developed and – importantly – underdeveloped countries for improvements in the healthcare sector.

AI-based research in the ophthalmology domain is progressing rapidly by providing tools to aid in the diagnosis of various ocular diseases such as diabetic retinopathy [35], age-related macular degeneration [54,55], cataract [56], pseudophakia/aphakia [57], and glaucoma [58,59]. Therefore, our proposed architecture can be helpful in

[2] https://aws.amazon.com/

integrating these tools for identifying ophthalmic disease. Overall, we envision an automated initial eye screening for the early detection of numerous ocular diseases. However, we also believe that robotic eye surgery and supervised surgical training will be ubiquitous in the immediate future, expanding the reach of expert and sub-specialized ophthalmology care.

8.4.4 Business Operations Model

Our proposed architecture would ensure multiple competitive service providers who maintain an agreement with a central regulatory body. Service providers could have either independent or shared cloud infrastructure to run DL and other statistical processes. The central regulatory body should perform periodic audits on the service providers to establish, monitor, and maintain data security, privacy, sharing, and storage governance. Most often the central body will likely be part of a government organization, also collecting periodic cost-covering fees based on several factors such as size of the company, user base, and the quality of services, for example.

The service providers would also collect fees from both consultant and user groups based on the type of services offered and received. Service providers might need to make additional arrangements with third-party medical data providers to optimize requested services for users who give consent to access their historical medical data.

Members of the consultant module could register with service providers using a fee-based plan. The users would be able to see the service charge before submitting requests at the kiosk, or at the beginning of their service agreements if they choose to upload the data periodically using their own devices.

8.5 PRIVACY, SECURITY, AND ETHICAL CONSIDERATIONS

The proposed architecture has potential to provide easy and affordable access to primary eye care for users from widespread geographical locations, including remote areas and underdeveloped countries. However, we should be aware and cautious of the risk inherent in possible data breaches, because the overall system will receive, store, and process sensitive health information. The proposed architecture would collect and process user information in cloud storage using DL and statistical analysis techniques. Although summarized results for each patient could exclude demographic data (later sent to consultants for additional adjudication) and the architecture would not voluntarily reveal user information to human experts within the system, unauthorized access to the cloud storage system or during user and cloud connectivity remains a real risk and concern. Unauthorized access and downloading of sensitive patient information could result in significant harms, including malicious use of demographic and health information, and potential for introduction of errors or modifications in data or models that could result in false diagnoses for users, in turn potentially inducing or exacerbating other health concerns such as hypertension, depression, and social anxiety. It is of critical importance that service providers implement the highest level of data security practices and ensure reliable secure communication between end-user devices such as a kiosk and cloud.

Broadly, two types of communication need to be secured in the proposed architecture. First, user and cloud communication should be encrypted so that transferred data can not be decoded by online attackers. The report from the cloud to the consultant group should also be encrypted. Second, the final report to the user should also be secured and properly authenticated. Overall, the service providers should follow HIPAA and GDPR while implementing all of the functions within individual modules of their systems.

Beyond these essential infrastructure elements, other ethical concerns are important to consider. For example, patients need to have the opportunity to provide fully informed consent for potential storage and use of their collected data, as well as interpretation and counseling on results. Some patients may not fully understand or be comfortable with data collection of implantable or attached wearable devices; this needs to be ensured. And, as technology progresses, models may identify new patterns, predictive conclusions, or changes to initial results.

Consideration needs to be given to how patients might be contacted with any new information, and patients would need to consent to this upfront as well.

8.6 RESEARCH CHALLENGES

Research in DL and IoT for ophthalmology is still in a nascent stage compared to other domains such as computer vision and sensor networks. Although security is critical, a major challenge of performing research in any healthcare sector – including ophthalmology – is the lack of access to data because of patient privacy and security concerns. The secondary challenge is ensuring reproducibility of evaluations from related literature, which may result from inaccessible medical data.

In addition to the lack of data access, initial data collection from users is also a challenge as the proposed architecture is assumed to be used in different geographic locations with people having limited education or technological sophistication. Therefore, educating people about the usage and benefits of such systems is also important. Cost, feasible deployment, and repair or maintenance are also real challenges that will limit adoption and quality in the near future.

Furthermore, DL models generate abstract representation of features in intermediate layers, which are difficult for humans to interpret. Although DL methods provide promising outcomes in solving disease classifications, expert knowledge remains necessary to augment and validate outcomes, to inform and counsel patients, and to lead complex decision-making and procedural interventions which are not feasible through technology.

8.7 CONCLUSIONS AND FUTURE DIRECTIONS

DL and IoT-based solutions can provide benefits for patients, ophthalmologists, and other providers to improve access and quality of eye care across diverse geographic locations. This proposed architecture for primary eye care services would also reduce repetitive tasks for clinicians and offload preliminary screening so that ophthalmologists could focus more on high-level patient care. Going forward, DL and IoT-based ophthalmology services will increasingly be automated to enable improved eye care for patients.

These considerations are especially pertinent in the setting of a pandemic (as with COVID-19) or other events that might restrict patients from seeking or obtaining face-to-face care. The evolution of COVID-19 morbidity and disease burden has led to increased adoption and utilization of video visits, asynchronous telemedicine exams, and remote monitoring technologies. It is likely that these trends will continue and persist even after COVID-19.

For future directions, we envision that ophthalmology care will incorporate the benefits of Internet of Everything [60], where people, data, and devices will be interconnected altogether to provide seamless eye care services. Similarly, ophthalmic surgeries may be assisted by robotic and other assistive technology in the near future, to improve access, outcomes, and safety with the aid of expert knowledge and human intervention.

REFERENCES

1. Omar Hamdan, Hassan Shanableh, Inas Zaki, AR Al-Ali, and Tamer Shanableh. IoT-based interactive dual mode smart home automation. In *2019 IEEE International Conference on Consumer Electronics (ICCE)*, pages 1–2. IEEE, 2019.
2. Rodolfo WL Coutinho and Azzedine Boukerche. Modeling and analysis of a shared edge caching system for connected cars and industrial IoT-based applications. *IEEE Transactions on Industrial Informatics*, 16(3): 2003–2012, 2019.
3. Abhishek Khanna and Sanmeet Kaur. Evolution of internet of things (IoT) and its significant impact in the field of precision agriculture. *Computers and Electronics in Agriculture*, 157: 218–231, 2019.
4. Luca Catarinucci, Danilo De Donno, Luca Mainetti, Luca Palano, Luigi Patrono, Maria Laura Stefanizzi, and Luciano Tarricone. An IoT-aware architecture for smart healthcare systems. *IEEE Internet of Things Journal*, 2(6): 515–526, 2015.
5. Fawzi Behmann and Kwok Wu. *Collaborative Internet of Things (C-IoT): For Future Smart Connected Life and Business.* John Wiley & Sons: Hoboken, New Jersey, 2015.
6. Yann LeCun, Yoshua Bengio, and Geoffrey Hinton. Deep learning. *Nature*, 521(7553): 436–444, 2015.
7. Kim Thuat Nguyen, Maryline Laurent, and Nouha Oualha. Survey on secure communication protocols for the internet of things. *Ad Hoc Networks*, 32: 17–31, 2015.
8. Dave Evans. The internet of things how the next evolution of the internet is changing everything (April 2011). *White Paper by Cisco Internet Business Solutions Group (IBSG)*, 2012.
9. Mohammed Ali Al-Garadi, Amr Mohamed, Abdulla Al-Ali, Xiao-Jiang Du, Ihsan Ali, and Mohsen Guizani. A survey of machine and deep learning methods for internet of things (IoT) security. *IEEE Communications Surveys & Tutorials*, 22: 1646–1685, 2020.
10. Ala Al-Fuqaha, Mohsen Guizani, Mehdi Mohammadi, Mohammed Aledhari, and Moussa Ayyash. Internet of things: A survey on enabling technologies, protocols, and applications. *IEEE Communications Surveys & Tutorials*, 17(4): 2347–2376, 2015.

11. Arthur Gatouillat, Youakim Badr, Bertrand Massot, and Ervin Sejdic. Internet of medical things: A review of recent contributions dealing with cyber-physical systems in medicine. *IEEE Internet of Things Journal*, 5(5): 3810–3822, 2018.
12. Priyanka Kakria, NK Tripathi, and Peerapong Kitipawang. A real-time health monitoring system for remote cardiac patients using smartphone and wearable sensors. *International Journal of Telemedicine and Applications*, 2015, 2015.
13. Wyss Institute. Human organs-on-chips: Microfluidic devices lined with living human cells for drug development, disease modeling, and personalized medicine. https://wyss.harvard.edu/technology/human-organs-on-chips/, 2020. Online.
14. N Shabana and G Velmathi. Advanced telesurgery with IoT approach. In *Intelligent Embedded Systems*, pages 17–24. Springer, 2018.
15. Rajesh Gupta, Sudeep Tanwar, Sudhanshu Tyagi, and Neeraj Kumar. Tactile-internet-based telesurgery system for healthcare 4.0: An architecture, research challenges, and future directions. *IEEE Network*, 33(6): 22–29, 2019.
16. Sohail Iqbal, Shahzad Farooq, Khuram Shahzad, Asad Waqar Malik, Mian M Hamayun, and Osman Hasan. Securesurginet: A framework for ensuring security in telesurgery. *International Journal of Distributed Sensor Networks*, 15: 1550147719873811, 2019.
17. Mohammad Wazid, Ashok Kumar Das, and Jong-Hyouk Lee. User authentication in a tactile internet based remote surgery environment: Security issues, challenges, and future research directions. *Pervasive and Mobile Computing*, 54: 71–85, 2019.
18. Najmul Hassan, Saira Gillani, Ejaz Ahmed, Ibrar Yaqoob, and Muhammad Imran. The role of edge computing in internet of things. *IEEE Communications Magazine*, 56(11): 110–115, 2018.
19. Babatunji Omoniwa, Riaz Hussain, Muhammad Awais Javed, Safdar Hussain Bouk, and Shahzad A Malik. Fog/edge computing- based IoT (feciot): Architecture, applications, and research issues. *IEEE Internet of Things Journal*, 6(3): 4118–4149, 2018.
20. Sapna Tyagi, Amit Agarwal, and Piyush Maheshwari. A conceptual framework for IoT-based healthcare system using cloud computing. In *2016 6th International Conference-Cloud System and Big Data Engineering (Confluence)*, pages 503–507. IEEE, 2016.
21. Felipe Fernandez and George C Pallis. Opportunities and challenges of the internet of things for healthcare: Systems engineering perspective. In *2014 4th International Conference on Wireless Mobile Communication and Healthcare-Transforming Healthcare Through Innovations in Mobile and Wireless Technologies (MOBIHEALTH)*, pages 263–266. IEEE, 2014.
22. Tuan Nguyen Gia, Mingzhe Jiang, Amir-Mohammad Rahmani, Tomi Westerlund, Pasi Liljeberg, and Hannu Tenhunen. Fog computing in healthcare internet of things: A case study on ECG feature extraction. In *2015 IEEE International Conference on Computer and Information Technology; Ubiquitous Computing and Communications; Dependable, Autonomic and Secure Computing; Pervasive Intelligence and Computing*, pages 356–363. IEEE, 2015.
23. David C Klonoff. Fog computing and edge computing architectures for processing data from diabetes devices connected to the medical internet of things. *Journal of Diabetes Science and Technology*, 11: 647–652, 2017.
24. Ian Goodfellow, Yoshua Bengio, and Aaron Courville. *Deep Learning*. MIT Press: Cambridge, MA, 2016.
25. Trang Pham, Truyen Tran, Dinh Phung, and Svetha Venkatesh. Predicting healthcare trajectories from medical records: A deep learning approach. *Journal of Biomedical Informatics*, 69: 218–229, 2017.
26. Andreas S Panayides, Marios S Pattichis, Stephanos Leandrou, Costas Pitris, Anastasia Constantinidou, and Constantinos S Pattichis. Radiogenomics for precision medicine with a big data analytics perspective. *IEEE Journal of Biomedical and Health Informatics*, 23(5): 2063–2079, 2018.

27. Xiao Xu, Ying Wang, Tao Jin, and Jianmin Wang. A deep predictive model in healthcare for inpatients. In *2018 IEEE International Conference on Bioinformatics and Biomedicine (BIBM)*, pages 1091–1098. IEEE, 2018.
28. Anirban Mitra, Priya Shankar Banerjee, Sudipta Roy, Somasis Roy, and Sanjit Kumar Setua. The region of interest localization for glaucoma analysis from retinal fundus image using deep learning. *Computer Methods and Programs in Biomedicine*, 165: 25–35, 2018.
29. Jen Hong Tan, Sulatha V Bhandary, Sobha Sivaprasad, Yuki Hagi- wara, Akanksha Bagchi, U Raghavendra, A Krishna Rao, Biju Raju, Nitin Shridhara Shetty, Arkadiusz Gertych, et al. Age- related macular degeneration detection using deep convolutional neural network. *Future Generation Computer Systems*, 87: 127– 135, 2018.
30. Daniel SW Ting, Lily Peng, Avinash V Varadarajan, Pearse A Keane, Philippe M Burlina, Michael F Chiang, Leopold Schmetterer, Louis R Pasquale, Neil M Bressler, Dale R Webster, et al. Deep learning in ophthalmology: The technical and clinical considerations. *Progress in Retinal and Eye Research*, 72: 100759, 2019.
31. Ryo Asaoka, Hiroshi Murata, Aiko Iwase, and Makoto Araie. Detecting preperimetric glaucoma with standard automated perimetry using a deep learning classifier. *Ophthalmology*, 123(9): 1974–1980, 2016.
32. Atalie C Thompson, Alessandro A Jammal, Samuel I Berchuck, Eduardo B Mariottoni, and Felipe A Medeiros. Assessment of a segmentation-free deep learning algorithm for diagnosing glaucoma from optical coherence tomography scans. *JAMA Ophthalmology*, 138(4): 333–339, 2020.
33. Felix Grassmann, Judith Mengelkamp, Caroline Brandl, Sebastian Harsch, Martina E Zimmermann, Birgit Linkohr, Annette Peters, Iris M Heid, Christoph Palm, and Bernhard HF Weber. A deep learning algorithm for prediction of age-related eye disease study severity scale for age-related macular degeneration from color fundus photography. *Ophthalmology*, 125(9): 1410–1420, 2018.
34. Maximilian Treder, Jost Lennart Lauermann, and Nicole Eter. Automated detection of exudative age-related macular degeneration in spectral domain optical coherence tomography using deep learning. *Graefe's Archive for Clinical and Experimental Ophthalmology*, 256(2): 259–265, 2018.
35. Shuqiang Wang, Xiangyu Wang, Yong Hu, Yanyan Shen, Zhile Yang, Min Gan, and Baiying Lei. Diabetic retinopathy diagnosis using multichannel generative adversarial network with semisupervision. *IEEE Transactions on Automation Science and Engineering*, 18(2): 574–585, 2020.
36. Kelvin H Wan, Suber S Huang, Alvin L Young, and Dennis Shun Chiu Lam. Precautionary measures needed for ophthalmologists during pandemic of the coronavirus disease 2019 (covid-19). *Acta Ophthalmologica*, 98(3): 221–222, 2020.
37. Dariusz Wroblewski, Brian A Francis, Alfredo Sadun, Ghazal Vakili, and Vikas Chopra. Testing of visual field with virtual reality goggles in manual and visual grasp modes. *BioMed Research International*, 2014, 2014.
38. Justus Thies, Michael Zollhofer, Marc Stamminger, Christian Theobalt, and Matthias Niener. Demo of FaceVR: Real-time facial reenactment and eye gaze control in virtual reality. In *ACM SIGGRAPH 2017 Emerging Technologies*, pages 1–2, 2017.
39. Xi Wang, Hao Chen, An-Ran Ran, Luyang Luo, Poemen P Chan, Clement C Tham, Robert T Chang, Suria S Mannil, Carol Y Cheung, and Pheng-Ann Heng. Towards multi-center glaucoma OCT image screening with semi-supervised joint structure and function multi-task learning. *Medical Image Analysis*, 63: 101695, 2020.
40. Ahmed M Sayed, Mostafa Abdel-Mottaleb, Rashed Kashem, Vatookarn Roongpoovapatr, Amr Elsawy, Mohamed Abdel-Mottaleb, Richard K Parrish II, and Mohamed Abou Shousha. Expansion of peripheral visual field with novel virtual reality digital spectacles. *American Journal of Ophthalmology*, 210: 125–135, 2020.

41. Jason Kugelman, David Alonso-Caneiro, Scott A Read, Jared Hamwood, Stephen J Vincent, Fred K Chen, and Michael J Collins. Automatic choroidal segmentation in OCT images using supervised deep learning methods. *Scientific Reports*, 9(1): 1–13, 2019.
42. Migel D Tissera and Mark D McDonnell. Deep extreme learning machines: Supervised autoencoding architecture for classification. *Neurocomputing*, 174: 42–49, 2016.
43. Bernhard M Ege, Ole K Hejlesen, Ole V Larsen, Karina Mller, Barry Jennings, David Kerr, and David A Cavan. Screening for diabetic retinopathy using computer based image analysis and statistical classification. *Computer Methods and Programs in Biomedicine*, 62(3): 165–175, 2000.
44. Zhiyong Xiao, Mouloud Adel, and Salah Bourennane. Bayesian method with spatial constraint for retinal vessel segmentation. *Computational and Mathematical Methods in Medicine*, 2013, 2013.
45. M Victoria Ibanez and Amelia Simo. Bayesian detection of the fovea in eye fundus angiographies. *Pattern Recognition Letters*, 20(2): 229–240, 1999.
46. Simon P Kelly, Judith Thornton, Richard Edwards, Anjana Sahu, and Roger Harrison. Smoking and cataract: Review of causal association. *Journal of Cataract & Refractive Surgery*, 31(12): 2395–2404, 2005.
47. Daphne Koller and Nir Friedman. *Probabilistic Graphical Models: Principles and Techniques*. MIT Press: Cambridge, MA, 2009.
48. Jessica Loo, Matthias F Kriegel, Megan M Tuohy, Kyeong Hwan Kim, Venkatesh Prajna, Maria A Woodward, and Sina Farsiu. Open-source automatic segmentation of ocular structures and biomarkers of microbial keratitis on slit-lamp photography images using deep learning. *IEEE Journal of Biomedical and Health Informatics*, 25: 88–99, 2020.
49. Md Haque, Abdullah Al Kaisan, Mahmudur R Saniat, and Aminur Rahman. GPU accelerated fractal image compression for medical imaging in parallel computing platform. *arXiv preprint arXiv:1404.0774*, 2014.
50. Md Enamul Haque, KM Imtiaz-Ud-Din, Md Muntasir Rahman, and Aminur Rahman. Connected component based ROI selection to improve identification of microcalcification from mammogram images. *8th International Conference on Electrical and Computer Engineering*. Dhaka, Bangladesh, 2014.
51. Thomas Martin Lehmann, Claudia Gonner, and Klaus Spitzer. Survey: Interpolation methods in medical image processing. *IEEE Transactions on Medical Imaging*, 18(11): 1049–1075, 1999.
52. Anders Eklund, Paul Dufort, Daniel Forsberg, and Stephen M LaConte. Medical image processing on the GPU–past, present and future. *Medical Image Analysis*, 17(8): 1073–1094, 2013.
53. Parampal S Grewal, Faraz Oloumi, Uriel Rubin, and Matthew TS Tennant. Deep learning in ophthalmology: A review. *Canadian Journal of Ophthalmology*, 53(4): 309–313, 2018.
54. Avinash V Varadarajan, Pinal Bavishi, Paisan Ruamviboonsuk, Peranut Chotcomwongse, Subhashini Venugopalan, Arunachalam Narayanaswamy, Jorge Cuadros, Kuniyoshi Kanai, George Bresnick, Mongkol Tadarati, et al. Predicting optical coherence tomography-derived diabetic macular edema grades from fundus photographs using deep learning. *Nature Communications*, 11(1): 1–8, 2020.
55. Yifan Peng, Shazia Dharssi, Qingyu Chen, Tiarnan D Keenan, Elvira Agron, Wai T Wong, Emily Y Chew, and Zhiyong Lu. Deepseenet: A deep learning model for automated classification of patient-based age-related macular degeneration severity from color fundus photographs. *Ophthalmology*, 126(4): 565–575, 2019.
56. Xinting Gao, Stephen Lin, and Tien Yin Wong. Automatic feature learning to grade nuclear cataracts based on deep learning. *IEEE Transactions on Biomedical Engineering*, 62(11): 2693–2701, 2015.

57. Nathan Congdon, Johannes R Vingerling, BE Klein, Sheila West, David S Friedman, John Kempen, Benita O'Colmain, Suh-Yuh Wu, and Hugh R Taylor. Prevalence of cataract and pseudophakia/aphakia among adults in the United States. *Archives of ophthalmology (Chicago, Ill.: 1960)*, 122(4): 487–494, 2004.
58. Sonia Phene, R Carter Dunn, Naama Hammel, Yun Liu, Jonathan Krause, Naho Kitade, Mike Schaekermann, Rory Sayres, Derek J Wu, Ashish Bora, et al. Deep learning and glaucoma specialists: The relative importance of optic disc features to predict glaucoma referral in fundus photographs. *Ophthalmology*, 126(12): 1627–1639, 2019.
59. Ryo Asaoka, Hiroshi Murata, Kazunori Hirasawa, Yuri Fujino, Masato Matsuura, Atsuya Miki, Takashi Kanamoto, Yoko Ikeda, Kazuhiko Mori, Aiko Iwase, et al. Using deep learning and transfer learning to accurately diagnose early-onset glaucoma from macular optical coherence tomography images. *American Journal of Ophthalmology*, 198: 136–145, 2019.
60. Mahdi H Miraz, Maaruf Ali, Peter S Excell, and Rich Picking. A review on internet of things (IoT), internet of everything (IoE) and internet of nano things (IoNT). In *2015 Internet Technologies and Applications (ITA)*, pages 219–224. IEEE, 2015.

9 Deep Learning in IoT-Based Healthcare Applications

S.M. Zobaed, Mehedi Hassan, and Muhammad Usama Islam
University of Louisiana at Lafayette

Md Enamul Haque
Stanford University

CONTENTS

9.1 INTRODUCTION

Recently, the phrase coined as the Internet of Things (IoT) has gained immense popularity due to emergence as the fastest developing technology. IoT is a system of computing nodes that are interconnected to each other through the network and utilize embedded technologies to sense (when to collect) and collect data to share with others or host(s). Accessibility, adaptability, portability, and energy efficiency have made IoT applicable to various fields such as wearable devices, smart cities, smart homes, smart vehicles, supply chain, agriculture, and retail. It plays an important role in the healthcare domain as well due to the popularity and availability of its bonafide facilities such as wearable fitness tracker, real-time monitoring equipment, and other smart objects.

Subsequently, by utilizing these IoT devices, numerous E-health applications are being developed to track human activities and patient's medical conditions in real-time to facilitate good health and also on-demand emergency services. One of the successful outcomes of integrating IoT in several domains is the evolve of a smart healthcare system that communicates between network connected systems, applications, and devices [1]. Interactions among healthcare professionals, patients, and smart devices are established through a network.

Such IoT-based healthcare systems monitor, track, or record patient's vital, sensitive medical information by collecting real-time data for several types of patients from multiple sources over a predefined period of time. Usually, these data are complex and in an unstructured form where we require to apply machine learning (ML) to extract gist. However, traditional ML algorithms typically require feature engineering to obtain robust features or transformed feature space that depends on domain knowledge. Hence, obtaining optimal performance using traditional ML models that trained on complex data becomes challenging.

The latest advancements in the field of DL ease the limitation of traditional ML. This allows the system to work with complex data without performing exhaustive feature engineering. DL achieves much higher levels of accuracy in recent studies [1–8]. The advancement of DL is surprising as it outperforms humans in various tasks, most notably, classifying objects in images [9], language modeling, and video-to-video synthesis [10]. DL leverages a deeper neural network architecture that is capable to extract useful but complex hidden characteristics from the input data. Recent studies show that DL has the capability in generalizing the complicated inherent relationship in massive healthcare data obtained from various IoT devices.

In this chapter, we present a comprehensive discussion on IoT-based healthcare systems using different case studies with respect to deep learning models. More specifically, we discuss the recent trends, scopes, prospects, challenges, and limitations, and provide future research directions of deep learning in IoT-based healthcare systems.

9.2 HEALTHCARE AND INTERNET OF THINGS

At present, IoT is one of the promising technologies used in healthcare sectors. Owing to IoT, a number of smart healthcare applications are developed that provide solutions to keep our lives safe and sound. In this section, we describe medical IoT devices and IoT-aware healthcare systems that have added "smartness" to the healthcare sector.

9.2.1 IoT Medical Devices

Receiving alerts or notifications regarding health and fitness is the prime factor that retains users' attention toward IoT medical devices that is also known as the Internet of Medical Things (IoMT). IoMT is being widely adopted, and the global IoMT market was valued $44.5 billion in 2018. By 2026, it is expected to grow to $254.2 billion [11].

TABLE 9.1
List of Wearable IoMT Devices That Are Used in IoT Healthcare Applications

Types of Wearable IoMT	Collected Data
Smartwatches (e.g., Samsung, Apple, Tick, Fitbit)	Physical activity, heart rate
Smart cloths and shoes (e.g., Nike, Digitsole)	Physical activity, body temperature
Activity tracker (e.g., Garmin, Ticker, Fitbit)	Fitness data, heart rate, sleep

IoMT is a connected network infrastructure of medical devices, health system software, and services. IoMT provides personalized living environments to provide proper care and attention to the users (e.g., patients) through the data-driven-based treatment. IoMT also offers alternative feasible solutions to solve problems met by the traditional medical systems, such as the lack of nurses, doctors, healthcare resources, and research data.

There are two types of IoMT devices namely, wearable and implantable based on usage. Wearable devices can be attached to the human body in various items such as bracelets, wristwatches, glasses, and other accessories. Such wearable devices are capable of tracking illness, wellness of a person, and transmit the collected information to the central server for analysis. They provide biological information such as burned calories, heart rate, blood pressure, and time spent exercising, number of walked steps, and so on. In general, wearable IoMT comprises three components that are sensors, computing architecture, and displays. In Table 9.1, we provide the list of popular wearable devices and types of data they collect from users.

IoMT domain is assured for further growth as artificial intelligence is integrated to make them capable of the real-time measurement and analysis of patient data. Specifically, ML or big data analytic tools are implemented to process the data involved in IoMT for performing deeper medical analysis and disease diagnosis. The size of data gathered by IoMT is also growing massively in each second. Therefore, it is a challenge to perform deeper and accurate analysis over such types of big data while maintaining real timeliness.

9.2.2 IoT-Aware Healthcare System

In this section, we briefly depict end to end IoT-aware healthcare system that can collect data of physiological parameters from patients or users in real-time and outsource them to the cloud for processing. In Figure 9.1, we show an ideal three-tier architecture (client-edge-cloud) of such type of system. Various IoT devices such as heart bit sensor, smart wristwatch, smartwatch, and blood pressure sensors are worked at user premises. They collect various types of information and send the data to the nearest edge server. Edge server contains light-weight processing capability, and it removes possible noise from the data and combines the data obtained from similar sources. Later, the data is transmitted to the cloud server to be analyzed. As the edge server relentlessly transmits data, the cloud needs to store the data for performing analytics over that. The cloud server performs data analytics by leveraging ML (i.e., DL) to detect complex patterns or capture underlying information.

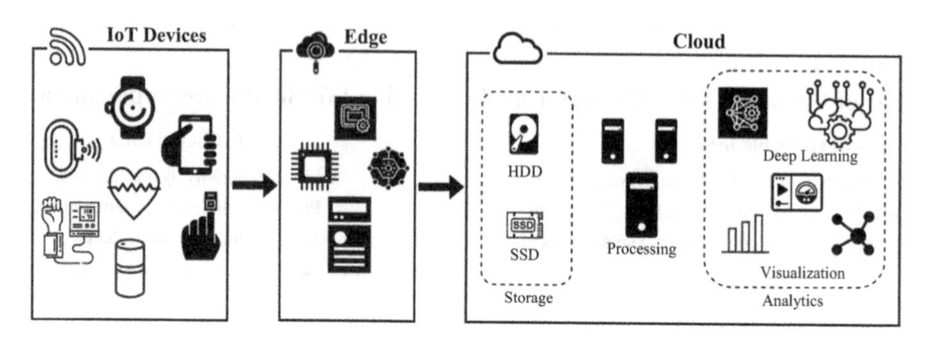

FIGURE 9.1 IoT healthcare architecture.

9.2.3 Emergence of IoMT in Healthcare

The IoMT-aware healthcare system has emerged over time with maturity in terms of diagnosis, prediction, and providing better healthcare solutions through aiding the healthcare professionals as well as all other stakeholders who are related to. Today's IoMT has insightful, intelligent, and astute features with tangible mobility through various connected devices that provide an extra edge over the predecessors that ruled the healthcare sector. In this section, we discuss some of the commercial applications and devices that have made promising changes in the healthcare sector with the aid of IoT.

The solution referred to as ColdTrace [12] offered a wire-free temperature monitoring system for the vaccines that are stored in refrigerators in remote healthcare facilities around the world. The technology focuses on the safekeeping of vaccines through IoT-enabled technology for least developed nations to focus on life-saving injections that are scarce and rare to find in remote regions.

SystemOne [13] utilized the Vodafone IoT network and connectivity to securely accumulate and transmit the medical diagnostic data in real-time to all stakeholders related to healthcare including the physician, nurse, health- care providers, and the patients. They have successfully developed several use cases for IoT IoT-enabled infrastructure and a data connectivity system through GxAlert [13] and tested for GeneXpert thus made an expansion to more than 35 countries with rural underdeveloped places with technical limitations.Application related to air quality monitoring emerged with industrialization in the developed world that led to the deterioration of the air quality.

The air we breathe is becoming toxic eventually due to massive industrialization which greatly affects the quality of life. Aclima [14] has provided an understanding of urban air quality with air quality index (AQI) visualization. The product focuses on the factors that are affecting the air quality.

QuiO [15] in a partnership with SHL has been providing an integrated platform with IoT support to connect various therapeutic devices. A dashboard provided by QuiO provides information of patients to caregivers and healthcare providers to effectively respond to situations. In addition, QuiO also works as a platform that provides opportunities for research studies participation with anonymity. Similar works are done using Azure IoT by Microsoft [16], Spry health [17], Dexcom [18], and

MediData [19]. Also, Philips E-Alert [20] used sensors to monitor the MRI performance and generate alerts through insight generated from device data [21].

Pfizer and IBM partnered up to assimilate the concept of IoT to provide a tracking solution [22] for the effectiveness of Parkinson's drugs on a patient and provided a real-time study to make dosage adjustments. A similar application is developed by Adheretech where it has provided an IoT-enabled technology to facilitate dosage reminders and personalized support for medication refills [23]. Similar work on dosage reminder is provided by Aeris communications [24] through the plug-and-play platform that they have in addition to a connected system with caregivers and health professionals for monitoring.

EightSleep [25] provided an intelligent mattress as a form of solution to sleep depravity. Apart from it being integrated into smart home systems, the intelligent sensing app senses the environment and determines the optimal sleep temperature. A similar application is also being developed by Apple-owned Beddit that used Bluetooth technology connected to a sleep monitor to track heart rate, snoring, and breathing patterns that provided an optimal solution to overall sleeping performance [26]. Furthermore, SNOO bassinet [27] has provided a similar solution for infants in aiding them to sleep longer through adjusting womb-like rocking, cry sensor which helps the parents to get much-needed rest in times of early parenting.

Amazon Care [28], an app developed by Amazon, has provided personalized healthcare with care courier services which delivered prescription drugs to its users or to a preferable local pharmacy securely. Furthermore, Alexa [29] symptom checker can check for mild fever from surrounding insights in smart home systems and provided basic medication guidelines which have added to the rich IoT-enabled services that are being provided by Amazon.

Abilify MyCite [30], a mood disorder treatment tablet, has an embedded marker which upon ingestion delivered data to the wearable sensor on mood and provided insights on the quality of rest via a secured portal with an additional feature that accumulated doctors and other healthcare stakeholders if required.

The AutoBed by GE healthcare [31] used deployable sensor technology to monitor bed requests in hospitals and tracked an insight into patient requirements such as nurse proximity [32] which determined the availability of nurse within a certain range for the request. Similar real-time patient location information using RFID [33] has been formulated by Stanley healthcare to provide a better experience to patients.

Genesis Touch [34,35] developed by Honeywell provided an integration with other devices such as oximeter and blood pressure monitor to subjugate data generated from the devices to provide health statistics. Furthermore, the dashboard's ability to host video visits enabled remote rehabilitation. Similar work was done by R-Style lab [36] and ENSA [37]. However, ENSA provided the additional syncing feature with Apple HealthKit and Dexcom CGM and provided a recommender system for advising wellness and supplements for the patients. Eversence [38,39], developed by Senseonics, used a biosensor for continuous glucose monitoring (CGM). It was aided by ML concepts in generating insights from the data for recommendations and actions related to glucose and blood sugar level of patients. Propeller [40,41], a device that has integrated sensors that are attached to asthma inhalers in its distributed enabled technology, is used to track medication usage and manage symptoms for asthma patients through ML-based predictive analytics.

9.3 PROSPECTS OF DEEP LEARNING

Deep learning is a subfield of ML that utilizes artificial neural networks to imitate the working procedures of the human brain [42]. DL models are built upon layers of neural networks (i.e., hidden layers) to perform unsupervised learning from unstructured data to extract the complex patterns. DL has offered substantial success in image classification, natural language processing, and other pattern recognition tasks [43]. Recently, DL has got attention for enormous success in IoT healthcare domain [32,43–45]. To perform DL comprehensively, a huge volume of data is required. Hence, utilizing DL is not possible in the domains where less data generation is involved [46–48]. On the contrary, the IoT healthcare domain gains immense benefit by utilizing DL as the domain generates a massive volume of data. Subsequently, DL has yielded promising results in complex diagnostics that help physicians by providing a substantial opinion regarding disease classification, marking problematic objects in medical imagery, and highlighting other critical findings [7,8,49].

9.4 CHALLENGES OF HEALTHCARE IoT ANALYTICS

Although DL-based and other IoT analytics techniques provide several benefits to many organizations, organizations face several critical issues that raise concerns about business growth and security. In this section, we discuss the challenges of IoT analytics that need to be resolved.

- **Privacy**

  To perform processing or analytics of big data collected from IoT healthcare devices, we require to shift to cloud services as local resources are not sufficient to perform. Healthcare organizations usually own patient data that are sensitive and required to maintain privacy. However, organizations dealing with big data are hesitant to migrate to the cloud because of the trustworthiness [50–52]. The massive volume of data is sorted, filtered, and processed in different servers that are situated globally. Hence, the data are required to be available for parallel processing on different clouds which increases the likelihood of being attacked [50,53].

  As an example, in 2018, more than 14 million Verizon customers' credential records were exposed from their cloud repository [54]. Similarly, confidential data of 3 billion Yahoo users were exposed [55] in 2013. In [56], 15 biggest data privacy violation incidents in the 21st century are reported. These incidents vividly highlight the trustworthiness of the cloud. Cloud business should give proper attention to their customers owning the healthcare data.
- **Real-Time Big Data Intelligence**

  Performing real-time analytics over the collected healthcare data is one of the prime requirements for IoT-based healthcare applications that yield the gist of the data and share it with the users in real-time. However, process such a massive volume of data in real-time is challenging as it requires large-scale computing devices. In addition, the framework or the algorithms that are implemented for data processing should be optimized to achieve timeliness.

- **Data Structure and Filtration**

  The collected data are not only in unstructured form but also fragmented, dispersed, and noisy but rarely standardized. In addition, because of using various types of IoT devices in healthcare, a researcher needs to merge, process, and analyze all data collected from the devices.

  Since raw data arrives continuously as a stream to the storage, data filtration is essential so that DL models can be trained efficiently. Because of enormous data volume and shorter processing time constraints, performing effective data filtering (pre-processing) is a significant challenge.

- **Scalability and On-demand Resource**

  In the healthcare sector, a large number of sensors are involved to collect data. Storing, filtering, and processing such data to their constant burst are challenging. The computational overhead in terms of processing data is indeterminable as the size of the data is grown continuously and requires to perform various operations over data. Especially, DL models require to exploit a large-scale computing resource to train their model. Besides, adding real-time constraints could increase resource demand. Therefore, on-demand resource allocation, heterogeneous computing capability is required to scale up the cloud computing infrastructure to perform analytics whenever they are asked for [57].

9.5 DEEP LEARNING TECHNIQUES IN HEALTHCARE APPLICATIONS

The massive volume of data generated by IoMT devices is utilized for further analysis. To this end, we can leverage DL to identify specific patterns in the data or to classify the data which is useful in healthcare sectors. In this section, we briefly discuss several IoT incorporated DL-based applications that are served for various purposes in the healthcare domain such as health monitoring, human activity recognition, disease analysis, and data security.

9.5.1 Health Monitoring

The pervasiveness of wearable devices allows us to monitor human health conditions at any time. Wearables along with smartphone sensors facilitate IoT-based applications in the field of health monitoring. Due to the limited resources of wearable devices, many implemented DL modules leveraging edge and cloud-based services. Azimi et al. [58] proposed a three-tier hierarchical edge-based IoT healthcare system using a DL-based model. The first tier consists of a wireless body area network (WBAN) of heterogeneous bio-sensors that collect the health data of a patient. The second tier consists of a gateway device enabling continuous communication with cloud servers and WBAN. The third-tier is a cloud-based server responsible for deploying a convolutional neural network (CNN) to perform data analysis. The proposed system can handle a large volume of data as it leverages the advantages of high-end machines such as cloud-based servers. To avoid any delay in decision-making after the analysis of the health data, the heavy computations of the classifying algorithm are performed in the cloud servers, and the final decision-making task is performed in the edge devices. The authors have used three layers of CNN architecture to extract features from the data and Multi-Layer Perceptron (MLP) model with one hidden layer to perform the classification task.

HealthFog [59] is a health monitoring system for heart patients. Health-Fog leverages FogBus [60] framework which enables any IoT-based system to integrate in fog and cloud infrastructures. Using an ensemble deep learning, HealthFog analyses heart diseases in edge devices with high accuracy and low latency. The system was trained using a deep neural network and performed MLP on the Cleveland dataset. The heavy and complex computation is performed on a cloud. The prediction task is performed in real-time using the edge devices.

Alhussein et al. [61] developed an EEG-based seizure detection and monitoring method named *CHIoT*. *CHIoT* uses a DL module: CNN and Stacked Autoencoder to extract features from patient's data and classify them as seizure or non-seizure with a probability score, respectively. The patient's data is collected using smart EEG sensors and other smart sensors that capture physiological and psychological signals. Along with EEG signals, other data such as movements, gestures, and facial expressions are analyzed in a cloud environment. Seizure detection is performed in the cloud, and the classifying results are sent to medical practitioners to monitor the patients.

Yu et al. proposed *EdgeCNN* model to diagnose atrial fibrillation by utilizing electrocardiogram recording [62]. The proposed EdgeCNN leverages edge and cloud computing to provide a solution for storing and processing IoT healthcare data in the cloud environment. EdgeCNN is a deep learning-based system that is integrated with smart devices to provide real-time diagnostic capabilities. To this end, they implement CNN on the ECG dataset that was published by PhysioNet. EdgeCNN outperforms the competitors in terms of network latency, privacy, and cost.

In Table 9.2, we provide a summary of the aforementioned applications that were proposed for health monitoring.

9.5.2 Human Activity Recognition

Activity recognition is fundamental in healthcare systems as it permits to monitor and analyze daily activities or actions. There exist several challenges in classifying activities. Due to inter-activity similarity, different activities may represent similar characteristics. Several studies have employed DL modules as a solution to mitigate such problems. Zhou et al. [48] designed a semi-supervised DL approach for accurate human activity recognition (HAR). Different types of wearable devices are used to capture motions and posture data from different parts of the body. A large

TABLE 9.2
Summary of IoT-Based Healthcare Applications Employing DL for Health Monitoring

Related Work	Addressed Problem	DL Used	Sensor	Dataset
[58]	Abnormality detection	MLP	Body sensors	MIT arrhythmia
[59]	Heart diseases analysis in edge devices	MLP	EEG, ECG, temperature, oxygen, glucose	Cleveland
[61]	Seizure detection and monitoring	CNN, stacked AE	EEG, smart sensors	CHB-MIT
[62]	Atrial fibrillation	CNN	ECG, BP meters, smartwatch	PhysioNet
[3]	EEG Sensor	CNN	EEG	TUH
[4]	Heart rate prediction	CNN	Heart rate, respiratory, oxygen	MIT-BIH

portion of these data is unlabeled. Combined with a small portion of labeled data, classification results become biased and inaccurate. As a result, deep neural network and Q-learning are combined with a distance-based reward function to transform unlabeled data into labeled data. An LSTM-based classifier is employed to extract high-level features from the multi-sensor-based fusion data.

Bianchi et al. [2] proposed a HAR system using CNN. The proposed system recognizes nine daily activities in an Active Assisted Living (AAL) environment. To train the system, the authors exploited a UCI dataset and a captured dataset using Wi-Fi wearable sensors (accelerometer, gyroscope, and magnetometer). The CNN architecture is deployed in the cloud to perform heavy computational tasks. The system goes through a retraining phase each time a user is added for tuning the system in order to provide the best classification performance.

Keshavarzian et al. [6] proposed a modified CNN-based HAR system on the IoT environment. Accelerometers and gyroscope of smartphones are used to collect data for different body movements. The authors also proposed a function-as-a-model to analyze human actions in the cloud infrastructure. The proposed system classifies human activities without any feature engineering. The system outperforms state-of-the-art methods on three datasets: HCI-HAR, HAPT, and UniMiB.

Ravi et al. [63] proposed a DL-based HAR model that incorporates features obtained from the sensor data with the complementary information obtained from a set of shallow features. The contribution of the proposed model was to resolve the constraints of low-power IoT devices, real-time computation, and activity classification. The model was utilized to classify time-series data and evaluated with five other datasets that are ActiveMiles, Daphnet FoG, Skoda, WISDM v1.1, and WISDM v2.0.

Gao et al. [45] proposed a CNN and LSTM-based aggregated model named R3D to remotely monitor the human activities for smart healthcare system. They extracted

TABLE 9.3
Summary of IoT-Based Healthcare Applications Employing DL for Human Activity Detection

Related Work	Addressed Problem	DL Used	Sensor	Dataset
[48]	Classify human activities	DNN, DQN, LSTM	On-body accelerometer	RealWorld HAR, UniMiB SHAR
[63]	Classify human activities	CNN	Accelerometer, gyroscope	ActiveMiles, Skoda, Daphnet FoG, WISDM v1.1, WISDM v2.0
[2]	Classify daily activities	CNN	Accelerometer, gyroscope	UCI
[6]	Classify human activities	CNN	Accelerometer, gyroscope	HCI HAR, HAPT, UniMiB
[45]	Monitor human activities	CNN, LSTM	N/A	UCF-101

spatial-temporal features and aggregate them using LSTM to obtain a high-level abstraction representation of each human action.

In Table 9.3, we provide a summary of the aforementioned applications that were proposed for human activity detection.

9.5.3 Disease Analysis

Medical image segmentation is often performed based on the pixel classification approach to perform disease analysis [64,65]. CNN is utilized to infer the hierarchical representation of the images that facilitates the segmentation. Prasoon et al. [49] explained about 3D image classification method to predict the risk of osteoarthritis. The authors criticized traditional CNN-based 3D image classification models as these required large memory and high training time. Their proposed classifier classified pixels from 3D images with high prediction accuracy and took less time to train.

Similar to previous work, Gulshan et al. [8] leveraged the CNN concept to identify diabetic retinopathy by analyzing retinal fundus images. Their proposed model provided high sensitivity and specificity to predict over about 10,000 test images with respect to certified ophthalmologist annotations.

Lopez et al. [7] proposed to classify dermotropic images that contain skin lesions as malignant or benign. The authors implemented a binary classifier that takes skin lesion images labeled as benign or malignant as input and built a model using deep CNN to predict whether a (previously unseen) image of a skin lesion is either benign or malignant. Their model outperformed other state-of-the-art by achieving 78.66% sensitivity and 79.74% precision.

In Table 9.4, we provide a summary of the aforementioned applications that were proposed for disease analysis.

TABLE 9.4
Summary of IoT-Based Healthcare Applications Employing DL for Disease Analysis

Related Work	Addressed Problem	DL Used	Sensor	Dataset
[49]	Knee cartilage segmentation	CNN	MRI scanner	N/A
[8]	Retinopathy detection	CNN	Camera	EyePACS-1
[7]	Skin lesion classification	CNN	Camera	ISBI-2016

9.5.4 Security

Healthcare data is sensitive as it contains personal information related to users. Cyber-attacks in IoT devices are alarming. To ensure a secure IoT environment and privacy of data, several studies have used DL techniques to authenticate devices, classify traffic, and detect anomalies. Shakeel et al. [66] developed a DL approach: Learning-based Deep-Q-Network (LDQN) for providing security and privacy in IoT-based healthcare systems. The system utilizes a DL framework to authenticate as well as to detect attacks present in the IoT devices using the Adaboost training process. After the initial authentication process, the traffic features are classified as malware or secured using the LQDN approach. The proposed system ensures the security of IoT-based devices with enhanced accuracy.

UbeHealth [67] used DL, big data, and IoT to address the challenges of latency, bandwidth, and reliability in the networking domain of healthcare. The framework ensures the quality of service by using four layers and three components. Mobile, Cloudlet, Network, and Cloud layers comprise the four layers whereas the three components include Network Traffic Analysis and Prediction (DLNTAP), Deep Learning Network Traffic Classification (DL-NTC), and Flow Clustering and Analysis (FCA). Cloudlet is a self-managing mobile edge device which reduces latency and jitter between mobile and cloud layers. DLNTAP component lies in the cloudlet layer in order to predict the network traffic to optimize data rates, caching, and routing. DLNTC component classifies the network traffic to detect any malware and provide high QoS. FCA component clusters data to identify the different kinds of data generated from the same application protocol. UbeHealth leverages two DL architectures: MLP and Long Short-Term Memory (LSTM). Finally, UbeHealth is evaluated on three widely used datasets named SPDSL-II, Waikato-VIII, and WIDE-18.

Balakrishnan et al. [44] proposed an intrusion detection scheme using a deep belief network (DBN) on IoT devices. Significant features are extracted, and a binary cross-entropy model is used to classify the output of DBN. The proposed system achieves significant performance compared with the state-of- the-art methods to provide privacy and security in IoT devices.

Thamilarasu et al. [68] developed an Integrated Intrusion Detection (IID) system using Deep Neural Network (DNN). The proposed system utilizes a pretrained deep belief network (DBN) to create the DNN model. The system is protocol-independent and can detect zero-day attacks as well as provide security as a service. The system

TABLE 9.5
Summary of IoT-Based Healthcare Applications Employing DL for Data Security

Related Work	Addressed Problem	DL Used	Sensor	Dataset
[66]	Authentication and malware detection on IoT devices	LDQN	RFID	Simulated dataset
[67]	Malicious traffic detection and data clustering	MLP, LSTM	N/A	SPDSL-II, Waikato VIII, WIDE-18
[44]	Preventing security breach on IoT devices	DBN	N/A	N/A
[68]	Intrusion detection on IoT devices	DNN	N/A	Simulated dataset

investigates both simulations and test-bed of IoT devices and achieves superior performance against the state-of-the-art techniques.

We summarize the aforementioned proposed applications in Table 9.5.

9.6 CONCLUSION AND FUTURE RESEARCH

The future of IoT research amalgamated with DL in the healthcare domain poses significant challenges due to issues related to scalability, security, and accessibility. The data that is accumulated from various sources of sensor-based applications and heterogeneous IoT devices are raw and unprocessed which is yet to be analyzed in order to generate actionable insights. The next question of challenge is the procedures for generating actionable insights.

Generating actionable insights on-the-go in real-time is gaining momentum among caregivers and healthcare sector professionals as this method provides more flexibility over offline insight generation.

However, generating actionable insights in real-time takes a substantial toll on processing power and time. For resource-constrained devices which is typically the case in personalized healthcare takes much longer time and power. So, the connected devices and their nature must be reexamined from scratch with a trade-off balance strategy to deploy DL architectures in resource-constrained devices as per requirement. A subproblem of DL architecture is the interpretability of the data. While the model result is well understood, the internal layers interpretation that leads to the result that has been achieved remains a big query to be explored. A hypothetical example to interpret this problem is as follows; a pneumonia image dataset generated from Electronic Medical Records (EMR) which can be used to detect pneumonia has its first layers of the CNN works on simple features such as edges while the deep hidden layers work to find more meaningful and insightful features. This example provides us with an insight that how a DL model works. However, apart from the simulation that leads to the accuracy of results, there is no other interpretable explanation other than reiterating with tuning the hyperparameters to generate a better result and avoid overfitting. Furthermore, another challenge that has embroiled the healthcare industry is the security of the data. While the data has significant value in terms of various paradigms, the most important of it all is that healthcare data is highly sensitive, personalized, and expected to be handled with the utmost care and discretion. Apart from investing the research on DL and its incorporation with IoT devices, much research work is needed to be carried out in designing secured hardware, the security of the network over which the dataflow shall be incorporated, and proper data encryption that shall store and analyze this sensitive information. Much recent research shows a significant promise of blockchain as a solution to data security and reliability through miner node and consensus-building. However, in terms of data handling more research needs to be carried out for trade-off between lightweight approaches. Lastly, one of the serious future challenges that lie ahead is the accessibility of data for research purposes. Healthcare data is more personalized and secretive in nature which makes it much more challenging to gather the data. While huge corporations have access to a big chunk of data that leads them to pursue cutting-edge scientific research, the academic research community is held back and lags behind with primitive forms of data. A proper policing and standards system can be thought of as a solution approach. Such standard systems authority is expected to guide, monitor, and certify academic researchers to collaborate with healthcare stakeholders, infrastructures, and organisms to properly handle misappropriation of sensitive data.

In this chapter, we have effectively summarized IoT and its application in the healthcare arena with challenges associated with the domain. We have addressed the deep learning concept and its effect on healthcare IoT to generate actionable insights and research works that have addressed the amalgamations of the aforementioned concepts. Lastly, we have analyzed the challenges in future directions and open-ended questions that still need to be addressed in this domain and conclude with the summary of our review.

REFERENCES

1. AKMI Iqtidar Newaz, Amit Kumar Sikder, Mohammad Ashiqur Rahman, and A Selcuk Uluagac. Healthguard: A machine learning-based security framework for smart healthcare systems. In *2019 Sixth International Conference on Social Networks Analysis, Management and Security (SNAMS)*, pages 389–396, 2019.
2. Valentina Bianchi, Marco Bassoli, Gianfranco Lombardo, Paolo Fornacciari, Monica Mordonini, and Ilaria De Munari. IoT wearable sensor and deep learning: An integrated approach for personalized human activity recognition in a smart home environment. *IEEE Internet of Things Journal*, 6(5):8553–8562, 2019.
3. Syed Umar Amin, M Shamim Hossain, Ghulam Muhammad, Musaed Alhussein, and Md Abdur Rahman. Cognitive smart healthcare for pathology detection and monitoring. *IEEE Access*, 7:10745–10753, 2019.
4. Zubair Md Fadlullah, Al-Sakib Khan Pathan, and Haris Gacanin. On delay-sensitive healthcare data analytics at the network edge based on deep learning. In *2018 14th International Wireless Communications & Mobile Computing Conference (IWCMC)*, pages 388–393, 2018.

5. Shanto Roy, Abdur Rahman, Masuk Helal, M Shamim Kaiser, and Zamshed Iqbal Chowdhury. Low cost rf based online patient monitoring using web and mobile applications. In *2016 5th International Conference on Informatics, Electronics and Vision (ICIEV)*, pages 869–874, 2016.
6. Alireza Keshavarzian, Saeed Sharifian, and Sanaz Seyedin. Modified deep residual network architecture deployed on serverless framework of iot platform based on human activity recognition application. *Future Generation Computer Systems*, 101:14–28, 2019.
7. Adria Romero Lopez, Xavier Giro-i Nieto, Jack Burdick, and Oge Marques. Skin lesion classification from dermoscopic images using deep learning techniques. In *2017 13th IASTED International Conference on Biomedical Engineering (BioMed)*, pages 49–54, 2017.
8. Varun Gulshan, Lily Peng, Marc Coram, Martin C Stumpe, Derek Wu, Arunachalam Narayanaswamy, Subhashini Venugopalan, Kasumi Widner, Tom Madams, Jorge Cuadros, et al. Development and validation of a deep learning algorithm for detection of diabetic retinopathy in retinal fundus photographs. *JAMA*, 316(22):2402–2410, 2016.
9. What is deep learning and Why is it more relevant than ever? https://interesting-engineering.com/what-is-deep-learning-and-why-is-it-more-relevant-than-ever, Accessed February 28, 2020.
10. The major advancements in Deep Learning in 2018. https://tryolabs.com/blog/2018/12/19/major-advancements-deep-learning-2018/, Accessed July 2, 2020.
11. Global Internet of Medical Things (IoMT) market- segment analysis, opportunity assessment, competitive intelligence, industry outlook 2016–2026. https://www.allthe-research.com/report/166/internet-of-medical-things-market, Accessed April 30, 2020.
12. S Balachandar and R Chinnaiyan. Reliable pharma cold chain monitoring and analytics through internet of things edge. In *Emergence of Pharmaceutical Industry Growth with Industrial IoT Approach*, pages 133–161. 2020.
13. Emmanuel Andre, Chris Isaacs, Dissou Affolabi, Riccardo Alagna, Dirk Brockmann, Bouke Catherine de Jong, Emmanuelle Cambau, Gavin Churchyard, Ted Cohen, Michel Delmée, et al. Connectivity of diagnostic technologies: Improving surveillance and accelerating tuberculosis elimination. *The International Journal of Tuberculosis and Lung Disease*, 20(8):999–1003, 2016.
14. Joshua S Apte, Kyle P Messier, Shahzad Gani, Michael Brauer, Thomas W Kirchstetter, Melissa M Lunden, Julian D Marshall, Christopher J Portier, Roel CH Vermeulen, and Steven P Hamburg. High- resolution air pollution mapping with google street view cars: Exploiting big data. *Environmental Science & Technology*, 51(12):6999–7008, 2017.
15. Joy Dutta, Chandreyee Chowdhury, Sarbani Roy, Asif Iqbal Middya, and Firoj Gazi. Towards smart city: Sensing air quality in city based on opportunistic crowd-sensing. In *Proceedings of the 18th International Conference on Distributed Computing and Networking*, pages 1–6, 2017.
16. Pasquale Pace, Gianluca Aloi, Raffaele Gravina, Giuseppe Caliciuri, Giancarlo Fortino, and Antonio Liotta. An edge-based architecture to support efficient applications for healthcare industry 4.0. *IEEE Transactions on Industrial Informatics*, 15(1):481–489, 2018.
17. Kathleen G Fan, Jess Mandel, Parag Agnihotri, and Ming Tai-Seale. Remote patient monitoring technologies for predicting chronic obstructive pulmonary disease exacerbations: Review and comparison. *JMIR mHealth and uHealth*, 8(5):e16147, 2020.
18. Rajiv B Kumar, Nira D Goren, David E Stark, Dennis P Wall, and Christopher A Longhurst. Automated integration of continuous glucose monitor data in the electronic health record using consumer technology. *Journal of the American Medical Informatics Association*, 23(3):532–537, 2016.

19. Michelle Crouthamel, Emilia Quattrocchi, Sarah Watts, Sherry Wang, Pamela Berry, Luis Garcia-Gancedo, Valentin Hamy, and Rachel E Williams. Using a researchkit smartphone app to collect rheumatoid arthritis symptoms from real-world participants: Feasibility study. *JMIR mHealth and uHealth*, 6(9):e177, 2018.
20. Philippe Lachance, Pierre-Marc Villeneuve, Francis P Wilson, Nicholas M Selby, Robin Featherstone, Oleksa Rewa, and Sean M Bagshaw. Impact of e-alert for detection of acute kidney injury on processes of care and outcomes: Protocol for a systematic review and meta-analysis. *BMJ Open*, 6(5):e011152, 2016.
21. Rinaldo Bellomo, John A Kellum, Claudio Ronco, Ron Wald, Johan Martensson, Matthew Maiden, Sean M Bagshaw, Neil J Glassford, Yugeesh Lankadeva, Suvi T Vaara, et al. Acute kidney injury in sepsis. *Intensive Care Medicine*, 43(6):816–828, 2017.
22. J Michael Ellis and Matthew J Fell. Current approaches to the treatment of Parkinson's disease. *Bioorganic & Medicinal Chemistry Letters*, 27(18):4247–4255, 2017.
23. Peter P Reese, Roy D Bloom, Jennifer Trofe-Clark, Adam Mussell, Daniel Leidy, Simona Levsky, Jingsan Zhu, Lin Yang, Wenli Wang, Andrea Troxel, et al. Automated reminders and physician notification to promote immunosuppression adherence among kidney transplant recipients: A randomized trial. *American Journal of Kidney Diseases*, 69(3):400–409, 2017.
24. Cornel Turcu and Cristina Turcu. Improving the quality of healthcare through internet of things. *arXiv preprint arXiv:1903.05221*, 2019.
25. Anita Valanju Shelgikar, Patricia F Anderson, and Marc R Stephens. Sleep tracking, wearable technology, and opportunities for research and clinical care. *Chest*, 150(3):732–743, 2016.
26. Jarno Tuominen, Karoliina Peltola, Tarja Saaresranta, and Katja Valli. Sleep parameter assessment accuracy of a consumer home sleep monitoring ballistocardiograph beddit sleep tracker: A validation study. *Journal of Clinical Sleep Medicine*, 15(03):483–487, 2019.
27. Anna Vest. Development of early social interactions in infants exposed to artificial intelligence from birth. Human Resources and Communication Disorders Undergraduate Honors Theses. https://scholarworks.uark.edu/rhrcuht/6, April 2020.
28. Timothy Bickmore, Ha Trinh, Reza Asadi, and Stefan Olafsson. Safety first: Conversational agents for health care. In *Studies in Conversational UX Design*, pages 33–57. 2018.
29. Veton Kepuska and Gamal Bohouta. Next-generation of virtual personal assistants (Microsoft Cortana, Apple Siri, Amazon Alexa and Google home). In *2018 IEEE 8th Annual Computing and Communication Workshop and Conference (CCWC)*, pages 99–103. IEEE, 2018.
30. D Papola, C Gastaldon, and G Ostuzzi. Can a digital medicine system improve adherence to antipsychotic treatment? *Epidemiology and Psychiatric Sciences*, 27(3):227–229, 2018.
31. Zuraida Abal Abas, Zaheera Zainal Abidin, AFNA Rahman, Hidayah Rahmalan, Gede Pramudya, and Mohd Hakim Abdul Hamid. Internet of things and healthcare analytics for better healthcare solution: Applications and challenges. *International Journal of Advanced Computer Science and Applications*, 9(9):446–450, 2018.
32. M Alper Akkaş, Radosveta Sokullu, and Hüseyin Ertürk Çetin. Healthcare and patient monitoring using IoT. *Internet of Things*, 11:100173, 2020.
33. Toni Adame, Albert Bel, Anna Carreras, Joan Melia-Segui, Miquel Oliver, and Rafael Pous. CUIDATS: An RFID–WSN hybrid monitoring system for smart health care environments. *Future Generation Computer Systems*, 78:602–615, 2018.

34. Kara K Hoppe, Makeba Williams, Nicole Thomas, Julia B Zella, Anna Drewry, KyungMann Kim, Thomas Havighurst, and Heather M Johnson. Telehealth with remote blood pressure monitoring for postpartum hypertension: A prospective single-cohort feasibility study. *Pregnancy Hypertension*, 15:171–176, 2019.
35. Ah-Lian Kor, Max Yanovsky, Colin Pattinson, and Vyacheslav Kharchenko. Smart-item: IoT-enabled smart living. In *2016 Future Technologies Conference (FTC)*, pages 739–749. IEEE, 2016.
36. Vlad Gapchinsky. Smart healthcare solutions - IoT medical software developer: R-style lab. https://www.slideshare.net/R-StyleLab/rstyle-lab-smart-solutions-for-healthcare-providers. Accessed May 30, 2021.
37. Ensa: Ensa app. Ensa - continuous, connected health - ensa connects its users' medical records and health and fitness apps to give personalized wellness recommendations on their mobile devices. https://www.welcome.ai/tech/healthcare/ensa-continuous-connected-health. Accessed March 15, 2020.
38. Rabab Jafri, Courtney Balliro, Firas El-Khatib, Michele Sullivan, Mallory Hillard, Alexander O'Donovan, Rajendranath Selagamsetty, Hui Zheng, Edward R Damiano, and Steven J Russell. A three-way accuracy comparison of the Dexcom G5, Abbott freestyle libre pro, and senseonics eversense CGM devices in an home-use study of subjects with type 1 diabetes. *Diabetes Technology and Therapeutics*, 22:846–852, 2020.
39. Carrie Lorenz, Wendolyn Sandoval, and Mark Mortellaro. Interference assessment of various endogenous and exogenous substances on the performance of the eversense long-term implantable continuous glucose monitoring system. *Diabetes Technology and Therapeutics*, 20(5):344–352, 2018.
40. Rajan Merchant, Rubina Inamdar, Kelly Henderson, Meredith Barrett, Jason G Su, Jesika Riley, David Van Sickle, and David Stempel. Digital health intervention for asthma: Patient-reported value and usability. *JMIR mHealth and uHealth*, 6(6):e133, 2018.
41. Rajan K Merchant, Rubina Inamdar, and Robert C Quade. Effectiveness of population health management using the propeller health asthma platform: A randomized clinical trial. *The Journal of Allergy and Clinical Immunology: In Practice*, 4(3):455–463, 2016.
42. Fatih Ertam and Galip Aydın. Data classification with deep learning using tensorflow. In *Proceedings of Computer Science and Engineering (UBMK)*, pages 755–758, 2017.
43. Zhi-Hua Zhou and Ji Feng. Deep forest: Towards an alternative to deep neural networks. In *Proceedings of the 26th International Joint Conference on Artificial Intelligence, (AAAI)*, pages 3553–3559, 2017.
44. Nagaraj Balakrishnan, Arunkumar Rajendran, Danilo Pelusi, and Vijayakumar Ponnusamy. Deep belief network enhanced intrusion detection system to prevent security breach in the internet of things. *Internet of Things*, 100112, 2019.
45. Yongbin Gao, Xuehao Xiang, Naixue Xiong, Bo Huang, Hyo Jong Lee, Rad Alrifai, Xiaoyan Jiang, and Zhijun Fang. Human action monitoring for healthcare based on deep learning. *IEEE Access*, 6:52277–52285, 2018.
46. Hai Wang and Hoifung Poon. Deep probabilistic logic: A unifying framework for indirect supervision. *arXiv preprint arXiv:1808.08485*, 2018.
47. Muhammad Imran Razzak, Saeeda Naz, and Ahmad Zaib. Deep learning for medical image processing: Overview, challenges and the future. In *Classification in BioApps*, pages 323–350. 2018.
48. Xiaokang Zhou, Wei Liang, I Kevin, Kai Wang, Hao Wang, Laurence T Yang, and Qun Jin. Deep learning enhanced human activity recognition for internet of healthcare things. *IEEE Internet of Things Journal*, 7:6429–6438, 2020.

49. Adhish Prasoon, Kersten Petersen, Christian Igel, François Lauze, Erik Dam, and Mads Nielsen. Deep feature learning for knee cartilage segmentation using a triplanar convolutional neural network. In *International Conference on Medical Image Computing and Computer-Assisted Intervention*, pages 246–253, 2013.
50. SM Zobaed and Mohsen Amini Salehi. Big data in the cloud. In Laurie A. Schintler and Connie L. McNeely, editors, *Encyclopedia of Big Data*. Cham: Springer, 2019, 308–315.
51. SM Zobaed, Sahan Ahmad, Raju Gottumukkala, and Mohsen Amini Salehi. Clustcrypt: Privacy-preserving clustering of unstructured big data in the cloud. In *2019 IEEE 21st International Conference on High Performance Computing and Communications; IEEE 17th International Conference on Smart City; IEEE 5th International Conference on Data Science and Systems (HPCC/SmartCity/DSS)*, pages 609–616. IEEE, 2019.
52. Chao Liang, Bharanidharan Shanmugam, Sami Azam, Asif Karim, Ashraful Islam, Mazdak Zamani, Sanaz Kavianpour, and Norbik Bashah Idris. Intrusion detection system for the internet of things based on blockchain and multi-agent systems. *Electronics*, 9(7):1120, 2020.
53. Gomotsegang Ntehelang, Bassey Isong, Francis Lugayizi, and Nosipho Dladlu. IoT-based big data analytics issues in healthcare. In *Proceedings of the 3rd International Conference on Telecommunications and Communication Engineering*, pages 16–21, 2019.
54. Cloud leak: How a Verizon partner exposed millions of customer accounts. https://www.upguard.com/breaches/verizon-cloud-leak, Accessed June 2, 2020.
55. Every single Yahoo account was hacked -3 billion in all. https://money.cnn.com/2017/10/03/technology/business/yahoo-breach-3-billion-accounts/index.html, Accessed June 2, 2020.
56. The 15 biggest data breaches of the 21st century. https://www.csoonline.com/article/2130877/the-biggest-data-breaches- of-the-21st-century.html, Accessed June 2, 2020.
57. Haneul Ko, Jaewook Lee, and Sangheon Pack. Cg-e2s2: Consistency- guaranteed and energy-efficient sleep scheduling algorithm with data aggregation for IoT. *Future Generation Computer Systems*, 92:1093–1102, 2019.
58. Iman Azimi, Janne Takalo-Mattila, Arman Anzanpour, Amir M Rahmani, Juha-Pekka Soininen, and Pasi Liljeberg. Empowering healthcare IoT systems with hierarchical edge-based deep learning. In *2018 IEEE/ACM International Conference on Connected Health: Applications, Systems and Engineering Technologies (CHASE)*, pages 63–68. IEEE, 2018.
59. Shreshth Tuli, Nipam Basumatary, Sukhpal Singh Gill, Mohsen Kahani, Rajesh Chand Arya, Gurpreet Singh Wander, and Rajkumar Buyya. Healthfog: An ensemble deep learning based smart healthcare system for automatic diagnosis of heart diseases in integrated IoT and fog computing environments. *Journal of Future Generation Computer Systems*, 104:187–200, 2020.
60. Shreshth Tuli, Redowan Mahmud, Shikhar Tuli, and Rajkumar Buyya. Fogbus: A blockchain-based lightweight framework for edge and fog computing. *Journal of Systems and Software*, 154:22–36, 2019.
61. Musaed Alhussein, Ghulam Muhammad, M Shamim Hossain, and Syed Umar Amin. Cognitive IoT-cloud integration for smart healthcare: Case study for epileptic seizure detection and monitoring. *Mobile Networks and Applications*, 23(6):1624–1635, 2018.
62. Jian Yu, Bin Fu, Ao Cao, Zhenqian He, and Di Wu. Edgecnn: A hybrid architecture for agile learning of healthcare data from IoT devices. In *Proceedings of 24th International Conference on Parallel and Distributed Systems (ICPADS)*, 2018.

63. Daniele Ravi, Charence Wong, Benny Lo, and Guang-Zhong Yang. Deep learning for human activity recognition: A resource efficient implementation on low-power devices. In *Proceedings of 13th International Conference on Wearable and Implantable Body Sensor Networks (BSN)*, pages 71–76, 2016.
64. Md Enamul Haque, KM Imtiaz-Ud-Din, Md Muntasir Rahman, and Aminur Rahman. Connected component based ROI selection to improve identification of microcalcification from mammogram images. In 8th International Conference on Electrical and Computer Engineering, 2014.
65. Md Haque, Abdullah Al Kaisan, Mahmudur R Saniat, and Aminur Rahman. GPU accelerated fractal image compression for medical imaging in parallel computing platform. *arXiv preprint arXiv:1404.0774*, 2014.
66. P Mohamed Shakeel, S Baskar, VR Sarma Dhulipala, Sukumar Mishra, and Mustafa Musa Jaber. Maintaining security and privacy in health care system using learning based deep-q-networks. *Journal of Medical Systems*, 42(10):186, 2018.
67. Thaha Muhammed, Rashid Mehmood, Aiiad Albeshri, and Iyad Katib. Ubehealth: A personalized ubiquitous cloud and edge-enabled networked healthcare system for smart cities. *IEEE Access*, 6:32258–32285, 2018.
68. Geethapriya Thamilarasu and Shiven Chawla. Towards deep-learning- driven intrusion detection for the internet of things. *Sensors*, 19(9):1977, 2019.

10 Authentication and Access Control for IoT Devices and Its Applications

Ananda Kumar S.
VIT University

Arthi B. and Aruna M.
SRM Institute of Science and Technology

Aaisha Makkar
Chandigarh University

Uttam Ghosh
Vanderbilt University

CONTENTS

10.1 INTRODUCTION

The uthentication methods for the Internet of Things (IoT). There is a variety of methods implemented in the software industry but not all are based on the concepts of IoT. Various lightweight and mutual authentication and access control methods are covered thoroughly in this chapter, and it establishes the superiority of one method to the other based on the constraints present in the working of IoT devices and the functioning in a particular range of frequencies and bandwidths. It will also discuss various drawbacks present in other authentication and access control processes and how these processes can be made more feasible for efficient working such as the lightweight process. This chapter also explores several gaps present in the IoT which are further dealt with to provide more and more feasible options pertaining to user authentication. The rapid increase in the number of IoT devices is the main reason for the concern of efficiency of present authentication methods [1].

IoT is a large-scale network of various devices. This network involves large amounts of data, which is collected from a variety of environments such as day-to-day lives of people and medical environments. In such cases, the security of private data is very important. Security usually requires authentication and access control. This chapter focuses on various techniques in authentication and access control in IoT for high-security processes. Such techniques have the following advantages:

a. It provides a method for reliable communication between edge devices and end-to-end devices.
b. It deals with end devices having a weaker identity, which is a very common scenario in an IoT system
c. The method for sharing the authentication key is minimal, flexible, convenient, and reliable.
d. It is extensible, as it allows for addition or removal of participants easy.
e. It protects the device from any unauthorized access and ensures protection and security against any replay attacks that might be committed.
f. It provides protection against attacks that might steal data during communication.

The chapter contains the following sections. Section 10.2 covers the various authentication methods in IoT, and Section 10.3 discusses more about access control in IoT.

10.2 AUTHENTICATION IN IoT

Any physical device can be a "thing" (in terms of IoT). The things are equipped with sensors and actuators that assist them with receiving and sending information. These devices are interconnected to the Internet and each other, which facilitates the IoT a giant network of associated "things" where people are the most important for the network as well. There are three sorts of connections in an IoT network: things-things, human-things, and human-human [2]. More devices/gadgets are present in comparison to individual count. IoT devices/gadgets stay associated with the Internet to gather, trade, dissect, and in any event, being engaged with dynamic response generation with no human intercession [3].

IoT architecture embraces a sensor layer [4] to detect information; a network layer to move information between IoT devices and cloud; a cloud layer to store information as discussed in Table 10.1. Authentication, confidentiality, privacy, and authorization of data are the major security issues in IoT layers.

The overall goal of IoT is to empower things to be related whenever, wherever, for esteem included administrations, over advancing interoperable data and correspondence advances [5]. IoT would be used in a wide range of spaces, for example, general wellbeing of public, brilliant urban communities, savvy homes, natural monitory, keen transportation framework, squander the board, military, and agriculture. The standard targets of data security are authentication, confidentiality; integrity; availability; and non-repudiation as shown in Figure 10.1.

TABLE 10.1
Description of IoT Architecture

Layer	Responsible for	Technologies
Perception/physical layer	Physical properties such as temperature, humidity, speed, and location perceived from an environment	RFID, GPS, NFC
Network layer	Obtain data from the perception layer and transmit it to the application layer and data management done using cloud	Bluetooth, Zig-Bee, 3G, 4G, 5G, Wi-Fi
Application layer	Application-specific services are delivered to users	Smart metering, smart homes, smart cities, health care, building automation

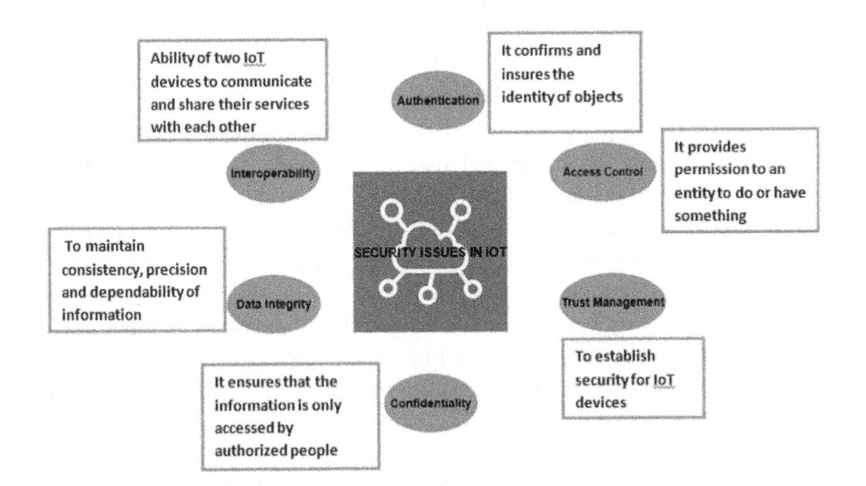

FIGURE 10.1 Security issues in IoT.

At the hardware layer, intruder gains access to IoT hardware and recover keys or security parameters inside IoT devices, and then the intruder could reproduce virtual IoT devices by utilizing the accessed security boundaries [6]. The virtual IoT devices can transfer counterfeit information to server and recover secured data about the client through server or network using a connected IoT device.

In IoT, billions of devices are connected using Internet, which gives an interloper a chance to control the IoT framework for huge scope. The IoT pools all the devices through sensors, and clients can get access to that IoT data wherein security and protection are significant for clients to keep them from enduring many known attacks in an open wireless platform. Authentication is the method for confirming entity's identity, and in IoT, it is one of the major challenging issues since the entities in the IoT environment cannot bear to incorporate cryptographic primitives which have high computational complexity as in customary Internet [7].

10.2.1 Authentication of IoT Devices

The IoT device collects the information using the sensors associated with it and transfers it to the server after processing the data if required. IoT devices communicate with the server using WAN by utilizing Internet and additionally a versatile mobile interface. IoT devices ought to be verified on the grounds that a huge variety of devices are connected through IoT gateway where authentication can guarantee that genuine information is available in networks, and in turn devices that demand that information are authentic. Most IoT devices are lightweight, and their functions are limited to save energy [8]. Using a cloud-based IoT environment, IoT users connect their IoT devices to the cloud platform but all information such as device ID, email, user contact, public IP address of IoT device, time in and time out of IoT device, source, and destination port are stored in the IoT platform itself. IoT devices generally have resource constraints so they outsource information to the cloud, which raises security-related issues such as authentication, integrity, access control, and confidentiality [9]. Using a secure authentication scheme, false identity of data, modification in data, and leakage of data during transmission can be avoided, and it also decreases computational and communication costs [10].

10.2.1.1 Security Issues in IoT Devices

Authentication is the initial phase in any network. Accordingly, being unauthenticated intrudes the whole framework and disregards the other security goals. To make sure about communication among users and servers and furthermore to confirm the identity of the server and user, Public Key Infrastructure (PKI) has been widely used as an authentication scheme [11]. In IoT networks, users interact with devices to access various services and to control IoT devices. An IoT gateway authenticates the entire IoT devices by verifying devices' secret key provided as input and also check their MAC addresses as shown in Figure 10.2. IoT gateway stores access log of Internet and also session log for traceability purposes. Basically, IoT devices are constrained devices with the following characteristics:

- Each IoT devices/gadget is allocated with a unique ID number.
- IoT devices have less computational roles and memory and low energy.

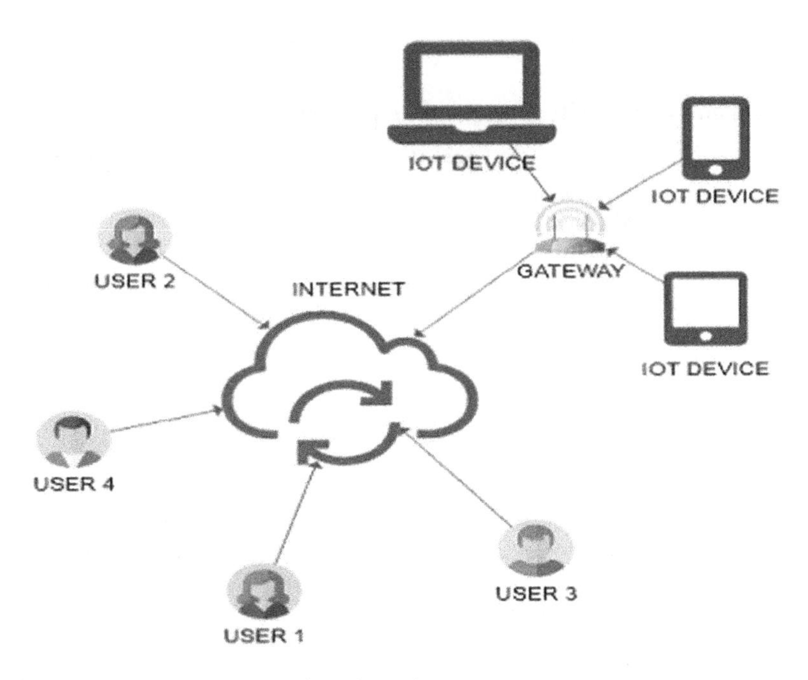

FIGURE 10.2 IoT device connections through gateway.

- An underlying secure vault is put away in IoT devices and is imparted to the server.
- The server has better-secured information base.
- Side-channel attacks performed by a foe.
- All information exchange is done between server and IoT devices over a trustworthy channel, so no information misfortune and each message is conveyed precisely.

During the authentication stage, IoT server and device build a common shared key, known as session key used to encode messages transferred between server and IoT device, and it acts as an encryption key for the message authentication code, which is utilized for message validation. The session key stays unaltered all through a solitary session; however, various session utilizes distinctive session keys. The security issues and requirements at each layer of IoT architecture are shown in Table 10.2.

10.2.1.2 Authentication Schemes in IoT

Generally, password-based authentication scheme is widely used to confirm the validity of a remote user using biometric authentication, smart cards, or one-way hash function [12]. Authentication scheme, avoids man-in-the-middle attacks, but cannot afford high authentication which is required to avoid attacks like DoS attack, stolen-verifier attack, and replay attacks. The three-way mutual authentication approach authenticates IoT server and device, in which devices send a connection request to the server to obtain a challenge [6]. Then the device sends back an authentication challenge to the IoT server for verification.

TABLE 10.2
IoT Layers and Its Security Issues and Requirements

IoT Layer	Security Issues	Security Requirement
Perception layer	Node capture	Lightweight encryption
	Mass node authentication	Authentication
	Routing threats	Key agreement
	Denial of Service (DoS) attack	Data confidentiality
	Distributed Denial of Service (DDoS) attack	
	Denial of sleep attack	
	Side-channel attack	
	Fake node/Sybil attack	
	Replay attack	
Network layer	Man-in-the-Middle (MITM)	Communication Security
	Denial of Service (DoS)	Routing security
	Eavesdropping/sniffing	Authentication
	Routing attacks	Key management
		Intrusion detection
Application layer	Data accessibility and authentication	Authentication
	Data privacy and identity	Privacy protection
	Availability of big data	Information security management

In a web-based authentication scheme or Extensible Authentication Protocol (EAP), users are identified using their usernames and passwords, but it is challenging for IoT devices to track all details since it is in large number [13]. In order to use a Wi-Fi Protected Access 2 (WPA2) approach, a shared secret key is used for authentication. As this approach, one secret key is used to access all devices, and it is less secured and more vulnerable to attacks, so it requires some security policy for user identification.

Continuous user authentication can be provided using contextual information like user's login details, their requests, user access durations and location, based on both IP and Bluetooth [14] shown in Figure 10.3.

In a group-based authentication scheme with access aggregation, a huge variety of devices form a device group and then group leader is selected among all the devices. At an instance of time, if any of the devices in the group wants to connect to the network, then each device has to send an access request message to the device group leader which aggregates all the access authentication request messages and sends them back to the network. Then device group leader generates an aggregate message authentication code (AMAC) for the verification of device group.

IoT systems use a light-weight user authentication scheme referred as Li-GAT (Lightweight Gait Authentication Technique) in which the authentication of users is done by extracting and identifying different walking patterns of users (gait) that build information about each user's behavior, analyze, and take actions accordingly [15,16]. Li-GAT exploits collected data from IoT devices, namely the intuitive level of user activities, authenticate the users with high accuracy and less resources [17].

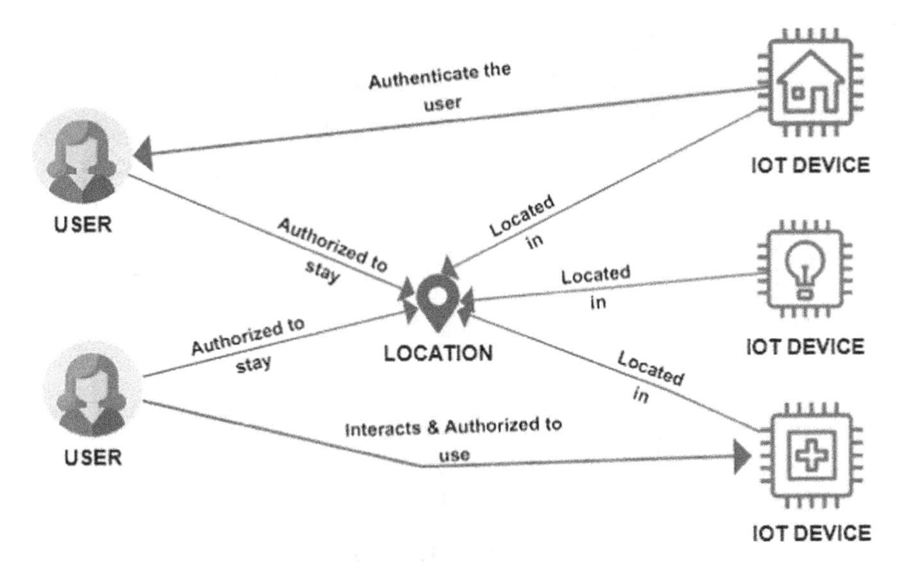

FIGURE 10.3 Continuous user authentication scheme.

PUF (Physical Unclonable Function) is a technology that allows each of the IoT devices to have distinct characteristics, such as human fingerprint or biometric information, which allows a device to have various characteristics even if it is made with a similar process [18]. A more secure device authentication done using PUF technology [19] should possess a unique identifier for identifying each device which can be generated by itself without being obtained from the outside.

10.2.2 Authentication in IoT-Based Applications

IoT devices coordinate and communicate with sensor devices by monitoring and modifying cyber and physical indicators, and actuators to provide authenticated access. A robust device authentication scheme is essential before establishing the connectivity with any IoT device which enhances the importance of exchange-to-exchange (E2E) data confidentiality and data privacy in industry, healthcare, and financial businesses [20]. Various IoT applications are shown in Figure 10.4. Cybercriminals access smart homes through router since it is more vulnerable to malware and harmful to the control of IoT devices. Secured smart home [21] can be built with Wi-Fi router that plays a significant role as it connects all connected IoT devices and helps them to operate. An efficient algorithm can be used to protect agricultural data such as weather condition, raindrop, temperature, humidity level of soil and atmosphere, soil chemical compositions, dioxide, and acidity from an unauthorized access. LPWANs in smart agriculture are ultimate in gathering agricultural data and environmental conditions with low energy consumption and cover long-range transmission.

10.2.2.1 Authentication in IoT-Based Smart Cities

Smart cities stay connected through IoT devices to manage traffic by smart parking and traffic management, vehicle diagnostic/monitoring, handling pollution through

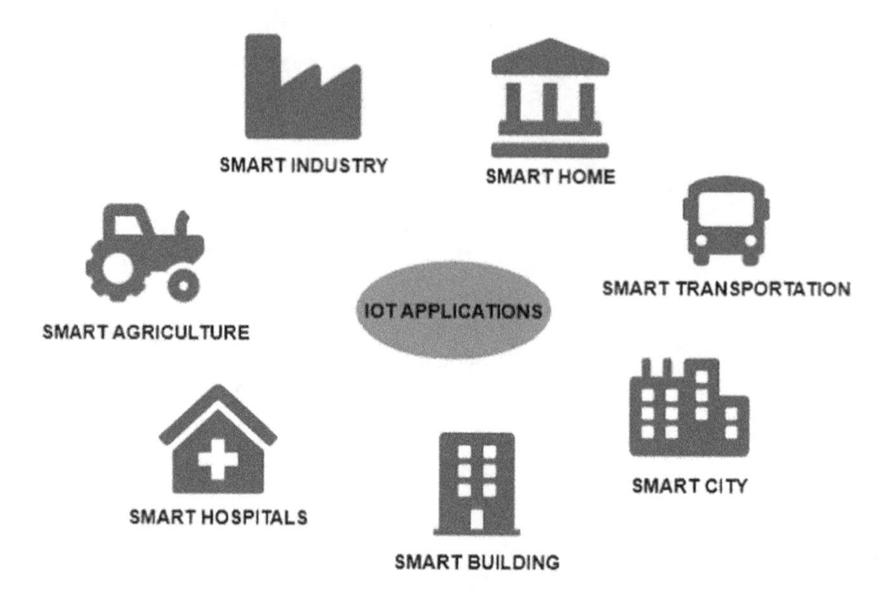

FIGURE 10.4 Applications of IoT.

environmental monitoring, enhancing protection by video surveillance, adopting smart wastage system, and many more advanced technologies. Fingerprint authentication in smart buildings acts as the chain of trust as the registration is done once a user logged in and provides facilities like lighting or temperature over interconnected IoT devices so that it can authenticate and identify each user based on personal information using a biometric ID or city card.

Behavior-based user authentication approaches can be used as smart home centralized hubs and commercial hubs which help in monitoring user's behavior that describes when, where, how, and with what the devices were used, and also help in authenticating user access. User behavior profiling can be categorized into network or host-based. Network-based scheme follows user calling pattern and migration behavior over the Internet service provider for developing a user behavior profile wherein host-based method captures hypothesis of mobile users and their apps' usage patterns at different periods in different locations.

10.2.2.2 Authentication in IoT-Based Smart Healthcare

Smart healthcare provides monitoring networked medical devices and health team management and enhancing routine healthcare workflow operations such as patient monitoring and medication management [22]. Medical devices are highly interconnected and automatically connected to other IoT devices for accessing patient sensitive data records that can be breached by malicious attackers by illegal access to medical system and patient data [23,24]. Consequently, access control in the smart healthcare should ensure authentication of medical representatives who handle medical devices to prevent unauthorized users access to avoid the formulation of critical medical system.

Objects Identification is the premier necessity of any sort of IoT innovation. Using object ID, IoT framework perceive individuals effectively and find them by perceiving their face, gadgets, and gestures with similar ID and can likewise know about insecure conditions [25]. Google maps observe the traffic on street using object identification by watching the time, spot, and number of gadgets accessible on a similar spot.

10.2.2.3 Authentication for IIoT

In the near future, 5G network that provides wide coverage, large capacity, low-power consumption, and low cost are used in Narrow Band Internet of Things (NB-IoT) system which become an significant part of the Internet of Everything [26]. Remote maintenance of machines or systems is done by developing an IIoT system using unique IoT device identifiers that avoid data breach. NB-IoT provides limited resources, data-centric and closely related applications, topology changes dynamically, and complex network environment that can be broadly used in smart agriculture, intelligent parking, remote meter reading, asset tracking, and so on. In order to ensure the security, it requires an effective access authentication and data distribution schemes. Smart maintenance is done using IIoT gateway for a secured communication requires more flexible and protected access to plant/machines/systems to which all devices are connected. Authentication of IoT devices are done using a Public Key Infrastructure (PKI) where all the machine certificates are managed.

10.2.3 Deep Learning–Based Authentication Techniques

10.2.3.1 Neural Networks for ECG-Based Authentication

In this model of deep learning, neural networks are used for ECG-based authentication. The author proposes the solution for authentication using the biometrics. The ECG-based authentication is very recent, and the ECG represents unique characteristics of heart of each user. The basic model that author proposed is first of all ECG must be preprocessed in order to remove noise or distortion and then features are extracted from it and these features are used as an input to deep learning model for classification. These results are compared to find the most effective transformation-classifier combination. Also, deep learning neural networks are utilized so as to make arrangement models that predict if an ECG has a place with a particular individual or not, that is basically to authenticate in light of the mix of the capabilities delivered in the change step. The authors have used deep learning model for the better accuracy [27].

The work states that the ECG is basically the electrical activity of heart over the period of time and records through the electrodes placed on the body surface. The authors have mentioned that the two approaches use ECG for authentication. First one is fiducial approach. This approach detects the process and classify those features for human identification. The other approach is non-fiducial. This method is mainly based on waveform and morphology and usually process signal in the frequency domain. The major consideration is choosing the data set [28]. The data set used in their work is Physikalisch-Technische Bundesanstalt (PTB) Diagnostic ECG

Database which contains 549 records from 290 subjects. Then next followed is preprocessing in order to remove noise and other anomalies in order to obtain accurate features. The feature extraction is done using various methods like Fourier transform and discrete cosine transform, followed by data classification. The most suitable and productive methodology is to utilize the ECG signal after the subject has been recognized so as to demonstrate the legitimacy of its character. So, in terms of characterization, the issue is to be understood as two classes grouping issue, with the two potential categories to determine whether the person is genuine or not. The four classification methods were used by the authors such as KNN, MLP, RBFN and random forest. The deep learning method is used for the speculation that the profound neural nets would have the option to recognize a noteworthy subset of the 60 coefficients that would empower the effective arrangement of ECG signals as per the individual they had a place with [29]. By using the deep learning, the author is able to achieve the accuracy from 86% to 97%. As far as grouping strategies, random forest classifier appears to perform in a way that is better than any of different techniques. The accuracy seems to be somewhat less as the size of the data set used is very less [30].

10.2.3.2 Convolutional Neural Networks – Deep Face Biometric Authentication

Authentication is a very important factor for any type of security in network. Authentication enforces access control, and the work is based on facial recognition authentication which is based on deep learning. In this research project, the authors propose on using Convolutional Neural Networks (CNN) [31]. They propose a CNN-based facial recognition which uses the component of Viola Jones face detector for detecting facial features from the provided input, and this uses already existing features of the faces which are stored in the database of the CNN to match and cross reference. This database contains images of the authorized personnel in different and various illumination conditions and face postures, and this proposed system for automated facial recognition authentication provides a very high success rate of 98.7%. The existing factors in this authentication systems are as follows:

A. **Processing Speed and Storage**: These systems require high storage size as it needs to store a vast number of images in the dataset and surfing through them also requires a certified amount of time and this is being dealt with using cluster computing.
B. **Image Size and Quality**: This is another important factor that affects the success rate is the matching ratio of the input quality with the dataset images.
C. **Surveillance Angle**: The angle of capturing the input image might vary with the dataset images.
D. **Light Variations**: The light variations must be accounted for while building a dataset as it might be different in the stored images and might cause a failure.
E. **Inter Class Variability**: This basically means that the resemblance among the dataset images might trigger false positives and this factor also must be accounted for.

There are multiple methods and approaches for this particular problem, and we are focusing on the deep learning methodology in this proposed solution. The proposed solution comprises the following:

- **Data Specifications**: This basically means building a dataset against which the software/system would cross reference with and verify with. This can be done by taking multiple images of the clints under different illuminations and postures and angles of their faces.
- **Face Detection**: Once the input has been provided to the software, the face detection is done using Viola Jones Detection system.
- **Data Augmentation**: For the purpose of increasing efficiency, dataset samples are mimicked and modified by using variables such as brightness and contracts so more dataset values to cross-reference to. This increases the dataset exponentially to increase success rate.
- **Facial Recognition**: This is obtained by using a trained CNN. A CNN has multiple layers, and the first layer, the input layer, takes care of the input size, height, and width.

The convolutional channel takes care of the feature extraction using filters. The batch layer focusses on speeding up learning. Pool layer performs down sampling. The connected layer identifies patterns, the last connected layer then fixes the output size, and the SoftMax layer converts the activations to probabilities. Finally, the classification layer uses this to provide as an input to a mutually exclusive class. Clearly, training a CNN is a rather time consuming and a complex process. They then trained multiple CNN models and had them all tested against a similar dataset, and the CNN model ResNet50 had the highest accuracy and efficiency. Each of the training models differs in their efficiency based on the application present at hand, and they could tailor the models based on their application that they faced and required such a data retrieval, attendance, etc.

We conclude that this research paper mainly focus on solving this growing problem at hand sing the deep learning solution which was to implement a Trained CNN model and use the CNNs for facial recognition and this has shown out to have a huge dataset for a small number of clients, therefore we sacrifice space and memory for the efficiency for the system here [32].

10.3 ACCESS CONTROL IN IoT

The field of IoT is growing rapidly day by day. Along with this swift evolution in IoT, ample Information Technology (IT)-based services and applications are also being developed which enhance the usage of IoT services and provide comfort to users. IoT has a great impact on our everyday life. It plays a vital role in improvising the day-to-day activities and hence has become an incipient topic of research. For enticing and achieving applications and services for the future, IoT uses the underlying proven fundamental principles and protocols. In addition, cloud computing technologies are progressing to support various applications and services that are developed on demand with data sharing competences, extreme flexibility both in storage and computation, etc. [9].

This assimilation of IoT with cloud computing technologies not only improves the performance, feasibility, integrity, and security of the existing legacy systems but also paves a way to develop framework and models for upcoming future applications.

IoT is an empowering factor that adds intelligence to many core functions of the contemporary world, and its features are used in all sectors like cities, hospitals, businesses, and constructions. The main challenge that is a constraint to the usage of IoT is the security and privacy. This is the major issues that hinders the extensive acceptance of the IoT. The storage and processing abilities of these devices are usually restricted, and their design is currently focused on ensuring functionality and to a large extent ignore other requirements, including security and privacy issues. Access control in IoT is an attractive research theme in computer security. Many research works are being conducted for this security challenges in IoT, and a few solutions involve designating encryption keys to users. However, the evolving IoT field and its unique applications require innovative and unique security solutions [33].

Most of the current security solutions involve too many calculations and are considered to be expensive. Conventional algorithms that are available are actually difficult to implement for many upcoming IoT devices or groups of IoT devices [34]. IoT requires an adaptable and reliable security access control structure to certify the security and privacy features of the IoT devices. A highly compatible, adaptable, and reliable trustworthy access control framework is essential to guarantee the privacy and security of trivial IoT devices. The standard access control model that has a centralized work structure cannot support the upcoming or evolving substantial and uncluttered IoT applications and devices [35].

One of the major challenges in IoT environment is maintaining data integrity and legitimacy of data sources. In addition, if the attribution metadata itself can be transmitted in a way that protects privacy, it is possible to extend the use of IoT systems to the territories of human society where privacy is critical. There are few investigations that provide a framework to merge data attribution and security and privacy solutions. Such research investigations have developed a few protocols for security and privacy preservation [36].

10.3.1 Blockchain-Based Access Control in IoT

With the swift development of smart devices, IoT has received enormous attention in recent years. The aim of the IoT is to efficiently integrate the existing physical world and the Internet through the available existing network infrastructure to promote data sharing between smart devices. However, the intricate and exhaustive network framework poses different security and privacy constraints to the IoT system. In order to guarantee data security, the conventional access control methods and frameworks that are available are not appropriate or adaptable for direct access control in IoT systems due to its complex access management schemes and absence of integrity [37,38].

Blockchain is another important subject that is constantly evolving among the technology and the entrepreneurial society. Blockchain is a freely accessible, visible, and distributed record that tracks all transactions between two groups very effectively which can be traceable, and which is permanent. All the transactions that are

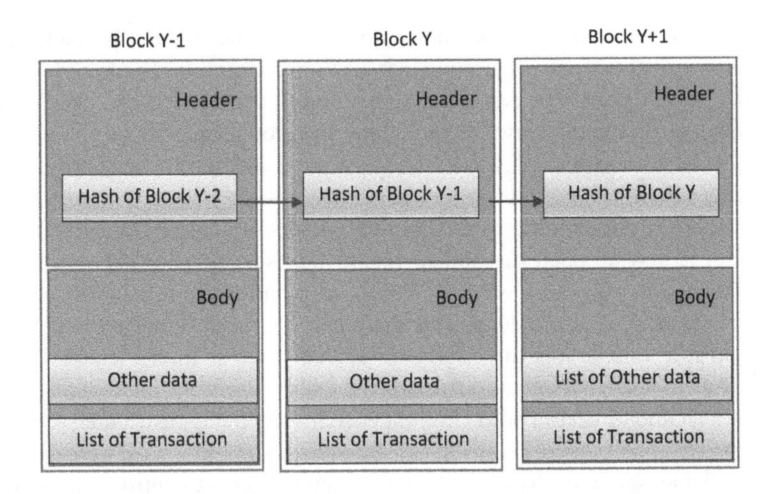

FIGURE 10.5 Blockchain structure.

recorded cannot be manipulated in blockchain unless there is a need [16,30,39,40]. Figure 10.5 represents the structure of blockchain. It is a series of blocks that are connected to each other by hash value. Each block is made up of two parts: the block header, and the block body part. The block body encompasses every transaction that are involved in each block.

Basically, blockchain can be categorized into three types: public blockchain, hybrid blockchain, and private blockchain. Table 10.3 shows the basic block chain

TABLE 10.3
Types of Blockchain and Its Characteristics

Types	Characteristics
Public	1. Completely decentralized 2. Transactions recorded in blockchain are not controlled by any entity 3. Order of the transaction recording is not controlled by any entity 4. Anyone can join the network 5. They have token associated with them
Private	1. More centralized 2. Not anyone can join the network, and they need proper consent 3. All the transactions are made private 4. The transaction is available to participants who are allowed to join the network 5. Also called permissioned blockchains
Hybrid	1. Hybrid in nature 2. Combines the features of both private and public blockchain 3. Combines transparency feature of public blockchain with privacy feature of private blockchain 4. Significantly flexible compared to the other two types 5. Easily compatible with the other blockchain protocols

categories and its characteristics. The combination of the IoT and blockchain technology is one of the promising trends that ensures the credibility of the IoT system and reduces its overall expenses. This helps the IoT build a decentralized, reliable, and widely certifiable database, so that many interconnected things can attain disseminated confidence out of it. Hence, as a favorable and optimistic distributed technology, blockchain can provide various solutions to solve many IoT challenges.

Figure 10.6 represents the blockchain-based access control framework for IoT. Blockchain is a wide-open distributed record that successfully logs exchanges in an irrefutable and perpetual way and keeps information from being messed with. A shrewd agreement is a mechanized exchange convention that is executed on a blockchain as a content. It permits authentic exchanges to be implemented without an outsider. Since blockchain gives a trustworthy environment for intelligent contracts, blockchain technology has a powerful influence in employing distributed responsible access control. Thus, the usage of these technologies for IoT access control means that a verifiable, secured, distributed, and ingenuous access control concept can be

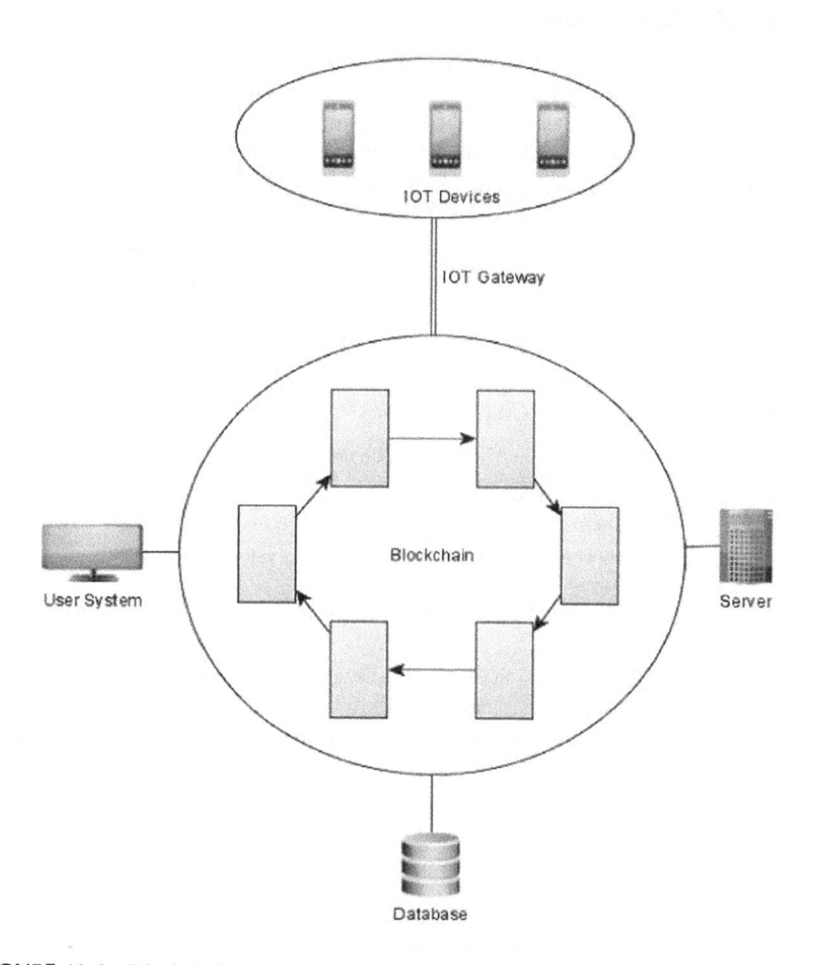

FIGURE 10.6 Blockchain-based access control for IOT.

implemented [35]. Many of the access control methods encounter a typical issue that a tenable focus is required for guaranteeing trust. As IoT devices and its applications are disseminated globally, it is not possible to manage them ventrally. Usually, each IoT device is administered by the specialists close to it [41,42].

Cao and Li proposed an innovative model for access control for IoT systems which is attribute-based using blockchain. With this model, there is no need to designate roles for all nodes in the system. Each device in the network is termed by a set of properties or attributes that are defined evidently in the system, and these attributes are published by the attribute authorities based on its identity or capability. The access control is based on the set attributes available that meet the access policy. If there are not enough attributes as per the access control policy, then the access is denied. The dissemination of attributes is taken care by blockchain technology. Being a smart technology, IoT needs an innovative access control framework that is compatible and adaptable to its precise requirements and characteristics. IoT access control should be so flexible that it should allow the users to control their security and privacy [37].

10.3.2 Access Control for IoT Applications

While we are considering security for IoT applications, we need to review and ensure it from various perspectives. The imminent threat that can be foreseen in large-scale infrastructure development such as smart cities, transport system, and power grids is the system being comprised of malicious intent by foreign agents.

The critical information in a system can be accessed by hacking IoT devices which are part of the system. Hence, with the expansion of the IoT-based system and application, the threat from hackers of critical information is also in rise [9].

A comprehensive approach is required toward protecting the IT application from possible intrusion and exploitation of data theft.

- Strong and unique passwords of all devices should be used to ensure security, and the same needs to change at frequent intervals.
- All the data before being transmitted between applications, the controllers, and the devices should be encrypted.
- Role-based access system to be provided for network and application access.

10.3.2.1 Access Control for IoT-Based Healthcare Applications

IoT devices play an important role in applications developed for healthcare. IoT devices collect data for the healthcare application which can be measured and analyzed. With IoT devices being vulnerable to data breach, protecting the data of the healthcare application is important.

With the advent of medical sensors and mobile devices, the healthcare industry is revolutionized and witnessing enormous growth in the medical and healthcare industry. IoT has been the game changer in the technological advancement witnessed in the healthcare industry.

The medical devices are interrelated and organized to exchange the data; these have brought in medical services like remote patient monitoring in reality. Healthcare organization is yet to allocate budgets to have enough resources to protect the privacy and security of patient information [9]. It is evident that in Medical Internet of Things (MIoT) data security and privacy are critical and play a key role. MIoT devices are generating a high volume of real-time data, which are very sensitive. Destroying such sensitive data could be disastrous. Managing the privacy of patient information at various stages like data collection, data transmission, and data storage is critical. Even though IoT is considered to be a paradigm shift in the healthcare sector, the privacy and security of the patient's information are considered as the most significant issue which needs to be addressed in a foolproof way. The best way to restrict the patient data is by restricting the user's access to such information. This can be achieved by providing access control-based usage by authorized users in the system. For IoT applications using a fog-based framework, access control functions are disseminated to make policy decision, and at the edges of the network near the end node, policy information mechanisms are used. By doing so, it increases the availability of policy decision and decreases the latency [24,43].

10.3.2.2 Access Control for IoT-Based Smart Home

Smart homes are the new norm in many countries. It has become a mandate to have security system and appliance management applications. This is achieved by having less power utilization and cost-effective embedded access control system. These access control systems enable end users to remotely monitor various home appliances and the security system like CCTV and intrusion alarms system. Today most of the appliance have a microprocessor in it, and each appliance has a user interface that allows users to manage appliances. In many cases, such interfaces are built on a complex framework, and users find difficulty in using appliances. A separate user interface device that a user can carry while on the move will help them in managing appliances. This can be achieved by having the interface on the smartphones which has the capability to run different applications on a single platform and communicate to various appliances and security systems. An abstract specification language for the description of home appliances can be used along with a two-way transmission protocol and a software that allows user interfaces to be customized with various home appliances and other systems that they are using. The essential or vital component of home security is ensuring all visitors at the door are detected and providing permission based on the approval provided by users using their smartphones. The records of visitors at the entrance are created by capturing the image and sending it to the owner to authorize the person's entry. Similarly, any intrusion is detected , then the alarm is triggered at the physical point and on the user mobile for them to action out on the security breach. Today various intelligent, low-cost, high-performance portable IoT devices are available in the market for smarter homes [44].

10.3.2.3 Access Control for IoT-Based Smart City

Smart cities emerged with the advancement of IoT devices and application in the space of infrastructure management. Using IoT devices, managing infrastructure, traffic, and pollution control has become a reality. In smart cities, an IoT network needs to relate to multiple devices with various storage and computing capabilities. During such a large scale of data gathering, a low latency network is required with access control to such systems. The data collected on these applications are heterogenous and are high in volume. This becomes a significant concern in managing the security and privacy of the data. This also brings in significant challenges to manage the access control that is provided to third party and other external organizations involved in the system.

In a large-scale project, the resource owner needs to keep full control of the system. This is achieved using structural relationships-based access control (SRBAC) model, where specific third party and external organization can be delegated access control rights. The delegation of the access control to the third party and external organization is done using smart contract and public blockchain, while the internal stakeholders will have full control of the system using local blockchain, thus providing a specific access control for resources [45].

10.3.2.4 Access Control for IoT-Based Vehicle Tracking

In the space of transportation, we have seen a greater influence of technology on the modes of transport. In this technological advancement, IoT has a greater role to play in monitoring and tracking the vehicle. One such application of IoT-based monitoring can be seen in fuel station where trucks entry and exit are monitored using RFID readers and tags. These IoT systems use Radio Frequency Identification (RFID) technologies [46]. The RFID tags act as the transponder and the reader as the receiver and work on radio frequency to send and receive information. The RFID tags are assigned to each truck, and during the entry and the exit, the reader captures the logs. This is considered better than barcode which works on optical AIDC technology [47]. Using the card-based access control system, vehicles entry and exit are restricted based on the access provided and also the logs of the vehicular movement are captured.

With the emergence of IoT, there has been more research from academia and industry on the secure localization of vehicles. Modern vehicles are equipped with microprocessors and devices that can connect with cellular networks, such as 5G network, which have the capability to communicate with other vehicles. Existing challenges in the secure localization and positions of the vehicle are further magnified by innovation such as driverless vehicles which use the 5G network, satellite, and IoT. IoT-based secure location techniques use machine learning, access control mechanism, and schematic segmentation. Access control providers grant the access to the secured information to the authorized user and restrict unauthorized access. This becomes vital to be implemented in military or government vehicles. Earlier the security in such establishment was managed through static Conflict of Interest (COI). However, these COI have administrative overloads leading to poor latency for execution which is not tolerable in IoT device communication. Considering the constraint, a hybrid access control system can be used which works on dynamic COI [48].

10.4 CONCLUSION

We have done an in-depth study on the importance of security in IoT. This chapter concludes with various authentication and access control protocols used in IoT and its applications. The performance of authentication protocol is varying for IoT-based applications. The extensive authentication protocols are studied for IoT healthcare applications, smart home applications, and industrial applications. The idea is to apply lightweight mechanism authentication for the communication between machines. The devices that are made to communicate are smart sensors and routers with a cryptography process that is secure. The chapter discussed the advanced authentication protocols to avoid the various attacks. Another major challenge in IoT is access control due to the restricted working of traditional access control in IoT applications because of a low computational process. The latest technique blockchain-based access control contains more security features. Attributed-based access control is more flexible in IoT applications. Dynamic-based access control is used for highly dynamic applications in IoT environment which attains faster than other methods.

REFERENCES

1. Mahmud Hossain and Ragib Hasan, "Boot-IoT: A Privacy-Aware Authentication Scheme for Secure Bootstrapping of IoT Nodes 2017," *IEEE International Congress on Internet of Things*, pp. 1–8, 2017.
2. Neeraj Kumar, and Aaisha Makkar. *Machine Learning in Cognitive IoT.* CRC Press, Boca Raton, 2020. doi: 10.1201/9780429342615.
3. M. El-hajj, A. Fadlallah, M. Chamoun, and A. Serhrouchni, "Ethereum for Secure Authentication of IoT Using Pre-Shared Keys (PSKs)," 2019 International Conference on Wireless Networks and Mobile Communications (WINCOM), pp. 1–7, 2019.
4. Guntuku, "Secure Authentication Scheme for Internet of Things in Cloud," 3rd International Conference On Internet of Things: Smart Innovation and Usages (IoT-SIU), pp. 1–7, 2018, doi: 10.1109/IoT-SIU.2018.8519890.
5. D. Wang, B. Da, J. Li, and R. Li, "IBS Enabled Authentication for IoT in ION Framework," Global Internet of Things Summit (GIoTS), pp. 1–6, 2017.
6. T. Shah and S. Venkatesan, "Authentication of IoT Device and IoT Server Using Secure Vaults," 17th IEEE International Conference On Trust, Security And Privacy In Computing And Communications/12th IEEE International Conference On Big Data Science And Engineering (TrustCom/BigDataSE) (IEEE), pp. 819–824, 2018.
7. K. S. Roy, "A Survey on Authentication Schemes in IoT," *2017 International Conference on Information Technology A Survey on Authentication Schemes in IoT*, pp. 2–7, 2017.
8. S. Choi, J.-S. Ko, and J. Kwak, "A Study on IoT Device Authentication Protocol for High Speed and Lightweight," International Conference on Platform Technology and Service (PlatCon), pp. 1–5, 2019.
9. B. Gupta and M. Quamara, "An Identity Based Access Control and Mutual Authentication Framework for Distributed Cloud Computing Services in IoT Environment Using Smart Cards," *Procedia Computer Science*, vol. 132, pp. 189–197, 2018.
10. Aaisha Makkar, Sahil Garg, Neeraj Kumar, M. Shamim Hossain, Ahmed Ghoneim, and Mubarak Alrashoud, "An Efficient Spam Detection Technique for IoT Devices using Machine Learning," *IEEE Transactions on Industrial Informatics*, vol. 17, pp. 903–912, 2020.
11. M. Loske, L. Rothe, and D. G. Gertler, "Context-Aware Authentication: State-of-the-Art Evaluation and Adaption to the IIoT," 2019 IEEE 5th World Forum on Internet of Things (WF-IoT), pp. 64–69, 2019.
12. J. Cui, Z. Zhang, H. Li, and R. Sui, "An Improved User Authentication Protocol for IoT," Presented at the International Conference on Cyber-enabled distributed computing and knowledge discovery (CyberC), pp. 2018–2021, 2018.

13. E. Rattanalerdnusorn, P. Thaenkaew, C. Vorakulpipat, "Security Implementation for Authentication in IoT Environments," 2019 IEEE 4th International Conference on Computer and Communication Systems (ICCCS), pp. 678–681, 2019.
14. Y. Ashibani, and Q.H. Mahmoud, "A Behavior Profiling Model For User Authentication in IoT Networks Based on App Usage Patterns," In Proceedings of the 44th Annual Conference of the IEEE Industrial Electronics Society (IECON), Washington, DC, USA, pp. 2841–2846, 21–23 October 2018.
15. Jin-Hee Han and JeongNyeo Kim, "A Lightweight Authentication Mechanism between IoT Devices," In *2017 International Conference on Information and Communication Technology Convergence (ICTC)*, IEEE, pp. 1153–1155, 2017.
16. P. Musale, D. Baek, and B. J. Choi, "Lightweight Gait based Authentication Technique for IoT using Subconscious Level Activities," *International Journal of Engineering Research & Technology (IJERT)*, vol. 5, pp. 564–567, 2020.
17. Z. Abbas, S. M. Sajjad, and H. J. Hadi, "Light Weight Secure Authentication for Accessing IoT Application Resources," 2019 22nd International Multitopic Conference (INMIC), pp. 1–5, 2019.
18. C. Lipps, A. Weinand, D. Krummacker, C. Fischer, and H. D. Schotten, "Proof of Concept for IoT Device Authentication Based on SRAM PUFs Using ATMEGA 2560-MCU," In Proceedings of 2018 1st International Conference on Data Intelligence and Security (ICDIS), South Padre Island, TX, 2018.
19. B. Kim, S. Yoon, Y. Kang, and D. Choi, "PUF based IoT Device Authentication Scheme," International Conference on Information and Communication Technology Convergence (ICTC), pp. 2019–2021, 2019.
20. P. Hao and X. Wang, "A Collaborative PHY-Aided Technique for End-to-End IoT Device Authentication," *IEEE Access*, vol. 6, pp. 42279–42293, 2018.
21. D. K. Sharma, N. Baghel, S. Agarwal, "Multiple Degree Authentication in Sensible Homes based on IoT Device Vulnerability," 2020 International Conference on Power Electronics & IoT Applications in Renewable Energy and its Control (PARC), pp. 539–543, 2020.
22. Rehman, N. A. Saqib, S. M. Danial, and S. H. Ahmed, "ECG Based Authentication for Remote Patient Monitoring in IoT by Wavelets and Template Matching," In Proceedings of the 2017 8th IEEE International Conference on Software Engineering and Service Science (ICSESS), Beijing, China, pp. 91–94, 24–26 November 2017.
23. M. Almulhim and N. Zaman, "Proposing Secure and Lightweight Authentication Scheme for IoT Based E-Health Applications," 20th International Conference on Advanced Communication Technology (ICACT), pp. 481–487, 2018. doi: 10.23919/ICACT.2018.8323802.
24. Akhilendra Pratap Singh, Nihar Ranjan Pradhan, Shivanshu Agnihotri, Nz Jhanjhi, Sahil Verma, Uttam Ghosh, and Ds Roy, "A Novel Patient-Centric Architectural Framework for Blockchain-Enabled Healthcare Applications," In *IEEE Transactions on Industrial Informatics*, 2020. doi: 10.1109/TII.2020.3037889.
25. K. Verma and N. Jain, "IoT Object Authentication for Cyber Security: Securing Internet with Artificial intelligence," IEEE International Students' Conference on Electrical, Electronics and Computer Science (SCEECS), Bhopal, pp. 1–3, 2018.
26. P. Yu, J. Cao, H. Li, B. Niu, and F. Li, "Quantum-Resistance Authentication and Data Transmission Scheme for NB-IoT in 3GPP 5G Networks," IEEE Wireless Communications and Networking Conference IEEE WCNC, pp. 1–7, 2019.
27. A. Makkar, N. Kumar, A. Y. Zomaya, and S. Dhiman, "SPAMI: A Cognitive Spam Protector for Advertisement Malicious Images," *Information Sciences*, vol. 540, pp. 17–37, 2020.
28. A. Makkar and N. Kumar, "Cognitive Spammer: A Framework for Pagerank Analysis with Split by Over-Sampling and Train by Under-Fitting," *Future Generation Computer Systems*, vol. 90, pp. 381–404, 2019.

29. A. Makkar and N. Kumar, "User Behavior Analysis-Based Smart Energy Management for Webpage Ranking: Learning Automata-Based Solution," *Sustainable Computing: Informatics and Systems*, vol. 20, pp. 174–191, 2018.
30. Ilias Chamatidis, Aggeliki Katsika, and Georgios Spathoulas, "Using Deep Learning Neural Networks for ECG Based Authentication," In *2017 International Carnahan Conference on Security Technology (ICCST)*, pp. 1–6. IEEE, 2017.
31. A. Makkar and N. Kumar, "An Efficient Deep Learning-Based Scheme for Web Spam Detection in IoT Environment," *Future Generation Computer Systems*, vol. 108, pp. 467–487, 2020.
32. Maheen Zulfiqar, Fatima Syed, Muhammad Jaleed Khan, and Khurram Khurshid, "Deep Face Recognition for Biometric Authentication," In *2019 International Conference on Electrical, Communication, and Computer Engineering (ICECCE)*, pp. 1–6. IEEE, 2019.
33. P. Z. Sotenga and K. Djouani, "Media Access Control in Large-Scale Internet of Things: A Review," *IEEE Access*, vol. 8, pp. 55834–55859, 2020.
34. H. Chen, C. Chang, and F. Leu, "Implement of Agent with Role-Based Hierarchy Access Control for Secure Grouping IoTs," In Proceedings of the 14th IEEE Annual Consumer Communications & Networking Conference (CCNC), Las Vegas, NV, USA, pp. 120–125, January 2017.
35. P. Wang, Y. Yue, W. Sun, and J. Liu, "An Attribute-Based Distributed Access Control for Blockchain-Enabled IoT," In Proceedings of the International Conference on Wireless and Mobile Computing, Networking and Communications (WiMob), Barcelona, Spain, pp. 1–6, 21–23 October 2019.
36. Hamadeh, "Privacy Preserving Data Provenance Model Based on PUF for Secure Internet of Things," 2019 IEEE International Symposium on Smart Electronic Systems (iSES), pp. 189–194, 2019.
37. J. I. N. Cao and C. Li, "A Novel Attribute-Based Access Control Scheme Using Blockchain for IoT," *IEEE Access*, vol. 7, pp. 38431–38441, 2019.
38. S. M. S. Hosen, Saurabh Singh, Pradip Kumar Sharma, Uttam Ghosh, Jin Wang, In-Ho Ra, and Gi Hwan Cho, "Blockchain-Based Transaction Validation Protocol for a Secure Distributed IoT Network," *IEEE Access*, vol. 8, pp. 117266–117277, 2020.
39. A. Ouaddah, A. Elkalam, and A. A. Ouahman, "FairAccess: A New Blockchain-Based Access Control Framework for the Internet of Things," *Security and Communication Networks*, 9, pp. 5943–5964, 2016.
40. L. Liu, H. Wang, and Y. Zhang, "Secure IoT Data Outsourcing with Aggregate Statistics and Fine-Grained Access Control," *IEEE Access*, vol. 8, pp. 95057–95067, 2020.
41. N. Liu, D. Han, and D. U. N. Li, "Fabric-IoT: A Blockchain-Based Access Control System in IoT," in IEEE Access, vol. 8, 2020.
42. A. Malik, D. K. Tosh, and U. Ghosh, "Non-Intrusive Deployment of Blockchain in Establishing Cyber-Infrastructure for Smart City," In *2019 16th Annual IEEE International Conference on Sensing, Communication, and Networking (SECON)*, Boston, MA, USA, 2019.
43. Wencheng Sun, Zhiping Cai, Yangyang Li, Fang Liu, Shengqun Fang, and Guoyan Wang, "Security and Privacy in the Medical Internet of Things: A Review," *Hindawi Security and Communication Networks*, vol. 2018, pp. 1–9, 2018.
44. Shaik Anwar and D. Kishore, "IOT Based Smart Home Security System with Alert and Door Access Control Using Smart Phone," *International Journal of Engineering Research & Technology*, vol. 5, Issue 12, pp. 504–509, 2016.
45. Sabrina, "Blockchain and Structural Relationship Based Access Control for IoT: A Smart City Use Case," 2019 IEEE 44th Conference on Local Computer Networks (LCN), pp. 137–140, 2019.

46. M. M. Bahgat, H. H. Farag, and B. Mokhtar, "IoT-Based Online Access Control System for Vehicles in Truck-Loading Fuels Terminals," In Proceedings of the 30th International Conference on Microelectronics (ICM), Sousse, Tunisia, pp. 2018–2021, 16–19 December 2018.
47. M. M. Bahgat, "Enhanced IoT-Based Online Access Control System for Vehicles in Truck-Loading Fuels Terminals," 2019 IEEE 6th International Conference on Industrial Engineering and Applications (ICIEA), pp. 765–769, 2019.
48. Muhammad Umar Aftab, Yasir Munir, Ariyo Oluwasanmi, Zhiguang Qin, Muhammad Haris Aziz, and Ngo Tung Son, "A Hybrid Access Control Model with Dynamic COI for Secure Localization of Satellite and IoT-Based Vehicles," *IEEE Access*, vol. 8, pp. 24196–24208, 2020.

11 Deep Neural Network–Based Security Model for IoT Device Network

Dukka Karun Kumar Reddy
Dr. L. Bullayya College of Engineering (for Women)

Janmenjoy Nayak
Aditya Institute of Technology and Management (AITAM)

Bighnaraj Naik
Veer Surendra Sai University of Technology

G. M. Sai Pratyusha
Dr. L. Bullayya College of Engineering (for Women)

CONTENTS

11.1 INTRODUCTION

'Internet' means a network of networks. It is a medium through which devices can connect, and the 'Thing' is any object, i.e., an object embedded with software and electronics with intelligence to connect to the internet. Internet of Things (IoT) is a platform where regular devices are connected to the internet, so they can interact, collaborate, and exchange data with each other. The term IoT was first used by 'Kevin Ashton' in 1999 at his presentation on Procter & Gamble and RFID (radio frequency identification). It relates to unique distinguishable objects (things) and their virtual portrayals on the internet-like framework. Before IoT existed, humans need to run from machine to machine telling them what to do. But in the case of IoT, things can interact, collaborate, and share experiences just like we humans do. With regard to IoT, machines are in continuous touch with each other and have been instructed what to do and when to do.

In modern days, cities are viewed as the reflecting face of a nation. A smart city is a multifaceted and modernistic urban area that addresses and serves the needs of inhabitants. With technological progress in computing and communication, people are moving to urban areas for improved amenities, healthcare, employment, and education. The metropolitan populace of the world has grown swiftly from 751 million in 1950 to 4.2 billion in 2018. As of 2018 human settlements and urbanization, 55% of the worldwide populace resides in metropolitan zones and the percentage expected to reach 68% by 2050 says the UN [1]. The latest projections indicate that in developing regions by 2030, the world is expected to have 43 megacities over 10 million inhabitants. For any successful development of a country, sustainable urbanization is also a key factor. As the world proceeds to urbanize, sustainable advancement relies progressively upon the effective management of urban development. Thereby, urban areas are turning out to be 'smart' to guarantee economically feasible and comfortable. Urbanization becomes a global phenomenon. Smart cities are enabled by cyber systems, which include connecting devices and systems through IoT technologies. IoT is a disruptive technology that enables the potential for significant improvements and innovations in business and societal environments [2]. The advancement of IoT implementations is drawn from the synergy of physical components and computational power, specifically by the use of critical infrastructure.

IoT is physical and engineered systems whose operations are integrated, coordinated, controlled, and monitored by computing and communication networks. Just as the internet transformed how people interact with each other. A smart city is an amalgamation of the technological platform to discourse the following challenges like public safety, health care, municipal corporation services, traffic and mobility, and transportation [3]. Designing and planning smart cities need specialists from different fields, including sociology, economics, information and communications technology, engineering, and policy regulations. Numerous outlines describing the framework of smart cities have been proposed by industry and scholastic sources. The most generally modified and appropriate reference model is projected by the U.S. National Institute of Standards and Technology [4]. Smart cities are composite systems, termed as systems-of-systems including process components, infrastructure, and people. Figure 11.1 shows the utmost smart cities model with essential components as environment, mobility, economy, government, people, and living.

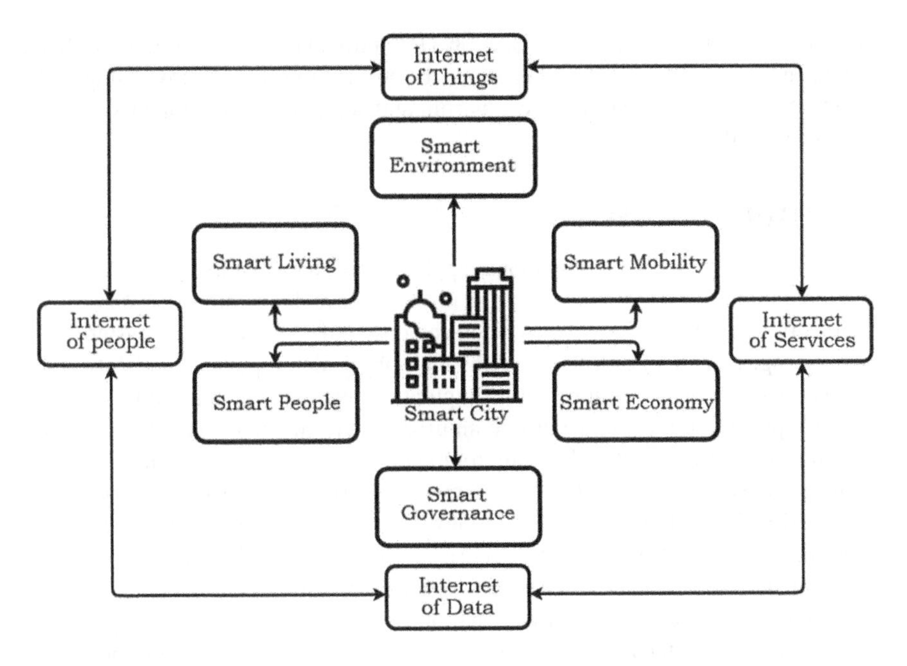

FIGURE 11.1 Essential components of smart city.

Security breach and abnormality have become general phenomena with lifting undesirable vulnerability to the systems due to the growing complication of the network infrastructure. Trust protection for efficient data mining, competent context-aware intelligence services, information security, and enhanced user confidentiality are the core feature of IoT. In order to gain people's trust, we must address the perception of uncertainty and risk in user utilization and acceptance of IoT applications and services [5]. Thus, the vision of a smart city can be boosted by IoTs using communication and information technologies for more effective and efficient resource management. The term smart city emphasizes delivering better quality and innovative services to its citizens by enhancing the infrastructure of the city and reducing the overall costs.

The main contributions of this chapter are as follows:

i. Design and implementation of Deep Neural Network (DNN) as a security model for IoT device networks.
ii. Understanding the influencing characteristics of IoT-DNL and OPCUA dataset for cyber system to recognize the feasible consequences and making it competent to detect various categorical attacks.
iii. Evaluating the performance using F1-score, recall, precision, accuracy values, ROC, and precision-recall curve.

Compared to different researchers' work, the proposed model has a lasting impact on the classification of different attacks and the accuracy of the solution. The following sections are structured in this chapter as follows: Section 11.2 provides a brief study of major-related work in this area. Section 11.3 summarizes the proposed DNN for the data analysis part. Section 11.4 provides the description and pre-processing of the datasets. Section 11.5 illustrates the experimental setup and result analysis. Section 11.6 concludes the chapter.

11.2 LITERATURE REVIEW

IoTs vary in their levels of operations, applications, and characteristics for smart cities operation. Different methods and designs have been proposed by several researchers to address these diverse requirements. IoT is a system of systems due to its heterogeneous and complex systems with physical components and networked systems through communication and computation interact in a continuous manner. The novel prototype for a smart city in an urban environment deals with new innovative services and amenities like environmental monitoring, emergency response, healthcare, business, transportation, energy distribution, and social activities. This augmented use of IoT brings various attacks and vulnerabilities that could lead to major consequences. Consequently, security problem is a global issue in this area that has made researchers design a secure, efficient, and robust IoT-based solution for a smart city. Here, we are presenting some of the methodology proposed by various researchers in characterizing anomaly detection techniques in different application domains related to a smart city.

Zhao et al. [6] proposed a system with principal component analysis to lessen dataset dimensions from a generous amount of features to a limited number of significant features. The developed classifier uses k-nearest neighbor and SoftMax regression algorithms. The KDD Cup 99 dataset's empirical findings indicate that the model worked optimally well in the marking of behaviors and classification of malicious behaviors.

Diro and Chilamkurti [7] proposed a distributed attack scheme of detection using the deep learning (DL) model for IoT. A comparative study is made between the proposed work and different machine learning (ML) approaches. The experimental results illustrate that the attack detection of a distributed system is fine and outstanding compared to attack detection of centralized systems using the DL model. It has been proved that the DL is well operative in attack detection and classification than compared to shallow counterparts. The evaluation of the model is performed on the NSL-KDD traffic dataset with binary and multi-classification. The DL model gives an accuracy of 99.2% for binary classification and an accuracy of 98.27 for multi-classification.

Baracaldo et al. [8] proposed an ML environment for the detection of poisoning attacks. The model is evaluated on two defense methods as a defense for partially trusted data and defense for fully untrusted data. The effectiveness of the proposed method is tested with two datasets namely poison 1(synthetic data) and poison 2(MNIST data) and compared with the support vector machine.

The solution proposed by Alrashdi et al. [9] can observe and recognize jeopardized IoT devices at distributed fog nodes. The work is evaluated on a part of the UNSW-NB15 dataset to demonstrate the model's accuracy. The AD-IoT (proposed model) achieved maximum accuracy of 99.34% for classification with the lowest

FPR of 0.02% to declare the percentage of true detection. The detection rate is 0.82% when compared to the Extra Trees Classifier, but precision is low with 0.79% and recall is better with 0.97%.

Hasan et al. [10] made a study analysis on a novel DS2OS dataset with various ML classifiers for detecting cyber-attacks on IoT networks. Among them, Decision Tree and Random Forest classifier predicted accurately for data probing, malicious operation, malicious control, spying, scan, and wrong setup classes. The normal and DoS attack classes are predicted more precisely than other classifier techniques. From the Random Forest confusion matrix, the DoS class is misclassified with 403 samples as Normal, from a total of 1178 DoS testing samples. The Normal class wrongly predicted 18 instances as DoS, from a total of 69,571 Normal testing samples. The Random Forest classifier results with an accuracy of 99.4% whereas precision, recall, and F1-score with 99%.

Rezvy et al. [11] projected a model for detection of attacks in IoT that shows 99.9% overall accuracy detection for flooding, injection, and impersonation type of attacks by applying a deep auto-encoded dense neural network approach. The intrusion detecting or attacks in 5G and IoT networks are evaluated by the algorithm on the benchmark Aegean Wi-Fi Intrusion dataset.

Anthi et al. [12] proposed a three-layer IDS approach with supervised learning to detect attacks in IoT-based networks. The proposed system comprises three key functions. First step involves the identification of normal behavior and type of each IoT device associated with the network. Secondly recognizes malicious packets in the network traffic when an attack is occurring, and finally categorizes the attack types that have been deployed on the network. The performance of the systems with three essential functions results in an F-measure of 96.2%, 90%, and 98% on the Benign Network Data.

Ullah and Mahmoud [13] proposed a hybrid anomalous IDS with a two-level detection approach for IoT networks. The first phase of the model classifies the network traffic as normal or anomalous by the flow-based anomaly detection approach. The second phase of the model uses Recursive Feature Elimination to detect significant and impacted features. The data cleaning and oversampling are carried by Edited Nearest Neighbors and Synthetic Minority Over-Sampling Technique approaches are applied on UNSW-15 and CICIDS2017 datasets. The proposed work derives the evaluation factors like F1-score, recall, and precision for level-1 with 99% on the UNSW-15 dataset and 100% for the CICIDS2017 dataset. The level-2 model derives F1-score, recall, and precision, which are measured as 97% for the UNSW-15 dataset and 100% for the CICIDS2017 dataset.

An automated framework for the secure and continuous availability of cloud service for smart connected vehicles was introduced by Aloqaily et al. [14]. The framework of the IDS mechanism enables to oppose security attacks and offers services to meet user's quality of experience and quality of service requirements. Intrusion detection is done through a three-phase approach with the analysis of data traffic, reduction of irrelevant data traffic features, and classification technique using Decision Tree ML mechanisms and Deep Belief network. The proposed work delivers a 99.92% detection rate and an overall accuracy of 99.43% on the NS-3 traffic and NSL-KDD datasets.

Li et al. [15] projected work deals with feature extraction of IoT network traffic and the IDS algorithm to oppose security attacks for a smart city with an intelligent-based learning model with deep migration. The intelligent-based learning model comprises DL model with intrusion detection technology. The experimental evaluation is performed on the KDD CUP 99 data set. It guaranteed a lower FPR and higher detection rate with improving efficiency compared to traditional methods.

Koumetio Tekouabou et al. [16] proposed a work system which is an integration of a predictive model based on ensemble learning methods with IoT to enhance the prediction of parking spaces available in smart parking. The parking dataset was used to carry out the testing process with the Bagging Regression algorithm and got a Mean Absolute Error of 0.06% on average.

A significant and critical problem is considered for the study of in-depth analysis for the selection of an effective ML algorithm when there are several ML algorithms for cyber-attack IDS for IoT security by Shafiq et al. [17]. A hybrid framework is developed, with a bijective soft set method on the BoT-IoT identification dataset to identify which ML algorithm is efficacious and considering intrusion traffic identification and IoT anomaly. The approach shows that the Naïve Bayes ML algorithm is efficacious for intrusion detection and anomaly detection within the IoT network.

Susilo and Sari [18] examined different ML and DL algorithms on an IoT network dataset from UNSW and concluded that Random Forests and Convolutional Neural Network presented the best outcome in terms of the AUC and accuracy for multiclass classification. The accumulation of epochs is experimented with 32 and 64 batches; as a result, there is a slender reduction in the accuracy. Similarly, with 128 batches, there was a small rise in accuracy. It is also noted that an increase in the batch size there is a speedup in the calculation process and doubling the variation in the batch size of Multi-Layer Perceptron could make the computation process 2.6 times faster, whereas Convolutional Neural Network could make the computation process 2.4 times faster.

11.3 PROPOSED METHOD

Due to the mitigation of a number of complex relationships and the acquisition of the most promising solutions through general evolution, ML models are popular in recent years. The Artificial Neural Network (ANN) is commonly used ML model of data handling inspired by normal apprehensive processes such as the organization of the human brain [19]. The ANN can discover complex and nonlinear relationship functions among dependent and independent variables based on processing and classifying the data through proper training. A DNN [20] consists of three-layer nodes, such as the input layer, the hidden layer(s), and the output layer. The input layer nodes take the input data and are linked to the hidden layer nodes, where the relation is represented by weights and the resulting link weights are computed with an activation function and are linked to the output layer nodes. By calculating the weighted sum and applying a bias, the activation function determines whether or not to activate a particular node.

The DL is the ML subdomain that is involved with networks that are able to learn unstructured or unlabeled knowledge. The computational ability of ANN is constrained in complexity. Developments in massive data analytics include advanced ANN, allowing machines to learn, analyze, and respond more quickly than humans to complex situations. Their efficiency continues to improve by building superior neural networks and training them with a massive amount of data or information, i.e., training larger neural networks with more data, making their output continue to grow. This is unlike other ML methods; the performance of traditional ML algorithms drops compared to the DL in performance as the amount of input data increases.

So, for revolutionary technology, DL offers truthful outcomes. More and more organizations are utilizing the DL technique to make novel action plans. The DL algorithm's most common structures are multiple layer neural networks.

Algorithm: DNN Training

Input: $X = X_i]_{i=1}^{N}$ is the dataset, ith = number of input vector, where $X_i = \{X_1, X_2 ... X_n\}$ is ith the input with n number of features

Repeat each epoch for

Repeat each layer for ℓ

Forward Phase:

The neuron net input Z^1 at layer l = 1 is computed as in Eq. 11.1 by using weight W^1, input X^1 and bias b^1. The net output A^1 has been calculated (Eq. 11.2) by using Relu activation function on Z^1.

$$Z^l = W^l \times X + b^l \tag{11.1}$$

$$A^l = \text{relu}(Z^l) \tag{11.2}$$

Forward propagation in the DNN (Figure 11.3), net output computation performance A^1 at layer 1 is propagated to layer l as in Eqs. 11.3 and 11.4.

$$A^{l-1} = \text{relu}(Z^{l-1}) \tag{11.3}$$

$$Z^l = W^l \times A^{l-1} + b^l \tag{11.4}$$

Output Prediction:

$$o^l = A^l = \sigma(Z^l) \tag{11.5}$$

Loss and Gradient Computation:

The DNN learning process is focused on the loss of the ℓ obtained from (Eq. 11.6) the network performance of output $o^l = A^l = \sigma(Z^l)$. Using backpropagation learning, the parameters of the DNN are updated by (Figure 11.4). The gradient of the loss functions where ℓ with respect to the constraint parameters (Eq. 11.7) gained. The gradient computation of ℓ regarding the parameters b^l, W^l, and A^{l-1} is derived as given in (Eqs. 11.8–11.10), where transpose matrix is represented by T.

$$\ell = -\frac{1}{m}\sum_{i=1}^{m}\left(y^i \log(o^i) + (1 - y^i)\log(1 - o^i)\right) \tag{11.6}$$

$$\Delta Z^{l} = \frac{\partial \ell}{\partial Z^{l}} \tag{11.7}$$

$$\Delta W^{l} = \frac{\partial \ell}{\partial W^{l}} = \frac{1}{m} \partial Z^{l} A^{l-1T} \tag{11.8}$$

$$\Delta b^{l} = \frac{\partial \ell}{\partial b^{l}} = \frac{1}{m} \sum_{i=1}^{m} \partial Z^{l,i} \tag{11.9}$$

$$\Delta A^{l-1} = \frac{\partial \ell}{\partial A^{l-1}} = W^{lT} \partial Z^{l,i} \tag{11.10}$$

Parameter Updation:

The parameter constraints b^l, W^l, and A^{l-1} and the obtained gradients are updated from ΔW^l, Δb^l, and ΔA^{l-1} as presented in Eqs. 11.11, 11.12, and 11.13, respectively.

$$W^{l} = W^{l} - \eta \times \Delta W^{l} \tag{11.11}$$

$$b^{l} = b^{l} - \eta \times \Delta b^{l} \tag{11.12}$$

$$A^{l-1} = A^{l-1} - \eta \times \Delta A^{l-1} \tag{11.13}$$

End of epoch

Return final $\left.o_i\right]_{i=1}^{N}$ for $X = \left.X_i\right]_{i=1}^{N}$

The output of each previous layer and bias is computed from the weighted inputs for the next layer of neural networks by a nonlinear activation function (Relu). In DL models, the cost function to be optimized is loss functions. The 'categorical cross-entropy' for multi-label categorization is the classic loss function. Activation functions are used to describe the neural network output that approximates yes or no. It maps the subsequent features between the activation function set, depending on the function. The model's final prediction is achieved via the feature of sigmoid activation, as seen in Figures 11.2 and 11.3. The dropout of nodes is used to prevent overfitting.

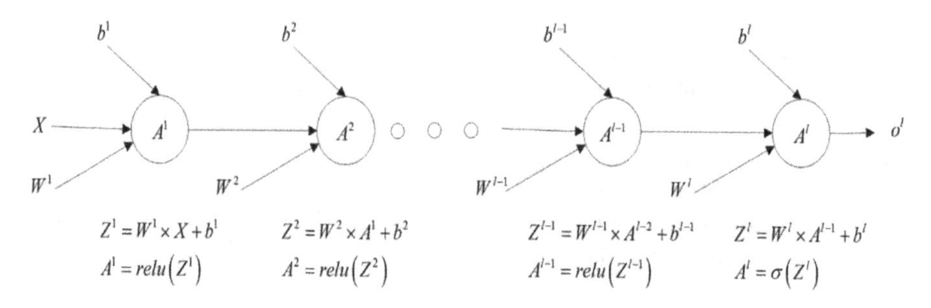

FIGURE 11.2 Forward phase.

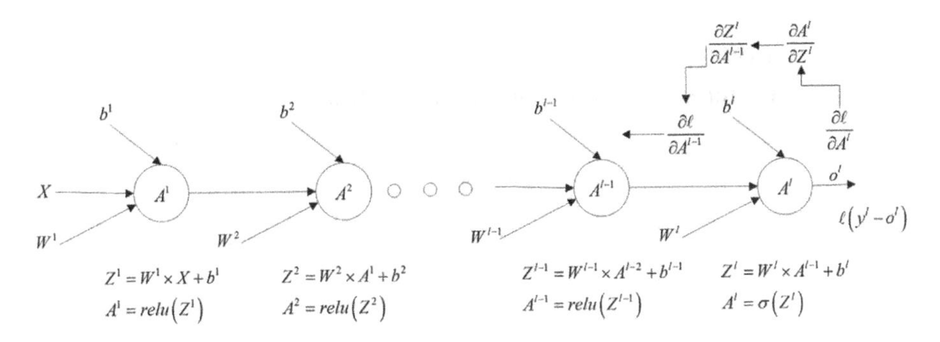

FIGURE 11.3 Backpropagation of error.

Dropout is a technique where, at certain stage in training, randomly selected neurons are ignored. They are arbitrarily 'dropped-out'. This represents that the presence in forward propagation of the activation mechanism of downstream neurons is temporarily impassive, and the weight update is not applied on the backward propagation of the neuron. This enables the DL network to train faster, reduce overfitting, and formulate enhanced forecasts. The dropout of nodes affects the simulation of a large number of neurons with a different network structure and the regular robustness of the nodes in the network structure to the specified inputs.

11.4 DATASET DESCRIPTION

This section illustrates the description about the dataset used in this experiment.

11.4.1 IoT-Device-Network-Logs Dataset

The dataset used in our experiment is IoT-Device-Network-Logs (IoT-DNL) dataset for network-based IDS, which is an open-source dataset collected from Kaggle and firstly introduced by Sahil Dixit [21]. It is a flow-based and labeled dataset specific to the wireless environment for the assessment of anomaly-based IDS. This is a preprocessed dataset for network-based IDS in IoT devices. The network logs are collected by monitoring the network through an ultrasonic sensor. The ultrasonic sensors used are Arduino and NodeMCU with the ESP8266 Wi-Fi module to send data to the server via Wi-Fi. The network traffic data furthermore contains 477,426 total instances with 14 features. The described dataset features are object type which is of float64 and int64 type as shown in Table 11.1. The dataset consists of 79,035 normal class and 398,391 anomaly instances and contains five classes that are classified. The normal class consists of 16.554% of total data, and the overall ratio between normal and anomalous class is 1:5.04. Table 11.2 gives a detailed representation of various attack distributions and anomaly in the IoT-DNL dataset. The dataset consists of 14 features with the target feature labeled as Normality. The Normality feature which consists of malicious traffic or attacks such as Wrong setup, DDoS, Data type probing (the ultrasonic sensor was used so in data type probing mostly string values are sent to the server), Man-in-the-Middle (MITM), and Scan attack was monitored within the network.

TABLE 11.1
Features Description of IoT-DNL Dataset

S. No.	Features	Data Type
1	Frame_number	Discrete
2	Frame_time	Discrete
3	Frame_length	Discrete
4	Ethernet_source	Discrete
5	Ethernet_destination	Discrete
6	IP_source	Discrete
7	IP_destination	Discrete
8	IP_protocol	Continuous
9	IP_length	Continuous
10	TCP_length	Continuous
11	TCP_source_port	Continuous
12	TCP_destination_port	Continuous
13	Value	Continuous
14	Normality	Discrete

TABLE 11.2
Distribution of Attacks in IoT-DNL Dataset

S. No.	Normality-Binary Classification	Normality-Multi Classification	Frequency Count	% of Total Data	% of Anomalous Data	Ratio between Normal and Anomalous Class
1	Normal	Normal	79035	16.554	--------	1:5.04
2	Anomaly	Wrong setup	82285	17.235	20.654	
		DDoS	79020	16.551	19.834	
		Data type probing	79002	16.547	19.83	
		Scan attack	79052	16.557	19.842	
		MITM	79032	16.553	19.837	

11.4.2 OPCUA Dataset

The dataset used in our experiment is the OPCUA for network-based IDS, which is an open-source dataset collected from IEEE Data Port and submitted by Rui Pinto [22]. The novel IDS dataset has a more comprehensive network, with a large number of flow-based and general features, and a labeled dataset specific to the wireless environment for the assessment of anomalous action or activity detection in IoT networks. The flow-based feature can be used to analyze and evaluate a flow-based IDS. The OPCUA dataset will give a reference point to distinguish anomalous activity over the IoT networks as intruders are aimed to deplete IoT network-based resources with malicious operations. The dataset is collected from a typical Cyber-Physical

TABLE 11.3
Distribution of Attacks in OPCUA Dataset

S. No.	Label Feature	Multi_Label Feature	Frequency Count	% of Total Data	% of Anomalous Data	Ratio between Normal and Anomalous Class
1	Normal	Normal	33566	31.18	--------	1:2.2
2	Anomaly	DoS	74012	68.76	99.92	
		Impersonation	49	0.045	0.066	
		MITM	6	0.005	0.008	

Production Systems (CPPS) configured testbed architecture. This dataset was generated by implementing and injecting various attacks on an OPCUA based CPPS testbed. The network traffic data, furthermore, contains 107,632 total instances with 32 features and target feature with binary classification labeled as label feature and for multi-classification labeled as Multi_Label feature. The described dataset features are with 5 object types, 11 float64 types, and 16 int64 types. The dataset with the Label feature consists of 33,566 normal class and 74,067 anomaly class instances. The overall ratio between normal and anomalous class is 1:2.2. The Multi_Label feature contains three attack classes with DoS, Impersonation, and MITM. Table 11.3 gives the attack taxonomy of the OPCUA dataset. OPCUA is an unbalanced dataset and may lead to have a negative influence on the performance functioning of the classification algorithm, so balancing of class distribution is made. SMOTE (synthetic minority oversampling technique) [23] is one of the oversampling techniques used to address the classification problem of imbalanced dataset. The class distribution is balanced by randomly increasing minority class instances by replicating them. Virtual records are generated for the minority class using linear interpolation by arbitrarily choosing one or more k-nearest neighbors for each instance in the minority class. The data is reconstructed after the oversampling process. As the OPCUA dataset is unbalanced because of the attack classes Impersonation and MITM with 49 and 6 instances in the overall dataset, we applied SMOTE. A brief explanation of IoT-DNL and OPCUA dataset categorical attacks is given in Table 11.4.

11.4.2.1 Data Preprocessing

In order to resolve and prioritize the features, data preprocessing is an essential activity because data processing improves efficiency with accurate and precise performance. The major requirements needed for ML research are exploratory data observation and analysis. The key task of a classification model is to supply the data with an acceptable classifier. Data cleaning is thus an initial step in dealing with an ML project. The OPCUA dataset consists of Proto, Service_errors, and Status_errors feature columns which are removed as they have a single value. The Src_IP, Dst_IP, and Service feature contains categorical values, which are encoded to numeric data.

TABLE 11.4
Attack Classes of IoT-DNL and OPCUA Dataset

Attacks	Definition
DDoS	The DDoS attack in which the offender attempts to make the network resources engaged to its proposed users by indefinitely distracting services of a host with a flood of internet traffic.
Scan	To identify the network users, establish the devices and state of systems, and take a list of network elements during the process of information acquisition by scanning. Where attacker crack to violate the applications and target systems.
MITM	A man in the middle attack permits a malicious actor to interrupt, send, and receive data meant for others.
Data type probing	Data probing is the practice of acquiring control over a data network or telecommunication without distracting the network from being altered or monitored from the structure of data.
Wrong setup	If the system is set up inappropriately, this will lead to other problems.
Impersonation	In impersonation attack, the opponent poses as a legitimate or trusted party to successfully assume in a communication protocols or in a system.

11.5 EXPERIMENTAL CONFIGURATION AND RESULT ANALYSIS

It is imperative that the method must be tested in order to assess the feasibility of the study and visualization of the dataset. Since any dataset does not reveal the appropriateness under practical noise structures and uncertainty. Undetected phenomenon cases from the datasets are revealed in the experimental setup and results review. The experimental setup utilized IdeaPad 330 (LENOVO) laptop. The operating system was Windows 10 Enterprise and i5-8250U CPU with 3.10GHz. The system memory contains 8 GB RAM. For data preprocessing, Pandas and NumPy framework were used. For data analysis Scikit-learn framework and for information visualization, Matplotlib framework was utilized through Spyder integrated development environment.

11.5.1 Performance Evaluation

The performance study of the proposed model in comparisons with other models is tested with several performance metrics. The metrics are a statistical determination of decision-making for the most appropriate model for this work. The confusion matrix illuminates the issue of classification by prediction results. It derives the errors acquired by a classifier and the forms of error analysis classified by true positive (TP), false positive (FP), true negative (TN), and false negative (FN), and correctness of the classifier through accuracy (Eq. 11.14), false-positive rate (FPR) (Eq. 11.15), true positive rate (TPR or also known as recall or sensitivity) (Eq. 11.16), F1-score (Eq. 11.17), precision (Eq. 11.18), ROC, and precision-recall curve.

$$\text{Accuracy} = \frac{TP + TN}{TP + TN + FP + FN} \tag{11.14}$$

$$\text{FPR} = \frac{\text{FP}}{\text{TN} + \text{FP}} \tag{11.15}$$

$$\text{TPR} = \frac{\text{TP}}{\text{TP} + \text{FN}} \tag{11.16}$$

$$\text{F1} - \text{score} = \frac{2 \times \text{TP}}{2 \times \text{TP} + \text{FP} + \text{FN}} \tag{11.17}$$

$$\text{Precision} = \frac{\text{TP}}{\text{TP} + \text{FP}} \tag{11.18}$$

11.5.2 Parameter Setting

Various DNN parameters are illustrated in Table 11.5.

11.5.3 Result Analysis

This section presents the result analysis of the proposed DNN algorithm with Adam optimizer, categorical cross entropy as its loss function, and with the number of epochs as 300. Table 11.5 shows the fine-tuned parameter setting of classifiers for better prediction. The proposed system is used to analyze two different datasets. The OPCUA dataset consists of 107,633 instances for multi-classification, where 207,233 (70%) instances are used for training and 88,815(30%) instances are used for testing. The IoT-DNL dataset consists of 477,426 instances for multi-classification, where 334,198 (70%) instances are used for training and 143,228 (30%) instances are used for testing.

Table 11.6 shows the evaluation metrics of the proposed method on the OPCUA dataset. Only 2 instances of Normal class are wrongly identified as MITM class. Also, the TPR is 1 for all classes except for the Normal class which has a TPR value of 0.9999 and so the FPR of MITM class is 0.0001, and others have an FPR value of 1. Except for the MITM class, the precision of all classes is 1. The F1-Score of MITM class is 0.0001, and the other classes is 0. The individual accuracy of DOS class is 1, Impersonation is 1, MITM is 0.9999, and Normal is 0.9999 which sums up to an overall accuracy of 99.998%.

Table 11.7 shows the evaluation metrics of the proposed method on the IoT-DNL dataset. All instances of Wrong Setup, DDOS, and MITM classes are

TABLE 11.5
Parameter Setting for DNN

S. No.	Dataset	DNN Parameter Setting
1	IoT-DNL, OPCUA	Epochs: 300 Optimizer: Adam Loss: Categorical Cross Entropy No of hidden layers: 4 Activation in hidden layers: Relu

TABLE 11.6
Results of OPCUA Dataset Using Proposed Method

DNN	DoS	Impersonation	MITM	Normal
TP	22167	22229	22135	22282
TN	66648	66586	66678	66531
FP	0	0	2	0
FN	0	0	0	2
TPR/Recall	1.0000	1.0000	1.0000	0.9999
FPR	0.0000	0.0000	0.0001	0.0000
F1 score	1.0000	1.0000	0.9999	0.9999
Precision	1.0000	1.0000	0.9999	1.0000
Accuracy	1.0000	1.0000	0.9999	0.9999
Overall accuracy	**99.998**			

TABLE 11.7
Results of IoT-DNL Dataset Using Proposed Method

DNN	Normal	Wrong Setup	DDoS	Data Type Probing	Scan	MITM
TP	229,755	24,749	23,733	23,723	23,640	23,582
TN	119,702	117,938	119,488	119,312	119,322	119,552
FP	0	541	7	11	173	94
FN	551	0	0	182	93	0
TPR / Recall	0.9765	1.0000	1.0000	0.9923	0.9960	1.0000
FPR	0.0000	0.0045	0.0001	0.0001	0.0014	0.0007
F1 Score	0.9881	0.9891	0.9998	0.9959	0.9944	0.9980
Precision	1.0000	0.9786	0.9997	0.99955	0.9927	0.9960
Accuracy	0.9961	0.9962	0.9996	0.9987	0.9981	0.9993
Overall Accuracy	**99.42**					

identified correctly. One hundred eighty-two instances of Data Type Probing are wrongly classified as Wrong Setup; 93 instances of Scan attack are identifying as MITM; and of all the instances of Normal class, 359 instances are misclassified as Wrong Setup, 7 as DDoS, 11 as Data type Probing, 173 as Scan, and 1 as MITM. The TPR of Wrong Setup, DDoS, MITM classes is 1. The FPR values of the classes Normal, DDOS, and Data type probing are least with values 0, 0.0001, and 0.0001 respectively. These lead to individual accuracies of 0.9961 for Normal, 0.9962 for Wrong Setup, 0.9996 for DDOS, 0.9987 for Data type probing, 0.9981 for Scan, and 0.9993 for the Man in the Middle classes respectively. This sums up to an overall accuracy of 99.42%.

Figures 11.4 and 11.5 show the confusion matrices of the proposed method over the OPCUA dataset and IoT-DNL dataset respectively. It is clearly evident that in OPCUA data only two instances of Normal class are wrongly identified as MITM class, and the rest of them are identified correctly. In the IoT-DNL data, all instances of

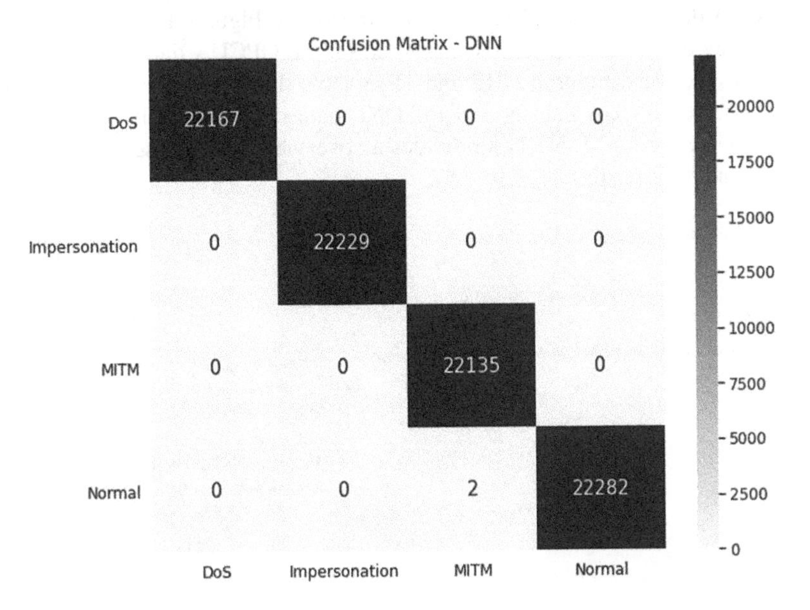

FIGURE 11.4 Confusion matrix of OPCUA data.

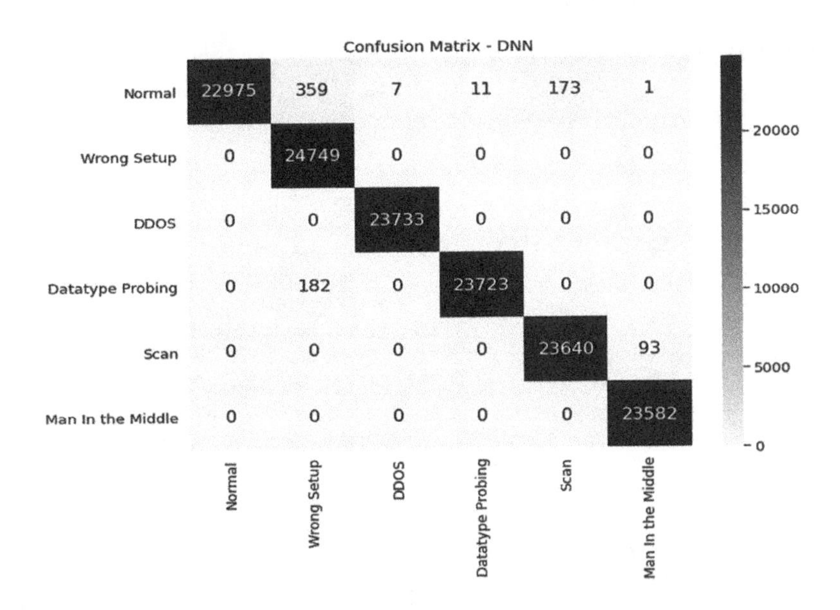

FIGURE 11.5 Confusion matrix of IoT-DNL data.

Wrong Setup, DDOS, and MITM classes are identified correctly, and a few instances of Normal class, Data type probing, and Scan attack classes are misclassified.

Figures 11.6 and 11.7 show the ROC curves of the proposed method over the OPCUA dataset and IoT-DNL dataset, respectively.

Figures 11.8 and 11.9 show the Precision-Recall curves of the proposed method on the OPCUA dataset and IoT-DNL dataset respectively. Figures 11.10 and 11.11 show the accuracy curves of the proposed method on the OPCUA dataset and IoT-DNL dataset, respectively. Figures 11.12 and 11.13 show the loss curves of the proposed method on the OPCUA dataset and IoT-DNL dataset, respectively. As per the literature review section, Table 11.8 provides an overview of all those early developed methods and their results.

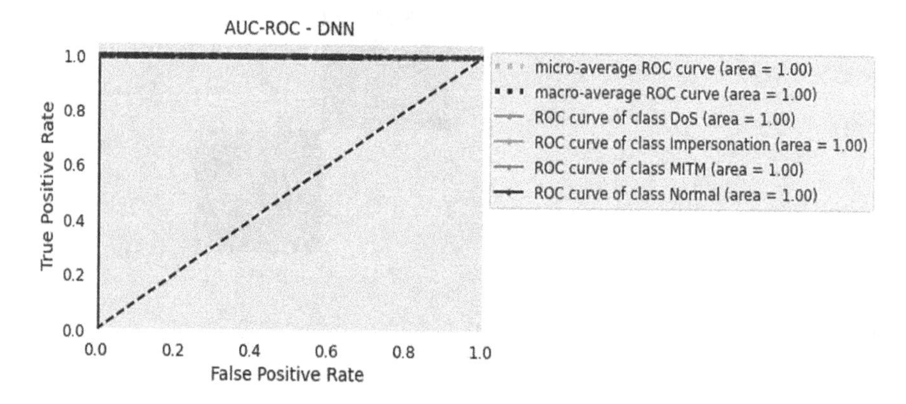

FIGURE 11.6 ROC curve of OPCUA data.

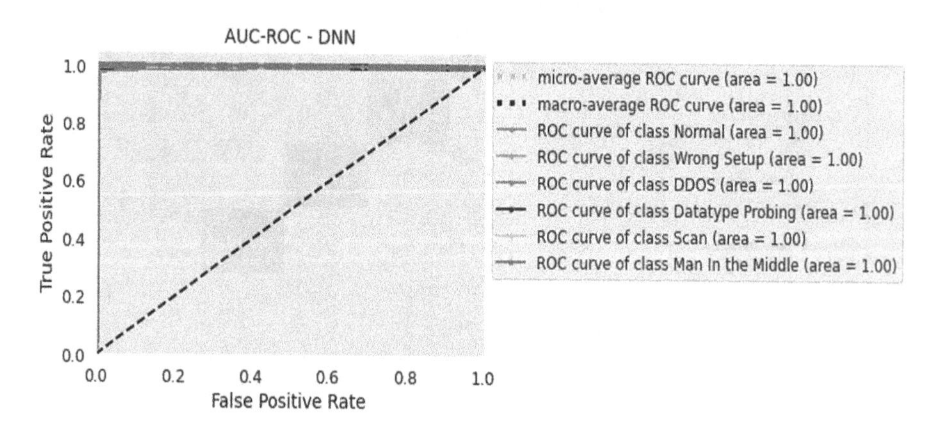

FIGURE 11.7 ROC curve of IoT-DNL data.

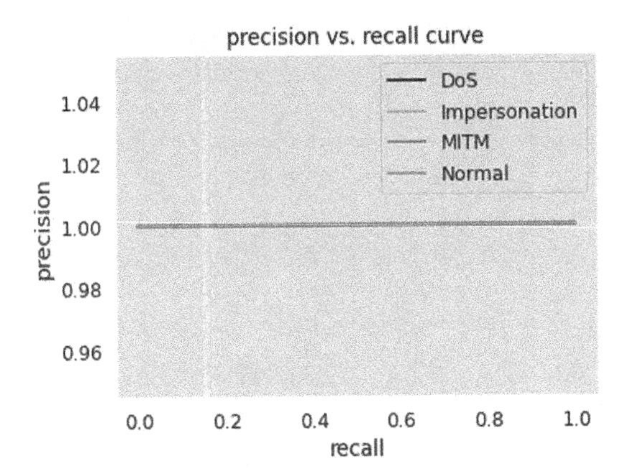

FIGURE 11.8 Precision-recall curve of OPCUA data.

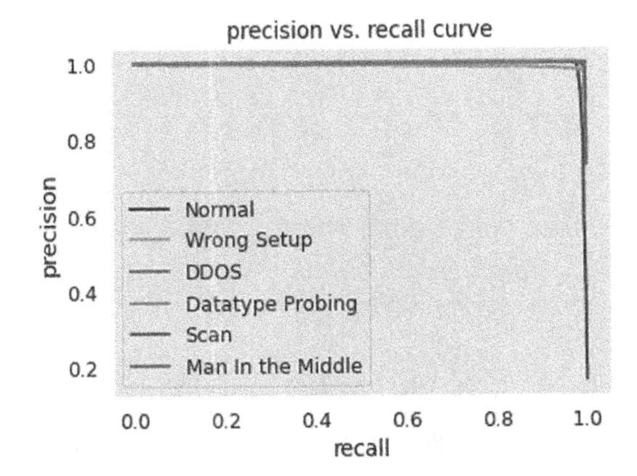

FIGURE 11.9 Precision-recall curve of IoT-DNL data.

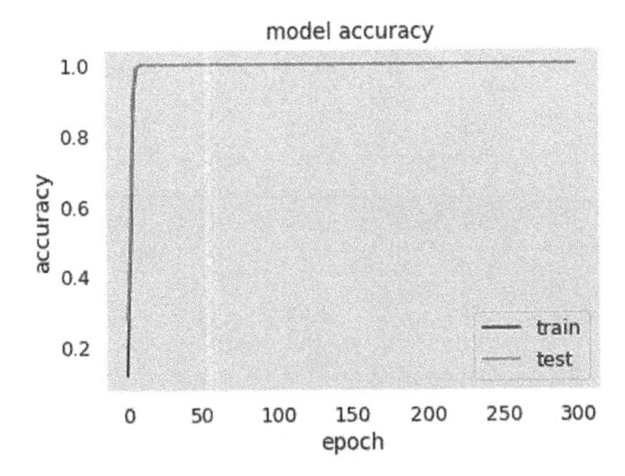

FIGURE 11.10 Accuracy curve of OPCUA data.

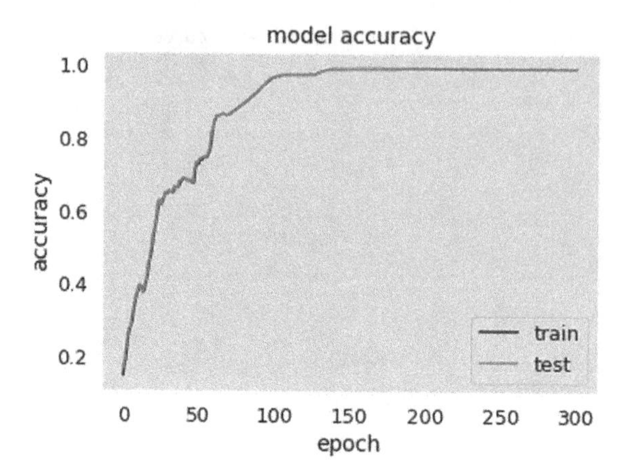

FIGURE 11.11 Accuracy curve of IoT-DNL data.

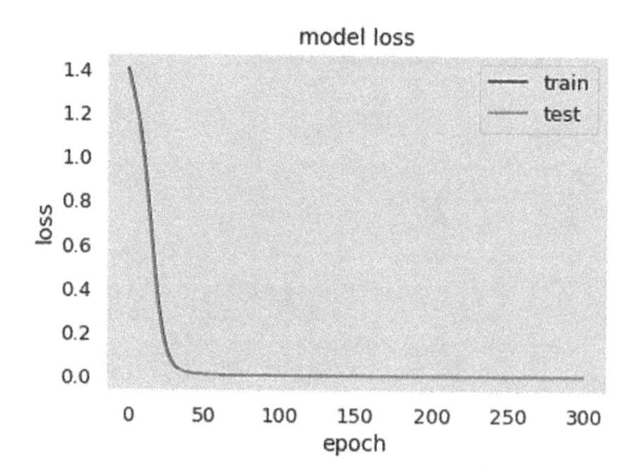

FIGURE 11.12 Loss curve of OPCUA data.

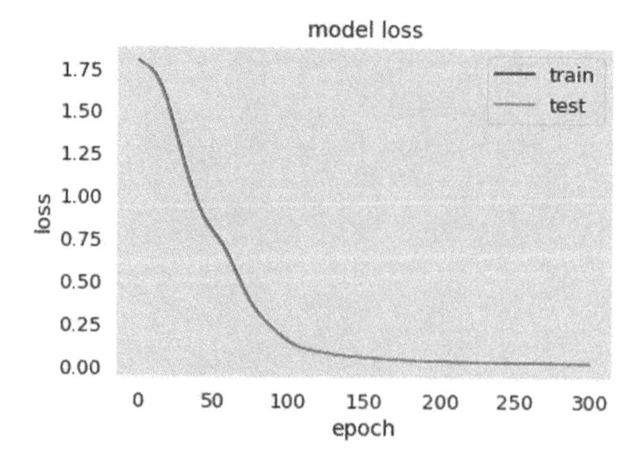

FIGURE 11.13 Loss curve of IoT-DNL data.

TABLE 11.8
The Accuracies of Proposed Method with Various IoT Models

Author	Dataset	Model Accuracy		Proposed Model Accuracy	
				IoT-DNL Dataset	OPCUA Dataset
[6]	KDD Cup 99	k-Nearest neighbor and SoftMax regression	84.9	99.42	99.99
[7]	NSL-KDD	Deep model	98.27		
[8]	MNIST data set	Gradient ascent with support vector machine	89		
[11]	Aegean Wi-Fi intrusion data	Auto-encoded DNN	99.9		
[10]	DS2OS	Random forest	99.4		
		Decision tree	99.4		
		ANN	99.4		
		Linear regression	98.3		
		Support vector machine	98.2		
[13]	CICIDS2017, UNSW-15 datasets	Two-level anomalous detection system	99.9		
[14]	NSL-KDD datasets.	Decision tree machine learning and deep belief mechanisms	99.43		

11.6 CONCLUSION

At present times, the technological evolution of IoT in smart cities is still in the conceptual stage which needs further study in comparison to information security, data security, and network security accompanied by the development of the IoT. The network traffic of a smart city operates in real-world time via IoT objects. Consequently, the objective is to make an essential cyber-physical system designed to alleviate IoT-related security threats and attacks that exploit security vulnerabilities. The objective of this chapter is to propose an intelligent DNN-based framework approach for classifying and distinguishing anomalies from normal behavior based. We conducted a distinct combination search to assess the best learning parameters for DNN on IoT-DNL and OPCUA dataset. The DNN achieved an overall accuracy of 99.42% and 99.99%. The experimental study and result analysis from modeling knowledge of DNN show a promising approach for anomaly detection in cyber systems with the categorizing of normal and attacks. The experimental observations of the analytical model report that the design of the deep neural network performs well in most categorical attacks through noticeable improvement.

REFERENCES

1. "UN," *Notes and Queries*, 2018. [Online]. Available: https://www.un.org/development/desa/en/news/population/2018-revision-of-world-urbanization-prospects.html.
2. P. B. Dash, J. Nayak, B. Naik, E. Oram, and S. H. Islam, "Model based IoT security framework using multiclass adaptive boosting with SMOTE," *Security and Privacy*, vol. 3, no. April, pp. 1–15, Jun. 2020.
3. U. Ghosh, P. Chatterjee, S. Shetty, and R. Datta, "An SDN-IoT-based framework for future smart cities: Addressing perspective," in *Internet of Things and Secure Smart Environments: Successes and Pitfalls*. Boca Raton, FL: CRC Press, pp. 442–463, 2020.
4. E. S. Madhan, U. Ghosh, D. K. Tosh, K. Mandal, E. Murali, and S. Ghosh, "An improved communications in cyber physical system architecture, protocols and applications," in *2019 16th Annual IEEE International Conference on Sensing, Communication, and Networking (SECON)*, Boston, MA, USA, 2019.
5. D. K. Reddy, H. S. Behera, J. Nayak, P. Vijayakumar, B. Naik, and P. K. Singh, "Deep neural network based anomaly detection in Internet of Things network traffic tracking for the applications of future smart cities," *Transactions on Emerging Telecommunications Technologies*, no. June, pp. 1–26, Oct. 2020.
6. S. Zhao, W. Li, T. Zia, and A. Y. Zomaya, "A dimension reduction model and classifier for anomaly-based intrusion detection in internet of things," in *2017 IEEE 15th Intl Conf on Dependable, Autonomic and Secure Computing, 15th Intl Conf on Pervasive Intelligence and Computing, 3rd Intl Conf on Big Data Intelligence and Computing and Cyber Science and Technology Congress(DASC/PiCom/DataCom/CyberSciTech)*, vol. 2018-Jan, pp. 836–843, 2017.
7. A. A. Diro and N. Chilamkurti, "Distributed attack detection scheme using deep learning approach for Internet of Things," *Future Generation Computer Systems*, vol. 82, pp. 761–768, May 2018.
8. N. Baracaldo, B. Chen, H. Ludwig, A. Safavi, and R. Zhang, "Detecting poisoning attacks on machine learning in IoT environments," in *2018 IEEE International Congress on Internet of Things (ICIOT)*, pp. 57–64, 2018.
9. I. Alrashdi, A. Alqazzaz, E. Aloufi, R. Alharthi, M. Zohdy, and H. Ming, "AD-IoT: Anomaly detection of IoT cyberattacks in smart city using machine learning," in *2019 IEEE 9th Annual Computing and Communication Workshop and Conference (CCWC)*, pp. 0305–0310, 2019.
10. M. Hasan, M. M. Islam, M. I. I. Zarif, and M. M. A. Hashem, "Attack and anomaly detection in IoT sensors in IoT sites using machine learning approaches," *Internet of Things*, vol. 7, p. 100059, Sep. 2019.
11. S. Rezvy, Y. Luo, M. Petridis, A. Lasebae, and T. Zebin, "An efficient deep learning model for intrusion classification and prediction in 5G and IoT networks," in *2019 53rd Annual Conference on Information Sciences and Systems (CISS)*, pp. 1–6, 2019.
12. E. Anthi, L. Williams, M. Slowinska, G. Theodorakopoulos, and P. Burnap, "A supervised intrusion detection system for smart home IoT devices," *IEEE Internet of Things Journal*, vol. 6, no. 5, pp. 9042–9053, Oct. 2019.
13. I. Ullah and Q. H. Mahmoud, "A two-level hybrid model for anomalous activity detection in IoT networks," in *2019 16th IEEE Annual Consumer Communications & Networking Conference (CCNC)*, pp. 1–6, 2019.
14. M. Aloqaily, S. Otoum, I. Al Ridhawi, and Y. Jararweh, "An intrusion detection system for connected vehicles in smart cities," *Ad Hoc Networks*, vol. 90, p. 101842, Jul. 2019.
15. D. Li, L. Deng, M. Lee, and H. Wang, "IoT data feature extraction and intrusion detection system for smart cities based on deep migration learning," *International Journal of Information Management*, vol. 49, no. March, pp. 533–545, Dec. 2019.

16. S. C. Koumetio Tekouabou, E. A. Abdellaoui Alaoui, W. Cherif, and H. Silkan, "Improving parking availability prediction in smart cities with IoT and ensemble-based model," *Journal of King Saud University-Computer and Information Sciences*, Feb. 2020. DOI:10.1016/j.jksuci.2020.01.008.
17. M. Shafiq, Z. Tian, Y. Sun, X. Du, and M. Guizani, "Selection of effective machine learning algorithm and Bot-IoT attacks traffic identification for internet of things in smart city," *Future Generation Computer Systems*, vol. 107, pp. 433–442, Jun. 2020.
18. B. Susilo and R. F. Sari, "Intrusion detection in IoT networks using deep learning algorithm," *Information*, vol. 11, no. 5, p. 279, May 2020.
19. Y. Bengio, P. Lamblin, D. Popovici, and H. Larochelle, "Greedy layer-wise training of deep networks," in *Proceedings of the 19th International Conference on Neural Information Processing Systems (NIPS'06)*, pp. 153–160, 2006.
20. Y. Bengio, "Learning deep architectures for AI," *Foundations and Trends® in Machine Learning*, vol. 2, no. 1, pp. 1–127, 2009.
21. "Iot Device Network Logs _ Kaggle," 2020. [Online]. Available: https://www.kaggle.com/speedwall10/iot-device-network-logs.
22. R. Pinto, "M2M USING OPC UA," 2020. [Online]. Available: https://ieee-dataport.org/open-access/m2m-using-opc-ua. [Accessed: 18-Sep-2020].
23. N. V. Chawla, K. W. Bowyer, L. O. Hall, and W. P. Kegelmeyer, "SMOTE: Synthetic minority over-sampling technique," *Journal of Artificial Intelligence Research*, vol. 16, no. 2, pp. 321–357, Jun. 2002.

Index

A

B

C

D

E

F

G

H

I

J

K

L

M

N

O

P

Q

R

S

T

U

V

W

X

Z